Multifunctional Hybrid Materials Based on Polymers: Design and Performance

Multifunctional Hybrid Materials Based on Polymers: Design and Performance

Editors

M. Ali Aboudzadeh

Shaghayegh Hamzehlou

MDPI • Basel • Beijing • Wuhan • Barcelona • Belgrade • Manchester • Tokyo • Cluj • Tianjin

Editors
M. Ali Aboudzadeh
CNRS, University Pau & Pays Adour
E2S UPPA, Institut des Sciences Analytiques et de Physico-Chimie pour l'Environnement et les Matériaux
IPREM, UMR5254
Pau
France

Shaghayegh Hamzehlou
POLYMAT and Kimika Aplikatua Saila
Kimika Fakultatea
University of the Basque Country UPV-EHU
Joxe Mari Korta Zentroa
Tolosa Hiribidea 72
Donostia-San Sebastian
Spain

Editorial Office
MDPI
St. Alban-Anlage 66
4052 Basel, Switzerland

This is a reprint of articles from the Special Issue published online in the open access journal *Processes* (ISSN 2227-9717) (available at: www.mdpi.com/journal/processes/special_issues/hybrid_nanomaterial).

For citation purposes, cite each article independently as indicated on the article page online and as indicated below:

LastName, A.A.; LastName, B.B.; LastName, C.C. Article Title. *Journal Name* **Year**, *Volume Number*, Page Range.

ISBN 978-3-0365-2034-6 (Hbk)
ISBN 978-3-0365-2033-9 (PDF)

Contents

About the Editors

M. Ali Aboudzadeh

M. Ali Aboudzadeh is currently a Marie Curie Fellow in IPREM, a joint research unit attached to the CNRS and the University of Pau and Pays de l'Adour, France. He received his B.Sc. (2003) and M.Sc. (2006) in Polymer Engineering from Amir Kabir University of Technology and Iran Polymer & Petrochemical Institute, respectively. He obtained his Ph.D. (2015) in Applied Chemistry and Polymer Materials from University of the Basque Country, Spain. Dr. Aboudzadeh is author/coauthor of more than 30 professional papers in different research areas, including polymer synthesis and characterization, supramolecular assemblies, rheology, DNA nanotechnology, and encapsulation via emulsion-based systems. He recently edited a book entitled "Emulsion-based Encapsulation of Antioxidants"under the publisher "Springer".

Shaghayegh Hamzehlou

Dr. Shaghayegh Hamzehlou earned her B.Sc. and M.Sc. in polymer engineering at Amir Kabir University of Technology on 2006. Later, she worked for 4 years as R&D Engineer in an industrial manufacturing company. In 2010, she moved to University of Basque country to do a Ph.D. in the framework of a European project, Marie Curie training network. In 2014, she joined Basque Center for Macromolecular Design and Engineering as a postdoctoral fellow, being involved in another European project for 3 years. Currently, she is working as a researcher at the chemistry faculty of University of Basque Country. Her research is focused on polymer reaction engineering, modelling and simulation of kinetics, topology, microstructure and morphology of complex polymerization systems. She has (co)authored more than 30 scientific articles, 4 book chapters and had oral presentations in more than 20 national and international conferences and was a keynote lecturer at the Polymer Reaction Engineering X (PRE 10) May 2018.

Preface to "Multifunctional Hybrid Materials Based on Polymers: Design and Performance"

Hybrids and composite materials offer a synergic combination of polymer and inorganic features. Over the past few decades, hybrid materials based on polymers have gained a lot of attention, not only for their interesting structural characterization, but also for their promising functional applications. Thus, investigations into multifunctional polymeric hybrid materials establish an essential task for polymer science. Tremendous progress in the design and preparation of new hybrid materials based on polymers has created a large number of solutions to the current challenges in many applications such asadsorption, separation, gas storage, catalysis, sensing, electronic devices, etc. In particular, the hybridization of these materials can bring about exceptional superior multifunctions, and thus presents the promise of application in the fields of chemical and biological sensing, heterogeneous catalysis, energy transformation and storage, and atmosphere and human health.

Moreover, polymer hybrid materials can be created via blending of functional polymers with other nanostructured compounds, with the latter displaying size-dependent physical and chemical features. This has become a considerable area for research and technological development due to the significant properties and multifunctionalities emanated from polymers'nanocomposite/nanohybrid structure. Therefore, scientists are attempting to incorporate different types of nanostructured compounds to adjust structures and enhance the properties of conventional polymers, which will have a tremendous impact in the field of polymer science. Nevertheless, the design and development of multifunctional hybrid nanomaterials also remain challenging, and their introduction into realistic applications is not yet acceptable. Therefore, it is highly advantageous to implement a progress on state-of-the-art nanomanufacturing and scale-up nanotechnology so as to design and synthesize progressive multifunctional hybrid nanomaterials with improved efficiency.

In this book, we introduce an elegant selection of first-rate reviews and original research articles that demonstrate the significance of developing multifunctional hybrid material based on polymers (including nanomaterials) for different applications. Deep understanding and appropriate theoretical calculations for analyzing the behaviors of these materials (involved in the formulations) at their interface have also been achieved through fundamental investigations.

The main aim of the book is to inspire and to guide scientists in this field. For the industrial establishments, the book also presents easy-to-achieve approaches that have been developed so far and could create a platform for industrial material production.

The Editors express their appreciation to all contributors from different parts of the world that have cooperated in the preparation of this book. In this context, this international book gives the active reader different perspectives on the subject and encourages him/her to read the entire book.

M. Ali Aboudzadeh, Shaghayegh Hamzehlou
Editors

processes

MDPI

Review

Onco-Receptors Targeting in Lung Cancer via Application of Surface-Modified and Hybrid Nanoparticles: A Cross-Disciplinary Review

Fakhara Sabir [1], Maimoona Qindeel [2,3], Mahira Zeeshan [3], Qurrat Ul Ain [4], Abbas Rahdar [5,*], Mahmood Barani [6], Edurne González [7] and M. Ali Aboudzadeh [8,9,*]

1 Faculty of Pharmacy, Institute of Pharmaceutical Technology and Regulatory Affairs, University of Szeged, 6720 Szeged, Hungary; fakhra.sabir@gmail.com
2 Hamdard Institute of Pharmaceutical Sciences, Hamdard University Islamabad Campus, Islamabad 76400, Pakistan; mqindeel81@gmail.com
3 Department of Pharmacy, Faculty of Biological Sciences, Quaid-i-Azam University, Islamabad 45320, Pakistan; mz1712@yahoo.com
4 Department of Pharmacology and Clinical Pharmacy, School of Pharmacy, Bandung Institute of Technology, Bandung 40132, Indonesia; aineevirk.av@gmail.com
5 Department of Physics, University of Zabol, Zabol 98613-35856, Iran
6 Department of Chemistry, Shahid Bahonar University of Kerman, Kerman 6169-14111, Iran; mahmoodbarani7@gmail.com
7 POLYMAT and Kimika Aplikatua Saila, Kimika Fakultatea, University of the Basque Country UPV/EHU, Joxe Mari Korta Zentroa, Tolosa Hiribidea 72, 20018 Donostia, San Sebastián, Spain; edurne.gonzalezg@ehu.eus
8 Centro de Física de Materiales, CSIC-UPV/EHU, Paseo Manuel Lardizábal 5, 20018 Donostia, San Sebastián, Spain
9 Donostia International Physics Center (DIPC), Paseo Manuel Lardizábal 4, 20018 Donostia, San Sebastián, Spain
* Correspondence: a.rahdar@uoz.ac.ir (A.R.); mohammadali.aboudzadeh@ehu.eus (M.A.A.)

Citation: Sabir, F.; Qindeel, M.; Zeeshan, M.; Ul Ain, Q.; Rahdar, A.; Barani, M.; González, E.; Aboudzadeh, M.A. Onco-Receptors Targeting in Lung Cancer via Application of Surface-Modified and Hybrid Nanoparticles: A Cross-Disciplinary Review. *Processes* **2021**, 9, 621. https://doi.org/10.3390/pr9040621

Academic Editor: Carla Vitorino

Received: 1 March 2021
Accepted: 29 March 2021
Published: 1 April 2021

Abstract: Lung cancer is among the most prevalent and leading causes of death worldwide. The major reason for high mortality is the late diagnosis of the disease, and in most cases, lung cancer is diagnosed at fourth stage in which the cancer has metastasized to almost all vital organs. The other reason for higher mortality is the uptake of the chemotherapeutic agents by the healthy cells, which in turn increases the chances of cytotoxicity to the healthy body cells. The complex pathophysiology of lung cancer provides various pathways to target the cancerous cells. In this regard, upregulated onco-receptors on the cell surface of tumor including epidermal growth factor receptor (EGFR), integrins, transferrin receptor (TFR), folate receptor (FR), cluster of differentiation 44 (CD44) receptor, etc. could be exploited for the inhibition of pathways and tumor-specific drug targeting. Further, cancer borne immunological targets like T-lymphocytes, myeloid-derived suppressor cells (MDSCs), tumor-associated macrophages (TAMs), and dendritic cells could serve as a target site to modulate tumor activity through targeting various surface-expressed receptors or interfering with immune cell-specific pathways. Hence, novel approaches are required for both the diagnosis and treatment of lung cancers. In this context, several researchers have employed various targeted delivery approaches to overcome the problems allied with the conventional diagnosis of and therapy methods used against lung cancer. Nanoparticles are cell nonspecific in biological systems, and may cause unwanted deleterious effects in the body. Therefore, nanodrug delivery systems (NDDSs) need further advancement to overcome the problem of toxicity in the treatment of lung cancer. Moreover, the route of nanomedicines' delivery to lungs plays a vital role in localizing the drug concentration to target the lung cancer. Surface-modified nanoparticles and hybrid nanoparticles have a wide range of applications in the field of theranostics. This cross-disciplinary review summarizes the current knowledge of the pathways implicated in the different classes of lung cancer with an emphasis on the clinical implications of the increasing number of actionable molecular targets. Furthermore, it focuses specifically on the significance and emerging role of surface functionalized and hybrid nanomaterials as drug delivery systems through citing recent examples targeted at lung cancer treatment.

Keywords: lung cancer; nanoparticles; toxicity; surface modification; hybrid nanocarriers

1. Introduction

Lung cancer is one of the most prevalent diseases and the leading causes of death worldwide [1]. It is more common in males than in females and based on an estimation, this type of cancer caused 154,050 deaths in 2018 [2]. One of the most common causes of this devastating disease is chronic tobacco usage. The major reason for its high mortality is the late diagnosis of the disease, and in most cases, lung cancer is diagnosed at the fourth stage when the cancer has already metastasized to the nearby organs [3]. Among lung cancer patients, 85% exhibit nonsmall cell lung cancer (NSCLC) while the rest (15%) of the patients have small cell lung cancer (SCLC). The survival of the patients suffering from lung cancer mainly depends upon the early diagnosis and efficient surgical removal of the tumor tissues. Among the different treatments, chemotherapy is the most recommended therapy to treat lung cancer. However, the major limitation of conventional chemotherapy is related to the presence of inefficient drugs at the target site, which ultimately compromises the therapeutic efficacy [4]. To reduce this problem, repeated administration of systemic chemotherapy at higher concentrations is required, which is allied with dose-related systemic toxicities. Moreover, in conventional therapies the uptake of the cytotoxic agents by the healthy cells can increase the chances of cytotoxicity in these normal cells. Hence, novel approaches are required for both the diagnosis and treatment of lung cancers [5].

Due to numerous limitations associated with these conventional methods, several researchers have exploited nanotechnology-based approaches for the efficient diagnosis and delivery of therapeutic agents [6]. Among various nanoparticle-mediated drug delivery systems, the most frequently used ones for lung cancer treatment include polymeric nanoparticles [7,8], liposomes [9,10], bionanoparticles [11,12] and metallic nanoparticles [13,14]. These nanoparticles have been very effective due to their small size, large surface area, high biocompatibility and reduced renal clearance. Although the use of nanoparticles has shown several advantages [15], their site-specific delivery is still a problem for which passive and active-targeting approaches are necessary [16,17].

The passive targeting approach utilizes the exploitation of the enhanced permeability and retention (EPR) effect. In many disease conditions, including lung cancers, the endothelial lining of the blood vessels exhibits higher permeability than in normal conditions [17,18]. The presence of this leaky vasculature allows the higher permeation of the nanoparticles into the target site [19]. Moreover, the lack of a normal lymphatic drainage system in the tumor site contributes to higher levels of retention of the nanoparticles. However, this idiosyncratic property cannot be applied to low molecular weight drugs which have a small residence time and rapid excretion from the tumorous cells. Low molecular weight drugs can be encapsulated in unionized drug carriers to improve their pharmacokinetics (elongated systematic circulation), increasing tumor selectivity and lowering side effects. This phenomenon of tumor-targeting is called "passive" and depends upon the properties of the carrier molecule (its molecular weight and residence time) and the tumor anatomy (vascularity, porosity, etc.), but does not have any ligands for specific cells' binding sites. The EPR effect provides a 20–30% higher concentration of the drug targeted delivery of the tumorous site compared to normal body tissues [20,21].

EPR effect is extremely dependent on the intrinsic pathways of tumor cells' growth and it is controlled specifically by the rate of angiogenesis and lymphangiogenesis, the rate of perivascular tumor development, stromal thickness response and the intratumor pressure. All these elements, along with the physicochemical properties of nanoparticles, can influence the efficiency of the drug's targeted delivery [22]. However, the extrusion properties of the newly formed tumor's blood vessels have an impact on the nanomedicine impregnation; it causes an increase in the interstitial pressure, which may hinder the retention of the drug carriers in the tumor tissues. Furthermore, due to the imbalance between the pro- and antiangiogenetic signaling in different points of the tumorous tissues, the blood vessels are deviant with enlarged, curvy and saccular pathways, unorganized interconnection processes and branching. This miscellaneous blood circulation causes an irregular growth of the tumor cells and those cells surrounding the blood vessel grow rapidly compared to those that are far away, because of low oxygen and nutrition supply. This explains why the outer sites of large tumorous tissues have less blood supply (i.e., 1–2 cm in diameter in mice) and why is most often difficult for nanomedicines to reach the cores of tumor cells. Although the interstitial pressure is high in the inner portion of the tumor, the extrusion rate is unexpectedly small. This pattern was observed in some different types of murine and human tumor cells. The increased interstitial pressure does not only hinder the drug supply to the core tumorous tissues but also retards the growth of new blood vessels. This causes a higher blood supply to flow towards the tumor cells' periphery, indicating that there is the possibility of modifying the EPR effect chemically or mechanically to improve the growth of the blood vessels for the retention of the drug-loaded nanocarriers. It is worth mentioning here that some types of EPR enhancers like bradykinin (kinin), nitric acid, peroxynitrite, prostaglandins, etc. may cause hypertension that could enhance tumor extrusion [23].

To further improve the targeted delivery of the imaging modalities and therapeutic agents against lung cancer, many researchers have also exploited the receptor-mediated delivery of theranostics [24]. Several receptors are overexpressed in lung cancer, like oxytocin, vasopressin, chemokine, epidermal growth factor, bradykinins, bombsein, folate and tyrosine receptors. The majority of the lung cancer receptors are categorized as G-protein coupled receptors. These receptors have a potential role in the formation, progression and metastasis of lung cancer and are involved in angiogenesis process during tumor development and also during the progression of the cancer to the nearby organs [25]. The overexpression of several kinds of receptors in lung cancer has been exploited by researchers for the site-specific delivery of theranostics. As compared with the passive targeted approach, a higher amount of the drug can be made to reach the target site through active targeted delivery of the imaging modalities and therapeutic agents.

Active targeting is compulsory for the proper distribution of drugs, genes and theranostics to the action site so the therapeutic effect on normal body tissue can be avoided. By using active targeting, a sufficient amount of drug is placed at the tumor site increasing the drug efficiency by many folds. Thus, active targeting nanosystems are more efficient than passive targeting ones. Active targeting is possible exclusively when the nanocarriers are enriched with ligands that are specific for the overexpressed receptors in lung cancers [26]. This phenomenon enhances the binding capacity of the drug and imaging modalities to the tumor tissues and thus increases the drug entrapment capacity at the tumor site. Hundreds of ligands and antibodies have been discovered against the abovementioned receptors and are exploited for targeted delivery of the drug cargoes to the target site. A strong ligand/receptor binding affinity serves as role model to promote active binding technology. This can improve the targeted delivery of theranostics and therapeutic agents on the one hand and overcome the problems allied with conventional approaches on the other hand [26]. The visual illustration of various nanotechnology-based theranostic delivery approaches are shown in Figure 1.

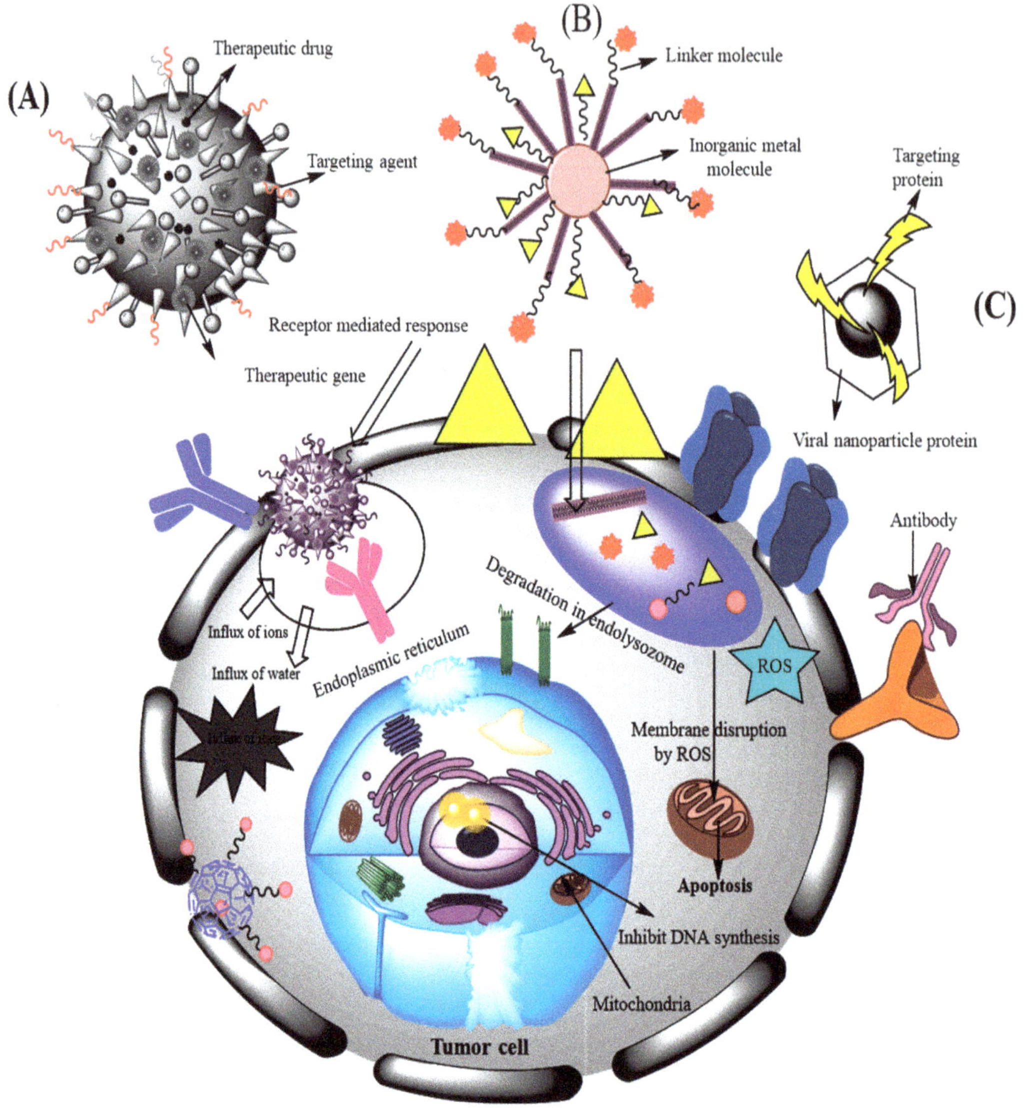

Figure 1. Schematic presentation of various methods for delivery of therapeutic agents against lung cancer including (**A**) polymeric nanoparticle-based approach, (**B**) metallic nanoparticle-based approach and (**C**) bioparticle-based approach.

This review first summarizes the current knowledge of the pathways implicated in the different types of lung cancer with an emphasis on the clinical implications of the increasing number of actionable molecular targets. The utilization of different targeting approaches to combat the toxicity of the chemotherapeutic agents is discussed here. The mechanism through which this targeted delivery is attained is also described. In this context, this review could attract the interest of medical scientists who are involved in biological systems. The second specific focus of this review is on the role of the surface-modified and hybrid nanomaterials as drug delivery systems in combating lung cancer. This spotlight was achieved through citing most the recent and representative examples targeted at lung cancer treatment. From this perspective, it could be highly interesting for material scientists.

The fields of biology and material science are traditionally rather separated, as much as they naturally rely on the very same basic principles. Through the novel cross-disciplinary focus of this review, we attempt to overcome this gap and create a more synergetic perspective on both areas, which will be highly beneficial for the scientific community given the plethora of discussions and discoveries that can be envisaged.

2. Pathways for Targeting Lung Cancer

Lung cancer is histologically classified into NSCLC and SCLC. The complex interplay between pathological changes and oncogenic mutations alters the signaling of multiple pathways and the expression of chemokines and various receptors. In turn, a modified tumor microenvironment facilitates the growth, proliferation, angiogenesis, metastasis and survival of the cancer cells. Traditional treatment strategies for the lung cancer include chemotherapy, radiotherapy and surgical excision. However, conventional chemotherapeutic agents have compromised therapeutic efficacy owing to pharmacokinetic issues, solubility problems and nonspecific action in normal cells with resultant toxicities. Moreover, high drug doses, tumor-associated alteration of pathways and subsequent treatment with multiple therapies will contribute to the occurrence of tumor resistance against chemotherapeutic agents [27] Therefore, the focus is now laid on the suppression of upregulated pathways including EGFR, RAS-RAF-MEK-ERK/MAPK, JAK-STAT, PI3K/AKT/mTOR through newly designed, specifically targeted small molecule inhibitors and antibodies (Figure 2). For instance, specific EGFR inhibitor (erlotinib) and PI3K/AKT/mTOR inhibitor (everolimius) replaced the first-line chemotherapy [28]. The most common genetic mutations in the lung cancer, along with their mode of aberration and the associated small molecule inhibitors to target specific pathways, are mentioned in Table 1. Nevertheless, the small-molecule-mediated targeted therapy is relatively successful and increases survival rates but is prone to therapeutic failure because of cancer relapse, and increased drug resistances due to targeting site mutations [29].

Hence, developing a highly targeted drug delivery system for specific action into the tumorous cells at an optimal dose is of great necessity. Broadly, lung cancer can be targeted through either passive or active targeting mechanisms or both. Passive drug delivery follows a certain principle to be deposited into the lung tissues under the EPR effect. EPR is attributed to leaky vasculature and deteriorative epithelial integrity that allows residence and accumulation of small sized particles into the lung tumorous tissue [30], which act either as a carrier to deliver the drug or act directly as a therapeutic moiety. In passive targeting, particle size is the main determinant for distribution and deposition in the lungs. For instance, large particles around >5 μm have fewer chances to concentrate and are mostly exhaled out of the lungs. Particles in the range of 1–5 μm are phagocytosed by the alveolar macrophages and particles with size <1 μm could be deposited in the alveolar cells with minimal clearance by the immune cells [31].

To achieve improved tumor-specific targeting and to avoid possible threats with dislocation and clearance of passively targeted delivery carriers, active targeting of overly expressed onco-receptors with specific ligands brings better outcomes [30]. The inhibition of overexpressed receptor functions through specifically targeting moieties modulates the expression of cancer projectors and improves drug action in the tumor-specific lung tissues. Various overexpressed receptors in the tumor microenvironment include EGFR, TFR, FR and CD44 receptor [32]. Tumor receptors and tumor-associated immune cells have a role in cancer growth, proliferation, metastasis and angiogenesis. Therefore, receptor-mediated targeting and immune cell targeting alter the onco-proteins' expression and inhibit oncogenic pathways to stop cancer growth and progression.

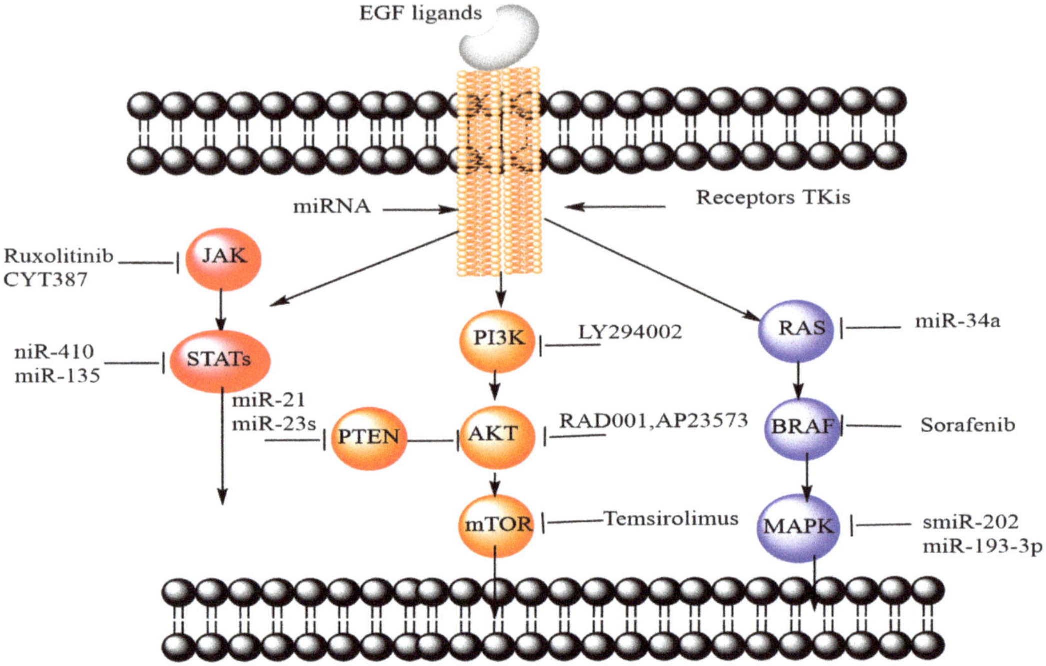

Figure 2. Oncogenic signaling pathways and drugs targeting abnormal signaling of EGFR, VEGFR, PI3/AKT/mTOR, RAS/BRAF/MAK, JAK/STAT pathways. Reproduced from the reference [27].

Table 1. Most common mutations in lung cancers and their relevant mechanisms.

Oncogene	Aberration	Activation Mechanism	Type of Lung Tumor	Targeted Drug Inhibitors	References
EGFR	Gatekeeper or oncogene mutation/ Amplification	Ligand binding → Activation of tyrosine kinase → phosphorylation of EGFR	NSCLC, ADC	Erlotinib, Gefitinib, Cetuximab,	[33–35]
EML/ALK	Fusion	Fusion of amino terminal of EML4 to intracellular kinase → ALK tyrosine kinase receptor rearrangement leads to activation	NSCLC, ADC	Lorlatinib, ensartinib, crizotinib, alectinib	[34,36,37]
BRAF	Mutation/fusion/kinase duplication	Autophosphorylation of kinase loop and MEK protein binding	NSCLC, ADC	Dabrafenib, Vemurafenib	[34,38]
PI3K	Modified/Activated	PIP2 and PIP3 phosphorylation → placement of serine threonine kinase AKT into membrane → PI3K phosphorylation	NSCLC, SCLC	LY294002, wortmannin	[39–41]

Table 1. *Cont.*

Oncogene	Aberration	Activation Mechanism	Type of Lung Tumor	Targeted Drug Inhibitors	References
mTOR	Activated	PIP2 and PIP3 phosphorylation → placement of serine threonine kinase AKT into membrane → mTOR phosphorylation	NSCLC, SCLC	Ridaforolimus, Rapamycin, sirolimus	[39,40]
RAS	Mutation	Conversion of GDP to GTP to activate G-protein (RAS) receptor	NSCLS, ADC	Tipifarinib, Lonafarinib, salirasib, sorafenib	[34,42]
p53	Mutation/Deletion	Inactivating of missense gene mutations	ADC, SCLC	Advexin (adenoviral vector)	[33,43,44]
MEK	Activated	RAS activation	NSCLS, ADC	Selumetinib, sorafenib, trametinib	[45,46]
c-KIT	Overexpression	Regulatory and functional c-KIT mutations → activation of protein kinase	SCLC	Imatinib, STI-571 (Gleevec)	[47,48]
VEGF	Overexpression	HIF-1 or EGR-1 upregulation → VEGF expression	SCLC, NSCLS	Bevacizumab	[33,49,50]
ROS1	Rearrangement	Autophosphorylation	NSCLS	Crizotinib	[51,52]

Epidermal growth factor receptor (EGFR); small cell lung cancer (SCLC); nonsmall cell lung cancer (NSCLC); adenocarcinoma (ADC); phosphatidylinositide-3 kinase (PI3K); hypoxia-inducible factor-1 (HIF-1); early growth response-1 (EGR-1); guanine diphosphate (GDP); guanine triphosphate (GTP); vascular endothelial growth factor (VEGF); echinoderm microtubule associated proteinlike-4 (EML4); phosphatidylinositol 4,5-bisphosphate (PIP2); phosphatidylinositol 3,4,5-bisphosphate (PIP3).

3. Onco-receptor Targets in Lung Tumors and Vasculature

3.1. Epidermal Growth Factor Receptor (EGFR)

The EGFR is a cell surface peptide receptor from the ErbB family of tyrosine kinase. It consists of the extracellular region with two homologous ligand-binding domains and two cysteine-rich domains, a single slanging transmembrane domain and an intracellular region comprising juxtamembrane, a tyrosine kinase domain and a regulatory region [53]. EGFR regulates growth, differentiation and migration of the alveolar and bronchial epithelial cells under normal conditions, while overfunctioning in cancer facilitates the proliferation, metastasis, and invasion of lung cancer cells [54]. EGFR is among the highly expressed onco-receptors in 85% of NSCLC, with negligible involvement in SCLC [55]. Various monoclonal antibodies (panitumumab, cetuximab) and tyrosine kinase inhibitors (erlotinib, gefitinib, lapatinib) are used to target EGFR to treat lung cancer [55]. Furthermore, anti-sense oligonucleotides, affibodies, peptides, and nanobodies worked to inhibit EGFR [56]. Recently, it has been observed that ligand anchored nanocarriers specifically bind to extracellular domains of EGFR to release the drugs intracellularly for the tumor-specific inhibition of the signaling pathway. Under this approach, biotinylated-EGF ligand-bound gelatin nanocarriers have delivered increased concentrations of cisplatin to the lung cancer cells and significantly reduced tumor volume via inhalation route [57]. Similarly, DNA aptamer conjugated chitosan-liposome complexes have delivered erlotinib specifically to the lung cancer cells via EGFR [58]. Additionally, monoclonal antibody linked polymeric nanoparticles have shown promising results against acquired EGFR-kinase resistance in cancer cell lines and could be designed to suppress EGFR resistant pathways in the lung tumor [59]. Ligand-bound nanocarriers favor site-specific tyrosine kinase inhibitors or monoclonal antibodies' delivery to the lung cancer cells, reduce off-site toxicities and endosomal clearance, and improve therapeutic efficacy with sustained drug release rate.

3.2. Transferrin Receptor (TFR)

Transferrin (TF) is a nonheme glycoprotein (~180 kDa), mainly responsible for iron (ferric ions) transport in the body [60]. Therefore, TFR (CD71) is expressed by the normal epithelial and immune cells. In tumors, the overexpression of TFR facilitates fast iron transport to accomplish the nutritional demand of the cancer cells. The expression of TFR in cancer cells is 10-fold higher than the expression in normal cells [54]. Overexpressed TFR can be targeted by the ligands including TF, ferritin and anti-TFR antibody, thus improving the tumor targeting efficiency of the carrier system. TFR is highly upregulated in lung cancer; about 88% of NSCLC cases have elevated TFR-1 levels [61]. In one study, the blocking of TFR through the anti-TFR antibody significantly retarded the cell proliferation of the lung adenocarcinoma cell lines [62]. Furthermore, TF conjugated doxorubicin (DOX) liposomes increased cellular internalization in A549 lung cancer cells compared to alveolar type *I* (AT*I*) and alveolar type *II* (AT*II*) cells [63]. Similarly, antibodies and peptides targeted TFR and inhibited tumor growth or induced apoptosis of the tumor cells [64].

3.3. α. vβ3 Integrin Receptor

Integrins belong to the transmembrane heterodimeric glycoproteins family, consisting of the 18 α and 8 β subunits [65]. Integrins are expressed in multiple forms in many tumor-associated cell types. In lung cancer, the integrins αv, α5, β1, β3 and β5 have been demonstrated to develop the survival and metastasis of cancer cells [66]. The role of integrins encompasses cell–matrix adhesion, the maintenance of cellular morphology, differentiation and proliferation [54]. About 82% of NSCLC cases have higher integrin expression, while only 13% of SCLC expressed integrins [67]. The widespread functions of integrins in lung cancer suggest that their inhibition could be beneficial in tumor targeting and therapy. It was demonstrated that the inhibition of the αvβ3 and αvβ5 integrins with targeted ligands can block the endothelial cell angiogenesis and tumor metastasis [66]. In this context, arginylglycylaspartic acid (RGD) peptide has the potential to target αvβ3 integrin, thus facilitating drug delivery to the lung cancer. In one study, RGD anchored poly(lactide-co-glycolide) (PLGA)-chitosan nanocarriers successfully delivered paclitaxel (PTX) specifically to lung cancer, while normal human bronchial epithelial cells with poor integrin expression had negligible cytotoxic effects of PTX [68]. Furthermore, cyclic peptide anchored formulation elevated the localized drug concentration and suppressed the tumor cells in the subcutaneous and orthotopic A549 xenograft mice models as compared to the free drug controls [69].

3.4. Folate Receptors (FRs)

FRs are from a family of glycoproteins (35–40 kDa) having a strong binding affinity for folic acid (FA). FRs are differentiated into four isoforms including FRα, FRβ, FRγ and FRδ [70]. Normal human cells have a very low content of FRs, whereas FRs are overexpressed in a variety of tumor cells—the first two isoforms (FRα, FRβ) are the most common [70,71]. In NSCLC, FRα is overly expressed especially in adenocarcinoma [54]. Therefore, FA or FR monoclonal antibodies could serve as a ligand to target lung cancer. Folate can be conjugated to chemotherapeutic agents, microparticles, nanocarriers, lipidic systems and oligonucleotides to directly target FR-positive tumor cells. Folate-PEG-modified cytochrome **c** nanomicelles have demonstrated selective targeting and internalization by FR expressed on the HeLa cells compared to FR negative cell lines [72]. Similarly, DOX and small interfering RNA (siRNA) were loaded into folate-biotin conjugated starch nanoparticles for codelivery into human lung cancer cells (A549). Folate-mediated codelivery has shown enhanced cytotoxicity and reduced proliferation of the A549 cells. The cytotoxicity was competitively inhibited in the presence of free folate; further, the expression of insulinlike growth factor 1 receptor (IGF1R) proteins was decreased through the treatment [73].

3.5. Cluster of Differentiation 44 (CD44)

CD44 is a cell-surface based glycoprotein receptor with a specific affinity for hyaluronic acid (HA). The binding of HA to the receptor regulates cell adhesion and the differentiation and migration of the normal cells [74]. In tumors, CD44 has the important functions of cell adhesion, growth, proliferation, metastasis and induction of the cancer cell resistance [74,75]. CD44 is highly upregulated in squamous cell metaplasia and NSCLC [76] and is involved in metastasis of NSCLC to the lymph node [77]. HA, as an anionic glycosaminoglycan and a polymeric ligand, can be anchored to the surface of the particles or itself is able to self-assemble to target the lung cancer [78,79]. For instance, HA anchored polyethyleneimine-PEG nanoparticles specifically delivered siRNA to lung cancer cells [78]. Furthermore, enzyme hyaluronidase-1 expressed heavily in the malignant tumors degraded HA, thus facilitating drug release from HA in the target cancer cells [80].

3.6. Other Onco-Receptors

Several other receptors are heavily expressed in the lung tumor microenvironment including luteinizing hormone-releasing hormone (LHRH) receptors [81], chemotactic chemokines receptor 4 (CXCR4) [82], fibroblast growth factor receptor [83], tyrosine kinase AXL receptor [84], vascular endothelial growth factor receptor (VEGFR) [85], death receptor/TNF-related apoptosis-inducing ligand-receptor (DR4/TRAIL-R1) [86], β2-adrenergic receptors (β2-AR) [87] and lectin receptors [88]. Targeting these receptors through specific ligands can inhibit lung cancer survival, growth and metastasis.

4. Extracellular Nanovesicles in Targeting Lung Cancer

The concept of applying nanoparticles for lung cancer targeting shares a lot of similarities with the function of extracellular vesicles (EVs). In this regard, we briefly review these particles in this section. EVs are cell-derived, membrane-bound particles known to mediate intercellular signaling and are sensitive in organ-specific metastasis. Depending on the biogenesis pathways or their subcellular origin and size, EVs are also referred to as apoptotic bodies, microvesicles or exosomes [89–91]. EVs confined from distinct body fluids transport immune response-related and immune-modulatory molecules. These molecules include proteins, lipids, and nucleic acids. Recent studies considered EVs as one of the main components in the tumor microenvironment. In the tumor microenvironment, the EVs are able to transport the biomolecules to the less malignant cells. As the result, the less malignant cells receiving the EVs may continue to show increased metastatic and migratory behavior [92].

Integrin receptors are enriched in small EVs and are major players in mediating EV functions. For example, αvβ3 integrin is upregulated during cancer progression and is known to account for the migration of cancer cells. These nanovesicles' signaling is capable of modifying the tumor cell's structure, characteristics and functionality, such as overcoming drug resistance [93,94]. In their study, Hoshino et al. demonstrated that the tumor-derived lung-tropic EVs carry integrins α6β1 and α6β4, which are favorably taken up by lung fibroblasts and surfactant protein C-positive epithelial cells. The authors demonstrated that the incorporation of EVs by lung resident cells enhanced the expression of the proinflammatory gene S100 and promoted the lung metastasis [95].

From the therapeutic perspective, EVs are novel drug delivery systems and have more biosafety and biocompatibility characteristics than other synthetic surface functionalized or hybrid nanoparticles. In this context, EVs can be divided into unmodified and modified EVs [96]. Similar to nonfunctionalized nanoparticles, unmodified EVs have shown less efficacy in various performed studies. Therefore, scientists are now developing modified EVs through the introduction of therapeutic molecules into EVs or modifying the surface components of EVs to enhance their efficacies in terms of tissue targeting and site specificity [97]. For example, Nakase et al. modified EVs with octaarginine peptide, which resulted in enhanced cellular EV uptake via the active induction of macropinocytosis without cytotoxicity. Additionally, the increased accumulation of EVs at the targeting site showed greater therapeutic effect [98]. Recent studies on EV-mediated lung cancer targeting at a specific site highlighted that there are many limitations involved in the modification of these nanovesicles, as the strategies adopted for modification may damage the EV membrane and consequently compromise the therapeutic efficacy of the EVs [99–103]. To overcome these limitations, surface modification of the synthetic nanoparticles has demonstrated more promising results. For instance in a recent report, $\alpha 3\beta 1$ integrins were targeted in NSCLC through cyclic peptide linked polymersome containing docetaxel. The results demonstrated better cellular uptake of cyclic peptide anchored formulation by A549 human lung cancer cells than by free docetaxel (DTX) and nontargeted polymersome.

5. Immunological Targets in Lung Cancer

Since the tumor is associated with pathophysiological, cellular and biochemical alterations, several immune cells like T-lymphocytes, macrophages, natural killer cells, B cells, MDSCs and dendritic cells infiltrate the lung tumor microenvironment. Hence, TAMs, MDSCs and regulator T-cells can be targeted through different ligands and strategies to modulate the tumor activity and reduce tumor progression [104].

5.1. *Tumor-Associated Macrophages (TAMs)*

Traditionally, activated macrophages of different phenotypes have commonly been categorized as M1 and M2 macrophages. M1 macrophages are activated through the classical pathway and are involved in proinflammatory response, while M2 macrophages are alternatively activated and associated with anti-inflammatory action. At first, macrophages polarize to M1 to assist the host immune response against an antigen, then they attain M2 phenotype to repair the damaged tissues. Macrophages linked with tumors are known as TAMs and are classified into two phenotypes—M1 and M2 (M1 type TAMs suppress cancer progression, while M2 type TAMs promote it). TAMs are characterized by increased M2/M1 ratio and play a crucial role in tumor progression, metastasis, matrix remodeling, angiogenesis and tumor resistance [105,106]. TAMs produce cytokines, growth factors like epithelial growth factor, matrix metalloproteinase-9, angiopoietin, etc. to assist tumor development. Therefore, TAM targeting can bring benefits to treat lung cancer. TAMs can be targeted through different ways including the repolarization of M2 into M1 cells, the prevention of macrophage recruitment into the tumor or the direct termination of M2 cells [104]. Moreover, several receptors such as C-type lectin, CD44, FRs have been expressed on the surface of TAMs, which can be specifically targeted for tumor eradication [107–109]. C-type lectin receptors are Ca^{2+} dependent carbohydrate recognition proteins and have multiple types including mannose receptor, macrophage galactose-type lectin-C and dectin receptor. Hence, different carbohydrate moieties are used to target C-type lectin receptors like mannose, glucose, D-galactose, N-acetyl-D-glucosamine (NAG) and maltose [110]. Recently, various mannose receptor targeting strategies involving mannose anchored liposomes, solid lipid nanoparticles, polymeric nanocarriers, niosomes, dendrimers and quantum dots have been fabricated to modulate macrophage function in the tumor [110]. In this quest, biotin and mannose conjugated lipid-coated calcium zoledronate nanocarriers have shown higher internalization in both TAMs and cancer cells, restraining tumor growth, progression and angiogenesis [111].

5.2. *Myeloid-Derived Suppressor Cells (MDSCs)*

MDSCs are the heterogeneous immature population of cells, comprised of myeloid progenitor cells, immature macrophages, immature dendritic cells and immature granulocytes [112]. MDSCs release immunosuppressive cytokines to retard immune system action against the tumor and thereby facilitating the tumor progression [113]. Thus, MDSCs are the major target cells in cancer immunotherapy. Several strategies have been designed to suppress MDSCs function in lung cancer including:

(a) Differentiation of MDSCs into mature myeloid cells [113];
(b) Suppression of MDSCs amplification through inhibition of stem cell function [114], VEGF [115] or STAT3 pathway [116];
(c) Direct elimination of MDSCs by antibodies or chemotherapeutic drugs like gemcitabine [117];
(d) Attenuating MDSCs functioning [118];
(e) Inhibition of the immune checkpoint to restore the antitumor immune response [119].

Nowadays, nanocarriers employing different options to target MDSCs, promote MDSCs maturation and modulate their function to regress tumor progression and angiogenesis [120]. Furthermore, MDSCs cell membrane coated iron oxide magnetic nanoparticles successfully evaded the immune system, actively targeted cancer cells along with magnetic and photothermal-induced ablation of the cancer cells [121].

5.3. Regulators T-Cells

Infiltrated regulatory T-cells in the tumors have downregulated the activation and response of cytotoxic T-cells against lung cancer [122,123]. Regulatory T-cells play an integral role in tumor development; thus, they could be targeted to suppress the tumor. For instance, glucocorticoid-induced tumor necrosis factor receptor-related protein (GITR) ligand was linked to PLGA nanoparticles for active targeting of regulator T-cells. The complex nanosystem, together with photothermal and photodynamic therapy, has remarkably reduced the tumor growth and recurrence [124,125].

6. Delivery Routes of Nanoparticles in Targeting Lung Cancer

The delivery of nanomedicines to the lungs increases the sustained local drug concentration to treat lung cancer [32]. Most chemotherapeutics act on normal tissues due to their nontargeting nature, leading to adverse effects [126]. Therefore, targeted drug delivery requires a low dose, which results in fewer systemic side effects. Although drugs can be administered through oral, intravenous or inhalational routes, the research on oral drug delivery to lungs has not shown promising results, as only a limited amount of the drug molecules are delivered to the lung tumors [127]. Additionally, the majority of the anticancer formulations are used as intravenous dosage forms. However, as most of the chemotherapeutic agents for lung cancer treatment are hydrophobic in nature, high doses and/or surface modification are needed to improve their systemic bioavailability [128]. On the other hand, the inhalational route is the most attractive option due to lower side effects and high biodistribution [129]. A safe and effective mean of lung cancer theranostics are the chemotherapeutic agents formulated by nanotechnology-based carriers, which are a novel targeted drug delivery "inhalational nanomedicine" that can be administered through the inhalational route [130]. In this context, the easiest way of drug delivery is inhalation by aerosols to target the cancerous tissues of the lung. The differential accumulation of drug particles or aerosol droplets in different regions of the lungs depends on their sizes. Drugs can be formulated as solutes or particles in aerosol droplets of appropriate size and used in drug delivery [131].

Okamoto et al. formulated gene powders with chitosan as a nonviral vector and mannitol as a dry powder carrier to compare their gene expression and therapeutic adequacy to intravenous or intratracheal gene solutions in mice having pulmonary metastasis prepared by injecting CT26 cells. In both normal and tumorous tissues, the genes expressed by intratracheal powder were higher than the one expressed by intravenous or intratracheal solutions, indicating that therapeutic gene powders are efficient for lung cancer treatment [132]. In another study, Dames et al. revealed that the targeted delivery of aerosols to the affected lung tissue might improve therapeutic efficiency and reduce undesired side effects. The authors showed theoretically that targeted aerosol delivery with superparamagnetic iron oxide nanoparticles along with a target-directed magnetic gradient field can be achieved to treat localized lung disease [133]. Ngwa et al. examine the potential of nanoparticle drones (smart nanomaterials) in targeting lung cancer. They compared and assessed inhalation (air) versus the traditional intravenous routes of navigating physiological barriers using such drones. They concluded that the inhalation route might be more promising for targeting tumor cells with radiosensitizers and cannabinoids in terms of maximizing the damage to lung tumor cells while minimizing any collateral damage or side effects [134].

7. Surface Modification of Nanoparticles to Combat Toxicity in Lungs Cancer

One of the challenging factors in the delivery of drugs to the lungs is to understand the interactions of the nanoparticles with the biological systems. The chemotherapeutic agents in the form of NDDSs are cell nonspecific, resulting in the undesired attack of healthy cells (an important factor in the failure of conventional nanotechnology cancer therapy). This is the reason why still further advancements need to be carried out in the field of NDDSs. The fast clearance of nanoparticles decreases the efficiency of sustained drug delivery and their translocation might bring nanoparticles to undesired areas of the body causing toxicity. Due to the complex nature of nanoparticles, research studies have led to different views of the nanomaterials' safety [135,136]. The physical properties of nanoparticles, such as morphology, geometry, dimensions and surface charge, have been found to change their therapeutic effect. Rod-shaped particles are more toxic than spherical particles. Long fibers cause inflammation because they are less likely to be engulfed by macrophages, thus minimizing their elimination from the system [136]. Nanoparticles produce pulmonary toxicity by oxidative stress because of the production of reactive oxygen species within the biological system [137]. It is evident from a research study that cytotoxicity occurs due to the production of free radicals after exposure to 3.5 to 23.3 μg/mL cerium oxide (CeO_2) nanoparticles. It causes oxidative stress in the cells by reducing glutathione and α-tocopherol levels and elevating the production of malondialdehyde and lactate dehydrogenase, which are indicators of lipid peroxidation and cell membrane damage, respectively [138]. The accumulation of nanoparticles in the tissue due to slow clearance produces potential free radicals as well as the prevalence of numerous phagocytic cells in the organs of the reticuloendothelial system (RES) making the lungs the main targets of oxidative stress [139].

According to a research study, 15 nm and 46 nm silicon dioxide (SiO_2) nanoparticles significantly reduced cell viability in a dose-dependent and time-dependent manner in bronchoalveolar carcinoma-derived cells at 10–100 μg/mL dosage. Both types of SiO_2 nanoparticles have higher cytotoxicity than the positive control material (Min-U-Sil 5). The reactive oxygen species (ROS) generated by exposure to 15 nm SiO_2 nanoparticles produces oxidative stress in these cells as reflected by reduced glutathione levels and the elevated production of malondialdehyde and lactate dehydrogenase, indicative of lipid peroxidation and membrane damage [140].

Surface functionalized nanoparticles have received tremendous importance as drug carriers. The physicochemical or biological properties of the nanoparticles can be altered by modifying their surfaces with different functional groups through covalent or non-covalent bonding, such as the adsorption of biologically active molecules (i.e., proteins, surfactants, enzymes, antibodies or nucleic acids) [141]. Nanoparticles functionalized with biodegradable polymers could be evaluated as the best chemotherapeutic delivery system. The surface chemistry of these nanoparticles must be carefully controlled as it is the shell of the nanoparticles that is in contact with body organs and fluids. As an example, nanoparticles have been coated with hydrophilic polymers or functionalized with ligands or proteins to enhance their circulation time or to achieve site specific delivery, respectively [142]. In another example, it was shown that the coating of nanoparticles with polymers could reduce their toxicity by changing their half-life distribution, disposition, stimuli reactivity and therapeutic application [30]. The oily nature of the nanocapsule's core can accommodate high loadings of lipophilic anticancer drugs [143]. Moreover, magnetic nanoparticles functionalized with polymers, monoclonal antibodies, peptides, heparin, hormones or other biologics are very effective and highly specific for cell biology and cancer therapeutic applications [144]. Surface modification of the NDDSs allows the targeted delivery of therapeutic agent such as antibodies and ligands (i.e., TF, FA, lactoferrins, lectins and mannose derivatives) into the tumors [135]. Additionally, surface PEGylation (the process by which polyethylene glycol chains are attached to biological molecules) does not only enhance the colloidal stability of nanoparticles but also increases their accumulation at the tumor site and decreases opsonization [145]. It was shown that surface-modified nanoparticles with 1, 2 dipalmitoylphosphatidylcholine (DPPC) are less prone to phagocytosis. The presence of phospholipids inhibits the adsorption of opsonic proteins on the inhaled nanoparticles, allowing them to escape phagocytosis [146]. In one research study, multiwalled carbon nanotubes (MWCNTs) were functionalized with amine-terminated poly (amidoamine) (PAMAM) dendrimers modified with fluorescein isothiocyanate (FI) and FA. This modified system acted as both a drug targeted system and a pH-responsive system for delivering DOX into cancerous cells [147]. Meenach et al. used an advanced organic spray-drying method to manufacture inhalable lung surfactant-based carriers comprising synthetic phospholipids, DPPC and dipalmitoylphosphatidylglycerol (DPPG), loaded with PTX, for targeted pulmonary delivery as high-performing nanoparticulate dry powder inhalers [148]. Li et al. suggested that a tumor-targeted PEGylated LPD formulation (liposome-polycation-DNA complex) enhanced cellular uptake by specific receptor-mediated pathways. They showed that the targeted drug delivery system caused a strong gene-silencing mediated by RNAi through delivering siRNA to the tumor cells after intravenous administration [149,150]. Grabowski et al. described the cytotoxicity and inflammatory action of nanoparticles made of PLGA through in vitro analysis on A549 human lung epithelial cells. Three different neutral, positively or negatively charged PLGA nanoparticles (230 nm) were obtained by using different types of stabilizers (polyvinyl alcohol, chitosan, or Pluronic® F68). For comparison, polystyrene nanoparticles were used as nonbiodegradable polymeric nanoparticles and titanium dioxide (anatase and rutile) as inorganic nanoparticles. As the result, the PLGA-based and polystyrene nanoparticles were less toxic than or equally toxic to titanium dioxide nanoparticles. On the contrary, the inflammatory response measured by the release of interleukin 6 (IL-6), IL-8, monocyte chemoattractant protein-1 (MCP-1) and tumor necrosis factor α (TNF-α) cytokines was low for all nanoparticles [151]. The PLGA-based nanoparticles led to a higher inflammatory response, which was correlated with a higher uptake of these nanoparticles. The authors claimed that both the coating of the PLGA nanoparticles and the nature of the core play a key role in the cell response.

8. Comparison between Surface-Modified Nanoparticles with Other Targeting Methods in Lung Cancer Treatment

There are many ongoing contributions of nanoparticles in the field of targeting lung cancer. However, there are a few limitations that inhibit amplifying their applications, including low stability, greater immunogenicity, nonuniform distribution, increased rate of clearance, poor ability in encapsulating imaging and targeting agents and unspecified internalization (via passive delivery method) at the malignant site. The morphology of the lung itself is a big barrier for the optimal transformation of agents into it. Therefore, it is important to design nanoparticulate systems that could reduce these complications. The functionalized nanoparticles can be used as powerful theranostic tools to enhance the delivery of drugs to the malignant site [34]. Many studies have been carried out comparing the performance of functionalized and nonfunctionalized nanoparticles in lung cancer targeting. For example, Chung et al. developed and compared the PLGA nanoparticles ligated with heparin, chitosan and pluronic with nonconjugated PLGA nanoparticles. The viability tests for both normal and tumor cells showed the less cytotoxic effect of the nanoparticles. The in vitro cellular uptake of the nanoparticles for both chitosan and heparin functionalization showed the desired effects. The in vivo tumor model study exhibited that there was a positive but insufficient effect of chitosan decorated nanoparticles, although it showed enhanced accumulation that was almost 2.4-fold higher than that of the control nanoparticles. The results concluded that the surface functionalization of the PLGA nanoparticles with chitosan and heparin may be an efficient strategy for the enhanced tumor theranostics [152]. Patil et al. performed a study with the aim of achieving targeted delivery through the single-step surface functionalization of nanoparticles with a tissue recognition ligand. They used biotin and a folic acid ligand to functionalize the PLA-PEG nanoparticles. The surface modification was confirmed through NMR, transmission electron microscopy and tumor cell uptake study. In comparison to the bare nanoparticles, the functionalized nanoparticles showed more precise and efficient results with greater binding affinity at the delivery site. The in vivo study result of the surface-modified PTX-loaded PLA-PEG nanoparticles showed an enhanced efficacy in comparison to the nonmodified nanoparticles [153]. The same authors in another study developed biotin functionalized PLGA nanoparticles encapsulating a combination of PTX and P-glycoprotein (P-gp) inhibitor tariquidar to overcome tumor drug resistance. The dual agent nanoparticles showed higher cell inhibition in the cell line study in comparison to only PTX-loaded ones. Additionally, performing in vivo studies in a mouse model, these nanoparticles demonstrated considerably enhanced inhibition of tumor growth. The authors concluded that these dual agent nanoparticles could be applied as an efficient system to overcome tumor drug resistance [154]. In another report, Xia et al. developed DOX-loaded selenium (Se) nanoparticles and functionalized them with cyclic peptide (Arg–Gly–Asp–$_D$-Phe–Cys [RGDfC]) to fabricate tumor targeting delivery. The aim of the study was to improve the antitumor efficacy of DOX in NSCLC. This nanodrug carrier displayed an efficient cellular uptake in A549 cells and entered the A549 cells mainly by clathrin-mediated endocytosis. Interestingly, comparing active targeting with the passive targeting delivery system, the authors concluded that the RGDfC functionalized DOX-loaded Se nanoparticles provide a promising approach for lung carcinoma therapy [155]. Perepelyuk et al. studied the therapeutic efficacy and in vivo efficacy of mucin1-aptamer-modified miRNA-29-loaded hybrid nanoparticles in a lung tumor model. The results displayed that the presence of MUC1-aptamer conjugates increase the delivery of miRNA-29b to the tumor cells. Moreover, the downregulation of DNMT3B by MAFMILHNs resulted in the inhibition of tumor growth in a mouse model [156]. Table 2 presents a summary of the recent studies on comparison between surface-modified and unmodified nanoparticles in lung cancer treatment.

Table 2. Summary of the recent studies on comparison between surface-modified and unmodified nanoparticles in lung cancer treatment.

Nanodrug Carrier Type	Encapsulated Drug	Ligand/Targeting Moiety	Outcomes	Reference
Cationic lipid nanosystems (CLNs)	Curcumin	Surface charged particle	Greater bioavailability pharmacokinetics inhibitory effect on cell growth and invasion, enhanced apoptosis in LL/2 cells, increased antitumor effect of curcumin loaded, CLNs in C57BL/6 J mice compared with control, reduced tumor volume and growth	[157]
Solid lipid nanoparticles (SLNPs)	Gemcitabine	Mannose	Reduced hemolysis due to the presence of cationic ammonium on the surface of SLNs, significant toxicity on A549 cells in vitro, greater uptake into A549 cells by receptor mediated endocytosis, enhanced concentration in lungs in in vivo studies	[158]

Table 2. *Cont.*

Nanodrug Carrier Type	Encapsulated Drug	Ligand/Targeting Moiety	Outcomes	Reference
Cationic liposomes	Vinblastine	Peptide nucleic acid (PNA)	Greater internalization of targeted liposome into LL/2 cells in vitro, inducing apoptosis in LL/2 cells, greater antitumor efficacy of PNA-modified vinblastine cationic liposome in tumor-bearing mice, increased survival rate of animals treated with PNA-modified liposomes	[159]
Polylactic acid (PLA)	Gemcitabine	Cetuximab	Greater uptake into A549 cells via EGFR mediated endocytosis, enhanced antiproliferative activity of targeted nanoparticles against lung cancer cells compared with nonmodified nanoparticles	[160]
Cationic liposome	Erlotinib/oxygen	Anti-EFGR aptamer	Greater cellular uptake, greater erlotinib resistance in vitro, inhibiting the tumor growth in xenograft model, accumulation of targeted liposomes at the site of tumor compared with other organs	[161]
Albumin self-assembly	Doxorubicin/ TRAIL protein	Not applicable (N/A)	Enhanced antiproliferative activity of doxorubicin and TRAIL protein on lung cancer (H226) cells, significant antitumor efficacy in BALB/c nu/nu mice having H226 cell induced tumor.	[162]
Multiwalled carbon nanotube (MWCNT)	Docetaxel/ curcumine-6	Transferrin	Greater uptake of targeted MWNT into A549 cells, cell cycle arrest in phase (sub-G1),significantly reduced lung toxicity of targeted MWNT.	[163]
pH sensitive liposomes	Afatinib	N/A	Enhanced stability of CL and PSL, induction of apoptosis in H-1975 cells	[164]
Silk fibroin	Gemcitabine	SP5–52 peptide	Increased potential in LL/2 cells targeting in both in vitro and in vivo studies, enhanced reduction in proliferation of tumor cells, greater accumulation of targeted nanoparticles at the site	[165]
Silica	10-Anthraquinone-2-carboxylic Acid (OCAq)/rose bengal (RB)	N/A	Enhanced efficacy of silica nanoparticles conjugated with dyes for photodynamic therapy, two folds phototoxicity on A549 cells by generating oxygen radicals	[166]
Thermally crosslinked supermagnetic iron oxide (TCL-SPION)	Cyanine/ Doxurubicin	N/A	Greater fluorescent intensity of TCL-SPION at tumor site compared to other tissues, greater accumulation of DOX encapsulated TCL-SPION at the tumor site.	[167]
Gold nanoparticles	N/A	N/A	For diagnosis of lung cancer by analyzing the volatile organic compounds in cancer patients	[168]
Polyamidoamine dendrimer	Cis-diamine platinum	Folate/HuR siRNA	Greater antiproliferative effect on H1299 cells by codelivery of anticancer drug and siRNA, enhanced toxicity of targeted formulation in comparison to nontargeted at tumor site	[169]

9. Different Types and Applications of Surface-Modified and Hybrid Nanoparticles for Targeting Lung Cancer

Nanoparticles are included in the drug delivery systems to overcome certain issues such as low solubility and permeability related to tumor targeting. The most significant advantages of nanoparticles are their excellent loading capacity and high surface to volume ratios. Various organic and inorganic nanomaterials have emerged as novel tools for cancer diagnosis and therapy due to their unique characteristics. In this review, based on the main structural moiety of nanoparticle we broadly divide them into three types: organic nanoparticles, inorganic nanoparticles and hybrid nanoparticles. The combinatorial therapeutic approach via hybrid nanoparticles is discussed in a separate subsection too.

9.1. Organic Nanoparticles

Organic nanoparticles can be defined as solid particles composed of organic compounds (mainly polymers, lipids or proteins). They have been widely studied for decades, presenting a large variety of materials and exciting applications in cancer therapies. There are many biopolymeric nanoparticles that are utilized in the drug delivery systems. For example, PLGA is a biodegradable copolymer approved by the US Food and Drug Administration (FDA) for use in distinct biological products. PLGA nanoparticles can be used to obtain extended and sustained delivery of therapeutic agents including protein, peptide, RNA, DNA and small molecules to their particular target sites [170,171]. As an example, Karra et al. developed cetuximab functionalized PLGA nanoparticles and loaded them with PTX. The results confirmed the in vitro targeting performance and enhanced the cellular internalization along with cytotoxicity of this targeted delivery system in lung cancer cells overexpressing EGFR. The intravenous administration of the nanoparticles to mice results in the considerable inhibition of tumor growth and the reduction of mortality rates. Pharmacokinetics studies results showed no increase in the aggregation of nanoparticles at the tumor tissue site. The authors concluded the promising potential of this system for enhanced efficacy against lung cancer [172]. In another report, Patil et al. compared YSA peptide functionalized and nonfunctionalized PLGA nanoparticles to improve delivery to bleomycin treated cultured endothelial cells in a bleomycin induced lung injury mouse model. When human umbilical vein endothelial cells (HUVEC) were treated with bleomycin, the 3 h uptake of both types of nanoparticles was increased up to 2-fold. The results showed that in mice the bleomycin injury led to 2.3 and 4.7 times increases in the lung concentrations of the nonfunctionalized and YSA-functionalized nanoparticles, respectively. The authors stated that PLGA nanoparticle delivery to cultured vascular endothelial cells and mouse lungs in vivo was higher directly after bleomycin treatment, with the delivery likely to be higher for YSA-functionalized nanoparticles [173].

Single chain technology is a new term developed in nanotechnology in order to broaden the functions of soft nano-objects through chain compaction. Using single chain technology, individual copolymer chains of different natures, compositions and molar masses have been folded intramolecularly to develop single chain nanoparticles (SCNPs). This leads to very small size polymer nanoparticles in the sub-20 nm size [174,175]. The folding is achieved by the self-assembly or crosslinking of functional groups on the precursor polymer, or rather moderated by the external cross-linker. There are several ways to develop SCNPs including dynamic and irreversible covalent crosslinking reactions such as cycloaddition. Moreover, there are huge number of SCNPs that have been introduced, from single and multiblock to star particles, hairpins and tadpole molecules. There are only a few examples present where a functionalized group has been incorporated into SCPNs [176]. However, these functionalized SCNPs still have not been used for lung cancer targeting. In this context, an insight was given by Benito et al. who evaluated the use of SCPNs based on poly(methacrylic acid) in targeting pancreatic adenocarcinoma. They functionalized SCPNs with somatostatin analogue PTR86 as a targeting moiety since these somatostatin receptors are overexpressed in pancreatic cancer. The imaging results showed a higher accumulation of targeted SCPNs in the tumor compared to the nontargeted nanoparticles, which was due to the enhanced retention in the tissues [177]. Later, Kröger et al. also reported the greater potential of these types of nanoparticles for cellular targeting [178,179].

Dendrimers are another class of polymers that are constructed by the stepwise addition of layers (generations) of molecules around a central core. This unique physicochemical properties of dendrimers enable a facile utilization of them as templates to funcionalize nanoparticles [180]. In this regard, a group of researchers reported greater penetration and higher stability of siRNA by implementation of surface-modified poly(propyleneimine) dendrimers. The siRNA nanoparticles were coated by a dithiol bearing cross linker that followed by a layer of PEG. In addition, a synthetic derivative of LHRH was linked at the end of the PEG polymer to conduct siRNA nanoparticles to the cancer cell. The developed system showed time- and concentration-dependent cellular uptake under in vitro conditions. It was proposed by the authors that this approach could be used for the in vivo systemic delivery of siRNA for efficient cancer therapy [181].

Solid lipid nanoparticles or lipid nanoparticles are nanoparticles composed of lipids as a matrix which are exceptionally biodegradable and biocompatible. They possess superior properties such as high drug payload, increased drug stability, large scale production and sterilization [182]. For instance, in one study Pooja et al. developed and evaluated TF conjugated and etoposide loaded solid lipid nanoparticles. The tissue distribution and pharmacokinetics were studied in Balb/c mice. The nanoparticles showed great anticancer activity of etoposide via antiproliferative assay and induced apoptosis in A549 cells. It was concluded that over expressed TF-receptors showed enhanced efficacy in NSCLC [183]. Liposomes are similar in design to lipid nanoparticles, but slightly different in composition and function. Riaz et al. developed the TF-7 surface functionalized liposomes loaded with quercetin (QR) for lung cancer therapy. These liposomes were evaluated for cellular uptake and in vitro cytotoxicity study and they exhibited higher cytotoxicity and S-phase cell cycle arrest. The in vivo study showed enhanced liposomes accumulation in the lungs and sustained release up to 96 h [184].

Considering that albumin has remarkable roles in human body, it can be used in the area of medicine and disease treatment. As an example, Yang et al. used hematoporphyrin (HP) functionalized albumin nanoparticles for cancer therapy. These nanoparticles further modified with gamma emitting nuclides (^{99m}Tc). HP-albumin nanoparticles showed improved accumulation in A549 and CT-26 cancer cell lines. The evaluation of the pharmacokinetics of ^{99m}Tc chelated HP-albumin nanoparticles via the scintigraphic imaging of rabbits resulted in acceptable imaging properties in the rabbit with a longer biological half-life compared to ^{99m}Tc-HP. The authors concluded these modified albumin nanoparticles could be applied as a diagnostic tool for cancer as well as the obvious application for photodynamic therapy [185].

9.2. Inorganic Nanoparticles

Inorganic nanoparticles including gold, silver, iron oxide and silica nanoparticles have been widely studied as therapeutic agents for cancer treatments in biomedical fields [186]. Among them, gold nanoparticles are attractive constituents for nanoparticle polymer hybrid materials as they support localized surface plasmon resonances, and the wavelength region of the surface plasmon resonance peak can be adjusted finely through the geometric parameters of the particles [187,188]. In one study, Heo et al. developed the gold nanoparticles surface-functionalized with PEG, biotin and rhodamine B and linked beta-cyclodextrin (β-CD). The specific interactions of these nanoparticles with cancer cells such as HeLa, A549 and MG63, as well as normal NIH3T3 cells, were evaluated. The authors observed that the modified nanoparticles were more effectively involved with the cancer cells. Confocal laser scanning microscopy (CLSM), fluorescence-activated cell-sorting (FACS) and cell viability analyses showed that the surface functionalized nanoparticles played a significant role in the diagnosis and treatment of the cancer cells, and could be used in theranostic agents [189]. Guo et al. developed a multifunctional nanocarrier encapsulated with methotrexate via electrostatic interaction between gold nanocluster conjugate chitosan and nucleolin targeting aptamer (AS1411). The in vivo study demonstrated that intravenous administration of nanodrug carrier systems into BALB/c mice caused the accumulation of methotrexate at the tumor site. The results suggested that the developed functionalized system can be applied for an effective delivery for anticancer agents and shows enhanced potential in clinical applications [190]. João Conde et al. fabricated the gold nanoparticles conjugated with siRNA/RGD and studied in a lung cancer murine model. The RGD treatment showed a significant downregulation followed by tumor growth inhibition and the increased survival of the tumor bearing transgenic mice. The results demonstrated that RGD gold nanoparticles stimulate the delivery by intratracheal application in mice that leads to the suppression of tumor cell proliferation. The enhanced targeted delivery of gold nanoparticles encapsulated with siRNA to cancer cells works towards effective silencing of the oncogene. The study showed gold nanoparticles stimulated the inflammatory and immune responses that can promote the therapeutic effect of the siRNA to reduce the tumor size at very low doses [191]. The schematic illustration of this study is described in Figure 3, which shows the enhanced efficacy of siRNA loaded into functionalized nanoparticles.

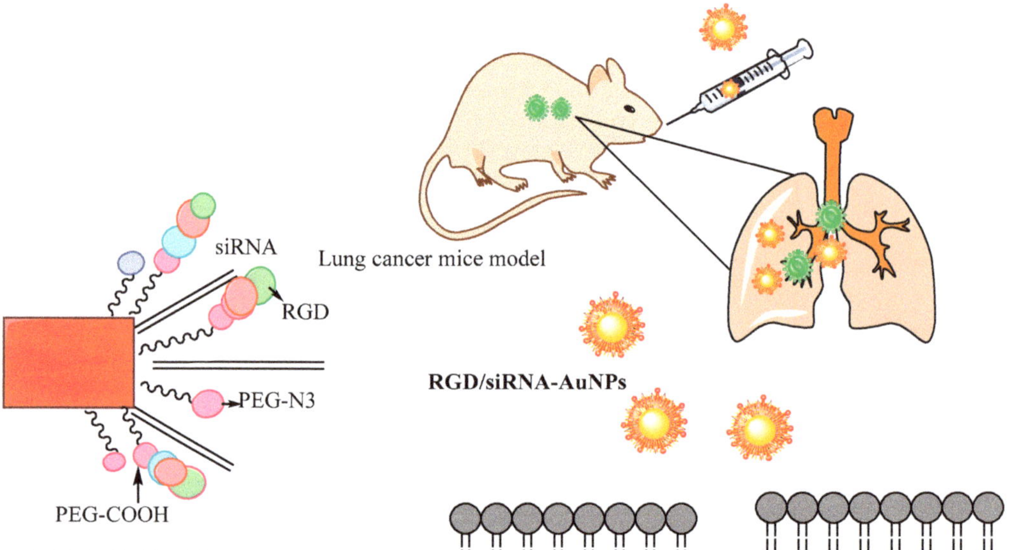

Figure 3. Inflammatory response and therapeutic siRNA silencing via RGD nanoparticles in a lung cancer mouse model.

Applications of silica or silicon dioxide (SiO_2) as another inorganic nanoparticle are broadly investigated in drug delivery. For example, Munaweera et al. prepared cisplatin and cisplatin/nitricoxide-loaded amine functionalized mesoporous silica nanoparticles for the treatment of lung cancer. The results demonstrated that for nonsmall lung cancer cell lines (i.e., H596 and A549), the toxicity of cisplatin/nitric oxide-loaded silica nanoparticles was higher than that of silica nanoparticles loaded with only cisplatin. The nitric oxide-activated sensitization of the tumor cell death, which showed that nitric oxide is a potential enhancer of platinum-based lung cancer therapy [192]. Another type of inorganic nanoparticles with biomedical applications is zirconium oxide (ZrO_2). In one study, ZrO_2 nanoparticles were coated with aminopropilsilane, tetraoxidecanoic acid or acrylic acid. The studied results showed dose-dependent signs of effectiveness. It was concluded that surface modifications of the ZrO_2 nanoparticles had very small effects on the inflammatory lungs of rats and mice but it had very clear efficacy in the allergic mouse model used. The results stated that the allergic mice are more responsive to exposure to surface-modified nanoparticles [193]. The unique properties of molybdenum disulfide (MoS_2) make it an attractive candidate for drug delivery applications [194]. In their study, Wei Zhang et al. developed the riboflavin 5′-monophosphate sodium salt functionalized 2D MoS_2 nanosheets prepared by the simple ultrasonication method, then they applied this nanocomposite having fine electrochemical redox activity as a platform to immobilize DNA probe. The results showed that the signal detection platform showed greater sensitivity with the limit of detection of 1.2×10^{-17} mol L^{-1} for PIK3CA gene from lung malignancy. The constructed biosensor was easy to achieve and could detect different pathogenic DNA without an intricate label process [195].

Magnetic nanoparticles can produce heat under the magnetic field and can also deliver drugs to the lung cancer site [196,197]. Among them, iron oxide nanoparticles are widely studied systems for biomedical applications [198,199]. In their study, Huang et al. reported the synergy effect of superparamagnetic iron oxide nanoparticles along with an anticancer drug (β-lapachone) for improved cancer therapy. The authors suggested that combination of superparamagnetic iron oxide nanoparticles with reactive oxygen species-producing drugs could conceivably enhance drug efficiency, thus presenting a synergistic strategy to integrate imaging and therapeutic functions in the discovery of theranostic nanomedicine [200]. In another study, dextran coated iron oxide nanoparticles were modified with the TAT peptide and they were used to improve the efficiency of the radiation. After performing the internalization study, it was revealed that TAT functionalized nanoparticles enhanced the generation of the reactive oxygen species in comparison to the nanoparticles without any surface modifications. These modified nanoparticles also affected the mitochondrial integrity of A549 cells in combination with the radiation, which resulted in a synergistic decrease in cell viability [201,202].

9.3. Hybrid Nanoparticles

In order to enhance the efficacy of the therapeutic regimen in lung cancer, it is necessary to develop new systems that can increase the survival rates. The development of hybrid nanoparticles (that could comprise both inorganic and organic structural moieties) in conjugation with other genes, biomolecules and other drugs are promising therapeutics systems for efficient targeting [203]. These types of nanoparticles are important for targeting the tumor site, for its early diagnosis and to measure the risk of malignancy in neighboring cells. These hybrid types are classified into diagnostics, therapeutic and theranostic nanoparticles. Figure 4 graphically shows the different types of hybrid nanoparticles that have been applied for lung cancer targeting.

In this regard, Sacko et al. studied anticancer effect of a combination therapy of miRNA-29b and genistein loaded in mucin-1 (MUC 1)-aptamer functionalized hybrid nanoparticles in NSCLC A549 cell line. This nanodrug carrier displayed a superior antiproliferative effect compared to individual genistein and miRNA-29b-loaded nanoparticles, thus, it can be considered a potential treatment modality for A549 cell line [204]. The same research group studied the pharmacokinetic response of novel antineurotensin receptor 1 monoclonal antibody (anti-NTSR1-mAb)-functionalized antimutant K-*ras* siRNA-loaded hybrid nanoparticles and compared it with that of naked siRNA formulation. As with the main findings, the plasma terminal half-life of the siRNA-loaded nanoparticle-delivered was 11 times higher than that of the naked siRNA formulation. In addition, high performance liquid chromatography (HPLC) analysis showed that the hybrid carrier system could protect the encapsulated siRNA against degradation in systemic circulation. The authors concluded that these hybrid nanoparticles can function as an effective nonviral vector for siRNA delivery for both experimental and clinical uses [205].

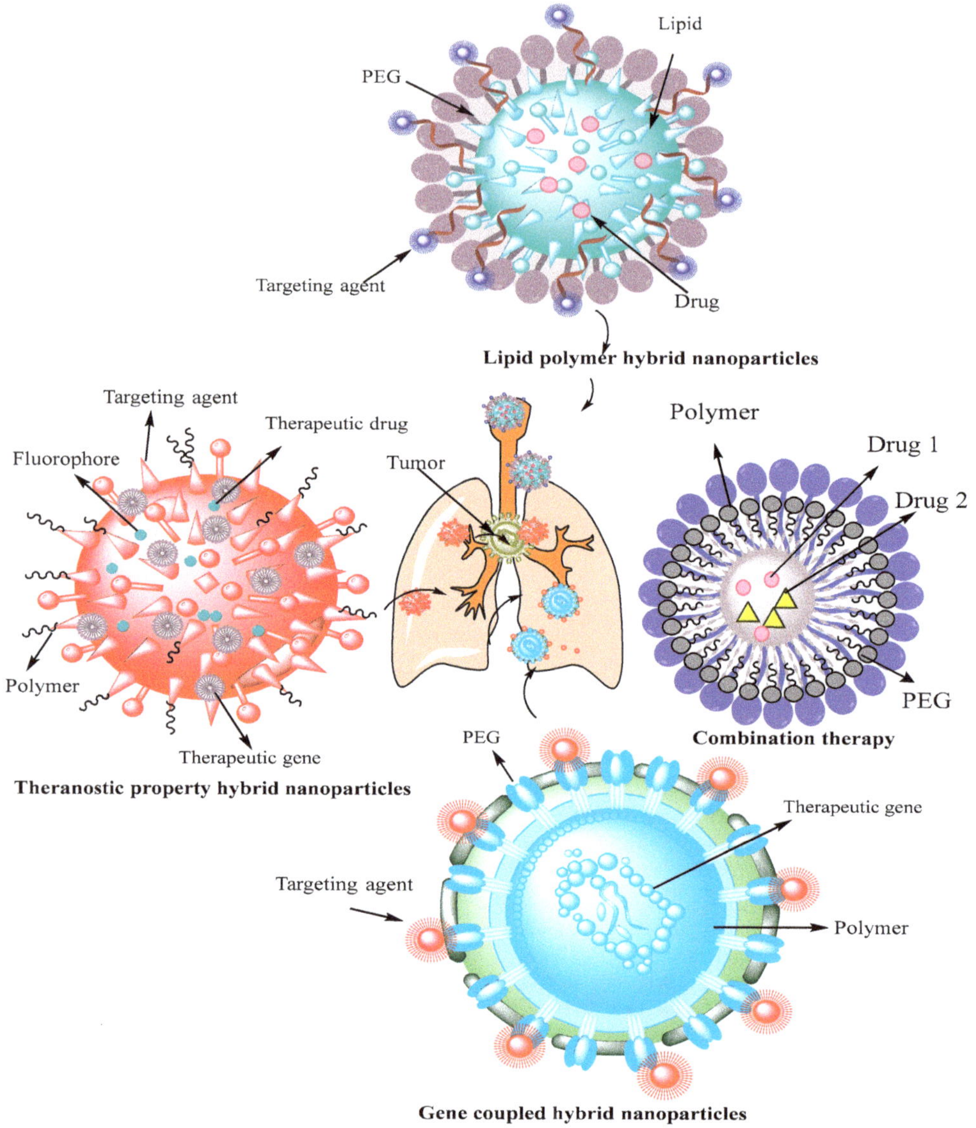

Figure 4. Schematic representations of different types of hybrid nanoparticles used for lung cancer targeting (i.e., theranostic hybrid nanoparticles, gene hybrid nanoparticles, combination therapy of hybrid nanoparticles and lipid polymer hybrid nanoparticles).

In another study, EGFR-targeted superaparamagnetic iron oxide nanoparticles (SPIONs) were conjugated to carboxy-terminated pluronic F127. The authors investigated the inhalation delivery of these nanoparticles as a potential approach for lung cancer treatment. As in the main findings, EGFR targeting enhanced tumor retention of SPIONs while minimizing systemic exposure. Additionally, magnetic hyperthermia using these nanoparticles resulted in a significant inhibition of in vivo tumor growth [206].

Another type of hybrid delivery system are lipid polymer nanoparticles or core shell lipid polymer nanoparticles, which combine the good biodegradability of polymeric nanoparticles with the excellent biomimetic characteristics of liposomes, and they are effective carrier systems for the delivery of anticancer drugs into the tumor site [207]. In this context, Bivash Mandal et al. showed that a hybrid system containing biodegradable polycaprolactone (as the core) and phospholipid-shell was able to deliver erlotinib into the lung cancer cells. Performing cell viability studies by this erlotinib-loaded hybrid system, a significant decrease in proliferation of A549 cells was observed, which affords this system a potential application to deliver erlotinib into lung cancer cells [208]. A similar study was carried out by Song et al. in which they showed the enhanced properties of EGFR-targeted lipid polymer hybrid nanoparticles in the codelivery of resveratrol and DTX. They developed this nanocarrier system by the conjugation of EGF and target the EGFR on the surface of the lung cancer cells in order to increase the endocytosis. Resveratrol (as an antioxidant) improved the production of reactive oxygen species through inducing cytotoxicity. The results exhibited the enhanced antitumor efficacy of resveratrol and DTX in both in vitro and in vivo studies [209]. Another related study reports the synthesis and characterization of lipid-coated poly D,L-lactic-co-glycolic acid nanoparticles that were modified with TF to deliver the DOX into A549 cells. These DOX-loaded hybrid nanoparticles exhibited higher cytotoxicity against lung cancer cells and showed an improved therapeutic effect in the lung cancer-bearing nude mice in comparison to their nontargeted counterparts. This finding marks this approach as an efficient targeted drug-delivery system for lung cancer therapy [210]. In another study, naturally occurring chitosan and hyaluronic acid were deposited on negatively charged hybrid solid lipid nanoparticles through layer-by-layer (LbL) assembly. Next, this hybrid system was loaded with DOX/dextran sulfate complex with the aim of tumor specific targeting. Employing this approach under in vivo studies, the DOX half-life was increased and its elimination rate was decreased compared to those measured for the uncoated solid lipid nanoparticles [211].

Another interesting type of nanocarriers in cancer therapy are hydrazine-based pH-sensitive nanoparticles. For example, Li et al. designed a dual-ligand lipid based nanoparticle system, in which TF conjugated PEG hydrazone nanoparticles were used for the codelivery of DTX and baicalein into A549 cells. Decorating the lipid nanoparticle with TF could internalize them into the cancer cells. Moreover, this hybrid nanocarrier achieved significant synergistic effects, the best tumor inhibition ability and the lowest systemic toxicity [212]. A recent progression in lung cancer therapy is gene therapy of lung cancer by applying siRNA hybrid nanoparticles [213]. Applying hybrid nanoparticles was useful to carefully transport the siRNA into the cytoplasm and cross the limitations of the traditional gene therapy [214]. For instance, the encapsulation of siRNA in calcium phosphate nanoparticles coated with DOPA (dioleoylphosphatydic acid) could target H460 lung cancer cells [215].

An overview of some other recent studies (that were not mentioned in this section) on developing nanoparticle-based delivery systems as a therapy against lung cancer is presented in Table 3.

Table 3. Summary of some recent studies on design of nanoparticle-based delivery approaches for therapy against lung cancer.

Nanoparticles Type	System Description and the Main Finding	Reference
Silver NPs [1] (AgNPs)	- Poly vinyl pyrrolidone coated AgNPs were used in this study. - After exposure to AgNPs, DNA damage induced by ROS was detected as an increase in bulky DNA adducts by ^{32}P postlabeling and these NPs were suggested as a mediator of ROS-induced genotoxicity.	[216]
Gold NPs (AuNPs)	- Glucose-bound AuNPs combined with radiation, can increase cytotoxicity on A549 cells not only by arresting the G2/M phase, but also by increasing apoptosis.	[217]

Table 3. *Cont.*

Nanoparticles Type	System Description and the Main Finding	Reference
Quantum dot (QD)	- The QD-pulsed dendritic cell vaccine was introduced as a new combination therapy to amplify antitumor immunity. - This combination boosts antigen-specific T-cell immunity and actively inhibits local tumor growth and tumor metastasis in vivo.	[218]
Liposome	- A liposomal curcumin dry powder inhaler for inhalation treatment of primary lung cancer was developed. - This liposomal system showed higher anticancer effects than the other medications regarding pathology and the expression of many cancers.	[219]
Graphene	- Graphene oxide/TiO_2/DOX loaded polymer composites were developed in the forms of nanofibers. - In the presence of magnetic field, these nanofibers showed higher proliferation inhibition effect on target lung cancer cells.	[220]
Carbon nanotubes (CNT)	- PEG-coated CNT nanodrugs were designed that improves the mitochondrial targeting of lung cancer cells. - This system increased the anticancer efficacy by increasing mitochondria accumulation rate of cytosol released anticancer nanodrugs.	[221]
Niosome	- A noisome-based formulation containing gemcitabine and cisplatin was presented for lung cancer treatment. - This system reduced cytotoxicity effects against both MRC5 and A549 comparing to with control (gemcitabine and cisplatin alone) after 72 h of treatment.	[222]
Solid lipid NPs (SLNPs)	- Sclareol-loaded SLNPs was formulated and tested for potential geno-cytotoxicity upon A549 lung cancer cells. - Flow cytometry analyses determined early and late apoptosis in sclareol and sclareol-loaded SLNPs treated cells.	[223]
Hydrogel	- A poloxamer-based thermoresponsive hydrogel was developed to exert local tumor control. - This hydrogel demonstrated a dose-dependent cancer cell-specific toxicity in vitro and was retained in situ for at least 14 days in the xenograft model.	[224]
Iron oxide magnetic NPs	- DOX and cetuximab were co-conjugated to dextran-coated Fe_3O_4 magnetic nanoparticles. - These NPs significantly suppress cell proliferation of A549 cells as compared with A549 cells treated with NPs only conjugated with DOX.	[225]
Nanoemulsion system	- Naringenin nanoemulsions for oral delivery were developed using employing a Box–Behnken design. - These nanoemulsion were more effective than the naringenin solution in reducing Bcl2 expression, while increasing proapoptotic Bax and caspase-3 activity.	[226]

Table 3. *Cont.*

Nanoparticles Type	System Description and the Main Finding	Reference
Porous Se@SiO_2 NPs	- Porous Se@SiO_2 NPs were fabricated with antioxidant properties. - These NPs significantly increased the resistance of airway epithelial cells under oxidative injury and shifted lipopolysaccharide-induced gene expression profile closer to the untreated controls.	[227]
Metal–organic frameworks (MOFs)	- A hydrolytically stable mesoporous gadolinium -MOF was prepared. - This nanostructure provided lewis basic sites for 5-Fu delivery and inhibition of human lung cancer cells in vivo and in vitro.	[228]

[1] NPs: Nanoparticles.

9.4. Combinatorial Therapeutic Approach via Hybrid Nanoparticles

The encapsulation of different drugs with multiple sites of action is considered to be an efficient method for targeting malignant cells. The application of combinatorial therapeutic approaches will reduce the dose and resistance of the applied anticancer drugs. However, each anticancer drug has a specific biochemical activity; therefore, combined administration would be inappropriate and ineffective in targeting lung cancers. Moreover, combining more drugs can cause harmful effects on healthy organs [229]. Various hybrid combination therapies have been designed for lung cancer targeting. Multifunctional hybrid nanoparticles have gained more recognition as a combinatorial therapeutic approach. For example, in one study an amphiphilic triblock copolymer functionalized with deoxycholate was synthesized and was used as a nanocarrier to codeliver DOX and PTX into lung cancer cells. Each residue of the copolymeric system comprises a unique property for the complex construction. The codelivery of DOX and PTX using hybrid nanovesicles exhibited an enhanced antitumor effect by reducing the growth of the A549 cells in lung cancer [230]. Another study demonstrated the enhanced efficacy of dual drug delivery in lung cancer therapy, in which PLGA/methacrylic acid copolymer nanoparticles were developed for the codelivery of DOX and chrysin. This nanoformulation significantly reduced the proliferation of A549 cells. The loaded agents showed higher antitumor activity under the in vitro cell line study [231]. Based on several studies, it is concluded that the combination therapies using two or more drugs in a single nanoparticle might be more useful for tumor targeting. Hybrid nanoparticle-based combination therapy, including chemotherapy with hyperthermia, can help to reduce the mortality rate in lung cancer patients. Through combining hyperthermia therapy with chemotherapy, upon increasing the temperature of tumor environment up to 40–45 °C, the malignant cells are killed but the healthy cells are not affected [232]. The hybrid magnetic nanoparticles can be used for combination therapy (chemo and hyperthermia) to achieve better antitumor efficacy. In one study, hydroxyapatite nanoparticles encapsulated with cisplatin were developed to target lung cancer by combining chemotherapy with hyperthermia. The results of this study showed a greater uptake of nanoparticles in A549 cells by activation of the (ERK) signaling pathway [233]. Figure 5 presents a schematic graphic design of combination therapy approaches via different types of nanoparticles.

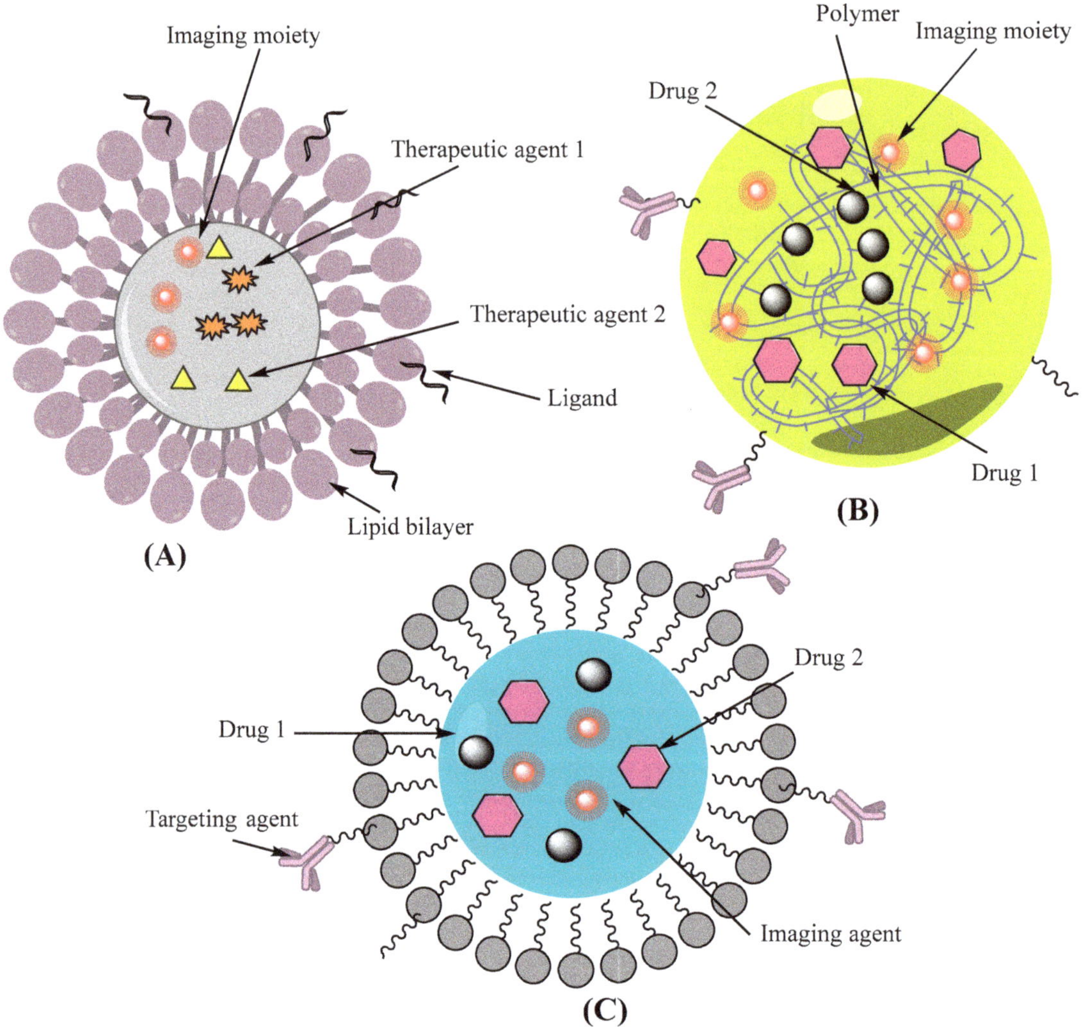

Figure 5. Schematic representation of combination therapy approaches in targeting lung cancer. (**A**) Combination lipid nanoparticles with imaging and therapeutic moiety (**B**) Polymer based nanocapsule hybrid therapy with imaging and targeting moiety along with therapeutic agent (**C**) Inorganic-polymer combination nanoparticle coencapsulated with drug molecule, targeting agent and imaging moiety.

10. Some Clinical Studies and Marketed Formulation of Nanoparticles for Lung Cancer Treatment

The application of nanoparticle-based drug delivery systems for lung cancer treatments is still in the development phase. Nevertheless, there are already some nanoparticle-based drugs in the market and various nanobased therapeutics are being used in clinical studies. Abraxane is the first nanotechnology-based drug for lung cancer therapy that has passed regulatory scrutiny and is already on the market and can be used to treat breast and pancreatic cancer as well. In this formulation, PTX is bonded to albumin nanoparticles as a delivery vehicle [234]. Genexol-PM is another PTX-loaded formulation, which is based on polymeric (PEG-PLA) nanoparticle micelles and has been approved in Europe and South Korea for the treatment of breast cancer and NSCLC [235,236].

In recent years, the US FDA has approbated numerous investigational new drug (IND) applications for nanoformulations, enabling clinical trials for lung cancer. Considering these developments, it seems that nanotechnology could improve the drugs' effects, overcoming their inherent conventional limits. In Table 4, a list of the FDA approved nanoparticle-based drug delivery systems and a list of INDs being tested in clinical trials are shown [237,238].

Table 4. List of FDA approved and under investigation nanoparticles encapsulated with anticancer drugs.

List of FDA Approved Anticancer Drug Loaded Nanoparticles			
Trade name	**Generic name**	**Benefits of encapsulation via nanoparticles**	**Reference**
Doxil (Janssen)	Doxorubicin HCl liposome injection	Increased delivery to disease site, decreased systemic toxicity of free drug	[239,240]
Marqibo (Spectrum Pharmaceuticals)	Liposomal vincristine	Increased delivery to tumor site, decreased systemic toxicity	[241]
Onivyde (Ipsen Biopharmaceuticals)	Liposomal irinotecan	Increased delivery to tumor site, decreased systemic toxicity	[241]
Vyxeos (Jazz Pharmaceuticals)	Liposomal daunorubicin and cytarabine	Increased efficacy through synergistic delivery of coencapsulated agents	[242]
List of INDs encapsulated by modified and unmodified nanoparticles under clinical trials			
Nanoparticle type (Investigation ID)	**Encapsulated Drug**	**Clinical trial phase**	**Reference**
Pegylated Liposomal MM-302	DOX	Phase 1	[243]
HER2-Targeted Liposomes	DOX	Phase 1	[244]
Thermo sensitive Liposome Thermodox (Celsion Corp.)	DOX	Phase 3	[244]
Conjugate Cyclodextran-PEG polymeric nanoparticle CRLX101	Camptothecin	phase1 and 2 clinical trials	[243]
Polymeric nanoparticle Conjugated CRLX301	Docetaxel	phase $\frac{1}{2}$a	[243]
Polyglutamic acid-conjugated nanoparticle (poliglumex) Opaxio	Paclitaxel	Phase3	[239]

11. Conclusions and Future Perspectives

In spite of the achieved advances in the drug delivery systems for lung cancer targeting, this cancer type is still the main cause of many deaths in the world. The major problem with the present treatment strategies is a lack of tools and smarter carrier systems for the drug targeting of malignant cells. It is evident from numerous research studies that the surface modification strategy reduces toxicity by changing the half-life, distribution, disposition, stimuli reactivity and therapeutic application. All the applications of surface-modified and hybrid nanoparticles paved the way for the clinical therapeutics through engineered and fine-tuned delivery to the lung tumor while reducing its side effects. These nanomaterials for lung cancer targeting are categorized into diagnostic, therapeutic, and theranostic multifunctional systems. The alteration of the onco-receptor function and modulating their pathways through specific strategies inhibit tumor growth and development through enhanced tumor-specific action. Surface-modified and hybrid nanoparticles can also help to advance the diagnosis process of lung cancer by stepping ahead from anatomical to molecular imaging for more precise diagnosis. However, the chemo-physiological aspects of these carriers should be carefully evaluated for optimal diagnosis. There are several conjugated nanodrugs recently filed in the list of the FDA's new drug applications. Camptothecin, a potent antineaoplastic agent being studied alone and in combination with other drugs, has been included in various phase 1 and phase 2 clinical trials for targeting lung cancer (SCLN and NSCLC) and other solid tumors. For evaluating the surface-modified and hybrid nanoparticles it is mandatory to evaluate the in vivo toxicity and biodistribution of the nanohybrid carriers. By addressing the challenges in the development of optimized modified/functionalized nanoparticles, it will be possible to translocate these systems for clinical applications. The current review described why and how hybrid or surface-modified nanocarrier systems have the potential to be the most insolent delivery system for targeting lung cancer. We anticipate that this cross-disciplinary review could inspire further research and discovery on the design and performance of surface-modified and hybrid nanoparticles for targeted drug delivery and lung cancer therapy.

Author Contributions: Conceptualization, A.R.; writing—original draft preparation, F.S., M.B., M.Z., M.Q., Q.U.A. and E.G.; writing—review and editing, A.R. and M.A.A.; supervision, A.R. and M.A.A. All authors have read and agreed to the published version of the manuscript.

Funding: The APC was funded through PHOTO-EMULSION project. Financing entity: European Union H2020-MSCA-ITN-2017.

Acknowledgments: M.A.A. would like to thank Radmila Tomovska for her great help in providing APC funding.

Conflicts of Interest: The authors declare no conflict of interest.

Abbreviations

AgNPs	silver nanoparticles
anti-NTSR1-mAb	antineurotensin receptor 1 monoclonal antibody
ATI	alveolar type I
ATII	alveolar type II
AuNPs	gold nanoparticles
β2-AR	β2-adrenergic receptors
β-CD	beta-cyclodextrin
CCL2	chemokine (C-C motif) ligand 2
CD44	cluster of differentiation 44
CeO2	cerium oxide
CLNs	cationic lipid nano-systems
CLSM	confocal laser scanning microscopy
CNT	carbon nanotubes
DNA	deoxyribonucleic acid
DPPC	1, 2 dipalmitoylphosphatidylcholine
DPPG	dipalmitoylphosphatidylglycerol
DOPA	dioleoylphosphatydic acid
DOX	doxorubicin
DR4/TRAIL-R	death receptor/TNF-related apoptosis inducing ligand-receptor
DTX	docetaxel
EGFR	epidermal growth factor receptor
EGR-1	early growth response-1
EML4	echinoderm microtubule-associated protein-like 4
EPR	enhanced permeability and retention
ERK	extracellular signal-regulated kinase
EVs	extracellular vesicles
FA	folic acid
FACS	fluorescence-activated cell-sorting
FDA	food and drug administration
FI	fluorescein isothiocyanate
FR	folate receptor
GDP	guanine diphosphate
GITR	glucocorticoid-induced tumor necrosis factor receptor-related

GTP	guanine triphosphate
HA	hyaluronic acid
HIF-1	hypoxia-inducible factor-1
HP	hemamiddleorphyrin
HPLC	high performance liquid chromatography
HUVEC	human umbilical vein endothelial cells
IL-6	interleukin 6
IND	investigational new drug
JAK-STAT	janus kinase-signal transducer and activator of transcription
KLF2	Krüppel-like Factor 2
KLF4	Krüppel-like Factor 4
LbL	layer-by-layer
LHRH	luteinizing hormone-releasing hormone
LPD	liposome-polycation-DNA complex
MAA	methacrylic acid
MAPK	mitogen-activated protein kinase
MCP-1	monocyte chemoattractant protein-1
MDSCs	myeloid-derived suppressor cells
MoS2	molybdenum disulfide
MWCNTs	multi-walled carbon nanotubes
NAG	N-acetyl-D-glucosamine
NDDS	nanodrug delivery systems
NPs	Nanoparticles
NSCLS	non-small cell lung cancer
OCAq	9,10-anthraquinone-2-carboxylic acid
PAMAM	poly(amidoamine)
PEG	polyethylene glycol
PI3K/AKT/mTOR	phosphatidylinositol 3-kinase/protein kinase B/mammalian target of rapamycin mTOR
PIP2	phosphatidylinositol 4,5-bisphosphate
PIP3	phosphatidylinositol 3,4,5-bisphosphate
PLA	polylactide
PLGA	poly(lactide-co-glycolide)
PNA	peptide nucleic acid
PNIPAAm	poly-(N-isopropylacrylamide)
PTX	paclitaxel
QD	quantum dot
(QR	quercetin
RAF	rapidly accelerated fibrosarcoma
RB	rose bengal
RES	reticuloendothelial system
RGD	arginylglycylaspartic acid
RNA	ribonucleic acid
RNAi	RNA interference
ROS	reactive oxygen species
SCNP	single chain nanoparticle
SCLC	small cell lung cancer
SiO2	silicon dioxide
siRNA	small interfering RNA
SLNPs	solid lipid nanoparticles
SPIONs	superaparamagnetic iron oxide nanoparticles
TAMs	tumor-associated macrophages
TCL-SPION	thermally crosslinked supermagnetic iron oxide
TF	transferrin
TFR	transferrin receptor
TLR3	toll-like receptor 3
TNFα	tumour necrosis factor α
VEGFR	vascular endothelial growth factor receptor
ZO1	zonula occludens 1
ZrO2	zirconium oxide

References

1. Lim, R.B.L. End-of-life care in patients with advanced lung cancer. *Ther. Adv. Respir. Dis.* **2016**, *10*, 455–467. [CrossRef]
2. Howlader, N.; Forjaz, G.; Mooradian, M.J.; Meza, R.; Kong, C.Y.; Cronin, K.A.; Mariotto, A.B.; Lowy, D.R.; Feuer, E.J. The effect of advances in lung-cancer treatment on population mortality. *New Engl. J. Med.* **2020**, *383*, 640–649. [CrossRef]
3. Bade, B.C.; Cruz, C.S.D. Lung cancer 2020: Epidemiology, etiology, and prevention. *Clin. Chest Med.* **2020**, *41*, 1–24. [CrossRef]
4. Rizvi, N.A.; Cho, B.C.; Reinmuth, N.; Lee, K.H.; Luft, A.; Ahn, M.-J.; van den Heuvel, M.M.; Cobo, M.; Vicente, D.; Smolin, A. Durvalumab with or without tremelimumab vs. standard chemotherapy in first-line treatment of metastatic non–small cell lung cancer: The MYSTIC phase 3 randomized clinical trial. *JAMA Oncol.* **2020**, *6*, 661–674. [CrossRef]
5. Subbiah, S.; Nam, A.; Garg, N.; Behal, A.; Kulkarni, P.; Salgia, R. Small cell lung cancer from traditional to innovative therapeutics: Building a comprehensive network to optimize clinical and translational research. *J. Clin. Med.* **2020**, *9*, 2433. [CrossRef]
6. Sukumar, U.K.; Bhushan, B.; Dubey, P.; Matai, I.; Sachdev, A.; Packirisamy, G. Emerging applications of nanoparticles for lung cancer diagnosis and therapy. *Int. Nano Lett.* **2013**, *3*, 45. [CrossRef]
7. Wang, X.; Chen, H.; Zeng, X.; Guo, W.; Jin, Y.; Wang, S.; Tian, R.; Han, Y.; Guo, L.; Han, J.; et al. Efficient lung cancer-targeted drug delivery via a nanoparticle/MSC system. *Acta Pharm. Sin. B* **2019**, *9*, 167–176. [CrossRef]
8. Hu, J.; Fu, S.; Peng, Q.; Han, Y.; Xie, J.; Zan, N.; Chen, Y.; Fan, J. Paclitaxel-loaded polymeric nanoparticles combined with chronomodulated chemotherapy on lung cancer: In vitro and in vivo evaluation. *Int J. Pharm.* **2017**, *516*, 313–322. [CrossRef]
9. Lin, C.; Zhang, X.; Chen, H.; Bian, Z.; Zhang, G.; Riaz, M.K.; Tyagi, D.; Lin, G.; Zhang, Y.; Wang, J.; et al. Dual-ligand modified liposomes provide effective local targeted delivery of lung-cancer drug by antibody and tumor lineage-homing cell-penetrating peptide. *Drug Deliv.* **2018**, *25*, 256–266. [CrossRef]
10. Song, X.L.; Ju, R.J.; Xiao, Y.; Wang, X.; Liu, S.; Fu, M.; Liu, J.J.; Gu, L.Y.; Li, X.T.; Cheng, L. Application of multifunctional targeting epirubicin liposomes in the treatment of non-small-cell lung cancer. *Int. J. Nanomed.* **2017**, *12*, 7433–7451. [CrossRef]
11. Perepelyuk, M.; Sacko, K.; Thangavel, K.; Shoyele, S.A. Evaluation of MUC1-Aptamer Functionalized Hybrid Nanoparticles for Targeted Delivery of miRNA-29b to Nonsmall Cell Lung Cancer. *Mol. Pharm.* **2018**, *15*, 985–993. [CrossRef]
12. Dostalova, S.; Vasickova, K.; Hynek, D.; Krizkova, S.; Richtera, L.; Vaculovicova, M.; Eckschlager, T.; Stiborova, M.; Heger, Z.; Adam, V. Apoferritin as an ubiquitous nanocarrier with excellent shelf life. *Int. J. Nanomed.* **2017**, *12*, 2265–2278. [CrossRef]
13. Ramalingam, V.; Varunkumar, K.; Ravikumar, V.; Rajaram, R. Target delivery of doxorubicin tethered with PVP stabilized gold nanoparticles for effective treatment of lung cancer. *Sci. Rep.* **2018**, *8*, 3815. [CrossRef] [PubMed]
14. Wang, Z.; Qiao, R.; Tang, N.; Lu, Z.; Wang, H.; Zhang, Z.; Xue, X.; Huang, Z.; Zhang, S.; Zhang, G.; et al. Active targeting theranostic iron oxide nanoparticles for MRI and magnetic resonance-guided focused ultrasound ablation of lung cancer. *Biomaterials* **2017**, *127*, 25–35. [CrossRef]
15. Chaturvedi, V.K.; Singh, A.; Singh, V.K.; Singh, M.P. Cancer nanotechnology: A new revolution for cancer diagnosis and therapy. *Curr. Drug Metab.* **2019**, *20*, 416–429. [CrossRef]
16. Iyer, S.; Prajapati, R.; Ramesh, A.; Basavalingegowda, M.; Todur, S.; Kavishvar, S.; Vijaykumar, R.; Naik, R.; Kulkarni, P.; Bhatt, A.D. The future of lung cancer therapy: Striding beyond conventional EGFR and ALK treatments. *Mol. Clin. Oncol.* **2019**, *10*, 469–475. [CrossRef]
17. Bertrand, N.; Wu, J.; Xu, X.; Kamaly, N.; Farokhzad, O.C. Cancer nanotechnology: The impact of passive and active targeting in the era of modern cancer biology. *Adv. Drug Deliv. Rev.* **2014**, *66*, 2–25. [CrossRef]
18. Kumar, M.; Jha, A.; Dr, M.; Mishra, B. Targeted drug nanocrystals for pulmonary delivery: A potential strategy for lung cancer therapy. *Expert Opin. Drug Deliv.* **2020**, *17*, 1459–1472. [CrossRef]
19. Vanza, J.D.; Patel, R.B.; Patel, M.R. Nanocarrier centered therapeutic approaches: Recent developments with insight towards the future in the management of lung cancer. *J. Drug Deliv. Sci. Technol.* **2020**, *60*, 102070. [CrossRef]
20. Patel, K.; Patel, K. Challenges and Recent Progress of Nano Sized Drug Delivery Systems for Lung Cancer Therapy: A Review. *Himal. J. Health Sci.* **2020**, *5*, 58–62.
21. Iyer, A.K.; Khaled, G.; Fang, J.; Maeda, H. Exploiting the enhanced permeability and retention effect for tumor targeting. *Drug Discov. Today* **2006**, *11*, 812–818. [CrossRef] [PubMed]
22. Haider, N.; Fatima, S.; Taha, M.; Rizwanullah, M.; Firdous, J.; Ahmad, R.; Mazhar, F.; Khan, M.A. Nanomedicines in diagnosis and treatment of cancer: An update. *Curr. Pharm. Des.* **2020**, *26*, 1216–1231. [CrossRef] [PubMed]
23. A Razak, S.A.; A Wahab, H.; Fisol, F.A.; Abdulbaqi, I.M.; Parumasivam, T.; Mohtar, N.; Mohd Gazzali, A. Advances in Nanocarriers for Effective Delivery of Docetaxel in the Treatment of Lung Cancer: An Overview. *Cancers* **2021**, *13*, 400. [CrossRef]
24. Behera, A.; Padhi, S. Passive and active targeting strategies for the delivery of the camptothecin anticancer drug: A review. *Environ. Chem. Lett.* **2020**, *18*, 1–11. [CrossRef]
25. Gorain, B.; Bhattamishra, S.K.; Choudhury, H.; Nandi, U.; Pandey, M.; Kesharwani, P. Chapter 3—Overexpressed Receptors and Proteins in Lung Cancer. In *Nanotechnology-Based Targeted Drug Delivery Systems for Lung Cancer*; Kesharwani, P., Ed.; Academic Press: Cambridge, MA, USA, 2019; pp. 39–75. [CrossRef]
26. Karpuz, M.; Silindir-Gunay, M.; Ozer, A.Y.; Ozturk, S.C.; Yanik, H.; Tuncel, M.; Aydin, C.; Esendagli, G. Diagnostic and therapeutic evaluation of folate-targeted paclitaxel and vinorelbine encapsulating theranostic liposomes for non-small cell lung cancer. *Eur. J. Pharm. Sci.* **2020**, *156*, 105576. [CrossRef]
27. Yuan, M.; Huang, L.-L.; Chen, J.-H.; Wu, J.; Xu, Q. The emerging treatment landscape of targeted therapy in non-small-cell lung cancer. *Signal. Transduct. Target. Ther.* **2019**, *4*, 1–14. [CrossRef]

28. Neel, D.S.; Bivona, T.G. Resistance is futile: Overcoming resistance to targeted therapies in lung adenocarcinoma. *NPJ Precis. Oncol.* **2017**, *1*, 1–6. [CrossRef]
29. Pathak, A.; Tanwar, S.; Kumar, V.; Banarjee, B.D. Present and future prospect of small molecule & related targeted therapy against human cancer. *Vivechan Int. J. Res.* **2018**, *9*, 36.
30. Nguyen, H.X. Targeted Delivery of Surface-Modified Nanoparticles: Modulation of Inflammation for Acute Lung Injury. In *Surface Modification of Nanoparticles for Targeted Drug Delivery*; Springer: Berlin/Heidelberg, Germany, 2019; pp. 331–353.
31. Carvalho, T.C.; Peters, J.I.; Williams, R.O., 3rd. Influence of particle size on regional lung deposition—What evidence is there? *Int. J. Pharm.* **2011**, *406*, 1–10. [CrossRef]
32. Ray, L. Polymeric Nanoparticle-Based Drug/Gene Delivery for Lung Cancer. In *Nanotechnology-Based Targeted Drug Delivery Systems for Lung Cancer*; Elsevier: Amsterdam, The Netherlands, 2019; pp. 77–93.
33. Palmer, J.D.; Zaorsky, N.G.; Witek, M.; Lu, B. Molecular markers to predict clinical outcome and radiation induced toxicity in lung cancer. *J. Thorac. Dis.* **2014**, *6*, 387.
34. Mottaghitalab, F.; Farokhi, M.; Fatahi, Y.; Atyabi, F.; Dinarvand, R. New insights into designing hybrid nanoparticles for lung cancer: Diagnosis and treatment. *J. Control. Release* **2019**, *295*, 250–267. [CrossRef] [PubMed]
35. Bartholomew, C.; Eastlake, L.; Dunn, P.; Yiannakis, D. EGFR targeted therapy in lung cancer; an evolving story. *Respir. Med. Case Rep.* **2017**, *20*, 137–140. [CrossRef] [PubMed]
36. Koivunen, J.P.; Mermel, C.; Zejnullahu, K.; Murphy, C.; Lifshits, E.; Holmes, A.J.; Choi, H.G.; Kim, J.; Chiang, D.; Thomas, R. EML4-ALK fusion gene and efficacy of an ALK kinase inhibitor in lung cancer. *Clin. Cancer Res.* **2008**, *14*, 4275–4283. [CrossRef] [PubMed]
37. Selinger, C.I.; Rogers, T.-M.; Russell, P.A.; O'toole, S.; Yip, P.; Wright, G.M.; Wainer, Z.; Horvath, L.G.; Boyer, M.; McCaughan, B. Testing for ALK rearrangement in lung adenocarcinoma: A multicenter comparison of immunohistochemistry and fluorescent in situ hybridization. *Mod. Pathol.* **2013**, *26*, 1545–1553. [CrossRef]
38. Cope, N.; Candelora, C.; Wong, K.; Kumar, S.; Nan, H.; Grasso, M.; Novak, B.; Li, Y.; Marmorstein, R.; Wang, Z. Mechanism of BRAF activation through biochemical characterization of the recombinant full-length protein. *Chembiochem A Eur. J. Chem. Biol.* **2018**, *19*, 1988. [CrossRef]
39. Papadimitrakopoulou, V. Development of PI3K/AKT/mTOR pathway inhibitors and their application in personalized therapy for non–small-cell lung cancer. *J. Thorac. Oncol.* **2012**, *7*, 1315–1326. [CrossRef]
40. Ji, M.; Guan, H.; Gao, C.; Shi, B.; Hou, P. Highly frequent promoter methylation and PIK3CA amplification in non-small cell lung cancer (NSCLC). *BMC Cancer* **2011**, *11*, 147. [CrossRef]
41. Tan, A.C. Targeting the PI3K/Akt/mTOR pathway in non-small cell lung cancer (NSCLC). *Thorac. Cancer* **2020**, *11*, 511–518. [CrossRef] [PubMed]
42. Pylayeva-Gupta, Y.; Grabocka, E.; Bar-Sagi, D. RAS oncogenes: Weaving a tumorigenic web. *Nat. Rev. Cancer* **2011**, *11*, 761–774. [CrossRef]
43. Gibbons, D.L.; Byers, L.A.; Kurie, J.M. Smoking, p53 mutation, and lung cancer. *Mol. Cancer Res.* **2014**, *12*, 3–13. [CrossRef]
44. Campling, B.G.; el-Deiry, W.S. Clinical implications of p53 mutations in lung cancer. *Methods Mol. Med.* **2003**, *75*, 53–77. [CrossRef]
45. Marks, J.L.; Gong, Y.; Chitale, D.; Golas, B.; McLellan, M.D.; Kasai, Y.; Ding, L.; Mardis, E.R.; Wilson, R.K.; Solit, D. Novel MEK1 mutation identified by mutational analysis of epidermal growth factor receptor signaling pathway genes in lung adenocarcinoma. *Cancer Res.* **2008**, *68*, 5524–5528. [CrossRef] [PubMed]
46. Heigener, D.F.; Gandara, D.R.; Reck, M. Targeting of MEK in lung cancer therapeutics. *Lancet Respir. Med.* **2015**, *3*, 319–327. [CrossRef]
47. Nakata, Y.; Kimura, A.; Katoh, O.; Kawaishi, K.; Hyodo, H.; Abe, K.; Kuramoto, A.; Satow, Y. c-kit point mutation of extracellular domain in patients with myeloproliferative disorders. *Br. J. Haematol.* **1995**, *91*, 661–663. [CrossRef]
48. Naeem, M.; Dahiya, M.; Clark, J.I.; Creech, S.D.; Alkan, S. Analysis of c-kit protein expression in small-cell lung carcinoma and its implication for prognosis. *Hum. Pathol.* **2002**, *33*, 1182–1187. [CrossRef]
49. Lai, Y.; Wang, X.; Zeng, T.; Xing, S.; Dai, S.; Wang, J.; Chen, S.; Li, X.; Xie, Y.; Zhu, Y.; et al. Serum VEGF levels in the early diagnosis and severity assessment of non-small cell lung cancer. *J. Cancer* **2018**, *9*, 1538–1547. [CrossRef]
50. Frezzetti, D.; Gallo, M.; Maiello, M.R.; D'Alessio, A.; Esposito, C.; Chicchinelli, N.; Normanno, N.; de Luca, A. VEGF as a potential target in lung cancer. *Expert Opin. Ther. Targets* **2017**, *21*, 959–966. [CrossRef]
51. Griffin, R.; Ramirez, R.A. Molecular targets in non–small cell lung cancer. *Ochsner J.* **2017**, *17*, 388–392.
52. Bergethon, K.; Shaw, A.T.; Ou, S.-H.I.; Katayama, R.; Lovly, C.M.; McDonald, N.T.; Massion, P.P.; Siwak-Tapp, C.; Gonzalez, A.; Fang, R. ROS1 rearrangements define a unique molecular class of lung cancers. *J. Clin. Oncol.* **2012**, *30*, 863. [CrossRef]
53. Ferguson, K.M. Structure-based view of epidermal growth factor receptor regulation. *Annu. Rev. Biophys.* **2008**, *37*, 353–373. [CrossRef]
54. Alhajj, N.; Chee, C.F.; Wong, T.W.; Rahman, N.A.; Abu Kasim, N.H.; Colombo, P. Lung cancer: Active therapeutic targeting and inhalational nanoproduct design. *Expert Opin. Drug Deliv.* **2018**, *15*, 1223–1247. [CrossRef]
55. Cadranel, J.; Ruppert, A.-M.; Beau-Faller, M.; Wislez, M. Therapeutic strategy for advanced EGFR mutant non-small-cell lung carcinoma. *Crit. Rev. Oncol. Hematol.* **2013**, *88*, 477–493. [CrossRef]
56. Yewale, C.; Baradia, D.; Vhora, I.; Patil, S.; Misra, A. Epidermal growth factor receptor targeting in cancer: A review of trends and strategies. *Biomaterials* **2013**, *34*, 8690–8707. [CrossRef]

57. Tseng, C.-L.; Su, W.-Y.; Yen, K.-C.; Yang, K.-C.; Lin, F.-H. The use of biotinylated-EGF-modified gelatin nanoparticle carrier to enhance cisplatin accumulation in cancerous lungs via inhalation. *Biomaterials* **2009**, *30*, 3476–3485. [CrossRef]
58. Li, F.; Mei, H.; Xie, X.; Zhang, H.; Liu, J.; Lv, T.; Nie, H.; Gao, Y.; Jia, L. Aptamer-conjugated chitosan-anchored liposomal complexes for targeted delivery of erlotinib to EGFR-mutated lung cancer cells. *AAPS J.* **2017**, *19*, 814–826. [CrossRef]
59. Zheng, Y.; Su, C.; Zhao, L.; Shi, Y. mAb MDR1-modified chitosan nanoparticles overcome acquired EGFR-TKI resistance through two potential therapeutic targets modulation of MDR1 and autophagy. *J. Nanobiotechnol.* **2017**, *15*, 66. [CrossRef]
60. Schieber, C.; Bestetti, A.; Lim, J.P.; Ryan, A.D.; Nguyen, T.L.; Eldridge, R.; White, A.R.; Gleeson, P.A.; Donnelly, P.S.; Williams, S.J. Conjugation of Transferrin to Azide-Modified CdSe/ZnS Core–Shell Quantum Dots using Cyclooctyne Click Chemistry. *Angew. Chem. Int. Ed.* **2012**, *51*, 10523–10527. [CrossRef]
61. Kukulj, S.; Jaganjac, M.; Boranic, M.; Krizanac, S.; Santic, Z.; Poljak-Blazi, M. Altered iron metabolism, inflammation, transferrin receptors, and ferritin expression in non-small-cell lung cancer. *Med. Oncol.* **2010**, *27*, 268–277. [CrossRef]
62. Wu, Y.; Xu, J.; Chen, J.; Zou, M.; Rusidanmu, A.; Yang, R. Blocking transferrin receptor inhibits the growth of lung adenocarcinoma cells in vitro. *Thorac. Cancer* **2018**, *9*, 253–261. [CrossRef]
63. Anabousi, S.; Bakowsky, U.; Schneider, M.; Huwer, H.; Lehr, C.-M.; Ehrhardt, C. In vitro assessment of transferrin-conjugated liposomes as drug delivery systems for inhalation therapy of lung cancer. *Eur. J. Pharm. Sci.* **2006**, *29*, 367–374. [CrossRef]
64. Daniels, T.R.; Bernabeu, E.; Rodríguez, J.A.; Patel, S.; Kozman, M.; Chiappetta, D.A.; Holler, E.; Ljubimova, J.Y.; Helguera, G.; Penichet, M.L. The transferrin receptor and the targeted delivery of therapeutic agents against cancer. *Biochim. Biophys. Acta BBA Gen. Subj.* **2012**, *1820*, 291–317. [CrossRef]
65. Hynes, R.O. Integrins: Bidirectional, allosteric signaling machines. *Cell* **2002**, *110*, 673–687. [CrossRef]
66. Aksorn, N.; Chanvorachote, P. Integrin as a molecular target for anti-cancer approaches in lung cancer. *Anticancer Res.* **2019**, *39*, 541–548. [CrossRef]
67. Bartolazzi, A.; Cerboni, C.; Flamini, G.; Bigotti, A.; Lauriola, L.; Natali, P.G. Expression of α3β1 integrin receptor and its ligands in human lung tumors. *Int. J. Cancer* **1995**, *64*, 248–252. [CrossRef]
68. Babu, A.; Amreddy, N.; Muralidharan, R.; Pathuri, G.; Gali, H.; Chen, A.; Zhao, Y.D.; Munshi, A.; Ramesh, R. Chemodrug delivery using integrin-targeted PLGA-Chitosan nanoparticle for lung cancer therapy. *Sci. Rep.* **2017**, *7*, 1–17. [CrossRef]
69. Zou, Y.; Sun, Y.; Guo, B.; Wei, Y.; Xia, Y.; Huangfu, Z.; Meng, F.; van Hest, J.C.; Yuan, J.; Zhong, Z. α3β1 Integrin-Targeting Polymersomal Docetaxel as an Advanced Nanotherapeutic for Nonsmall Cell Lung Cancer Treatment. *ACS Appl. Mater. Interfaces* **2020**, *12*, 14905–14913. [CrossRef]
70. Morales-Cruz, M.; Delgado, Y.; Castillo, B.; Figueroa, C.M.; Molina, A.M.; Torres, A.; Milián, M.; Griebenow, K. Smart Targeting to Improve Cancer Therapeutics. *Drug Des. Dev. Ther.* **2019**, *13*, 3753. [CrossRef]
71. Shen, J.; Hu, Y.; Putt, K.S.; Singhal, S.; Han, H.; Visscher, D.W.; Murphy, L.M.; Low, P.S. Assessment of folate receptor alpha and beta expression in selection of lung and pancreatic cancer patients for receptor targeted therapies. *Oncotarget* **2018**, *9*, 4485. [CrossRef]
72. Morales-Cruz, M.; Cruz-Montañez, A.; Figueroa, C.M.; González-Robles, T.; Davila, J.; Inyushin, M.; Loza-Rosas, S.A.; Molina, A.M.; Muñoz-Perez, L.; Kucheryavykh, L.Y. Combining stimulus-triggered release and active targeting strategies improves cytotoxicity of cytochrome c nanoparticles in tumor cells. *Mol. Pharm.* **2016**, *13*, 2844–2854. [CrossRef]
73. Li, L.; He, S.; Yu, L.; Elshazly, E.H.; Wang, H.; Chen, K.; Zhang, S.; Ke, L.; Gong, R. Codelivery of DOX and siRNA by folate-biotin-quaternized starch nanoparticles for promoting synergistic suppression of human lung cancer cells. *Drug Deliv.* **2019**, *26*, 499–508. [CrossRef]
74. Templeton, A.K.; Miyamoto, S.; Babu, A.; Munshi, A.; Ramesh, R. Cancer stem cells: Progress and challenges in lung cancer. *Stem Cell Investig.* **2014**, *1*, 9.
75. Misra, S.; Hascall, V.C.; Markwald, R.R.; Ghatak, S. Interactions between hyaluronan and its receptors (CD44, RHAMM) regulate the activities of inflammation and cancer. *Front. Immunol.* **2015**, *6*, 201. [CrossRef]
76. Penno, M.B.; August, J.T.; Baylin, S.B.; Mabry, M.; Linnoila, R.I.; Lee, V.S.; Croteau, D.; Yang, X.L.; Rosada, C. Expression of CD44 in human lung tumors. *Cancer Res.* **1994**, *54*, 1381–1387.
77. Zhao, S.; He, J.-L.; Qiu, Z.-X.; Chen, N.-Y.; Luo, Z.; Chen, B.-J.; Li, W.-M. Prognostic value of CD44 variant exon 6 expression in non-small cell lung cancer: A meta-analysis. *Asian Pac. J. Cancer Prev.* **2014**, *15*, 6761–6766. [CrossRef] [PubMed]
78. Ganesh, S.; Iyer, A.K.; Morrissey, D.V.; Amiji, M.M. Hyaluronic acid based self-assembling nanosystems for CD44 target mediated siRNA delivery to solid tumors. *Biomaterials* **2013**, *34*, 3489–3502. [CrossRef] [PubMed]
79. Cadete, A.; Olivera, A.; Besev, M.; Dhal, P.K.; Gonçalves, L.; Almeida, A.J.; Bastiat, G.; Benoit, J.-P.; de la Fuente, M.; Garcia-Fuentes, M. Self-assembled hyaluronan nanocapsules for the intracellular delivery of anticancer drugs. *Sci. Rep.* **2019**, *9*, 1–11. [CrossRef]
80. Yoo, J.; Park, C.; Yi, G.; Lee, D.; Koo, H. Active targeting strategies using biological ligands for nanoparticle drug delivery systems. *Cancers* **2019**, *11*, 640. [CrossRef]
81. Li, X.; Taratula, O.; Taratula, O.; Schumann, C.; Minko, T. LHRH-targeted drug delivery systems for cancer therapy. *Mini Rev. Med. Chem.* **2017**, *17*, 258–267. [CrossRef]
82. Meng, W.; Xue, S.; Chen, Y. The role of CXCL12 in tumor microenvironment. *Gene* **2018**, *641*, 105–110. [CrossRef]
83. Hallinan, N.; Finn, S.; Cuffe, S.; Rafee, S.; O'Byrne, K.; Gately, K. Targeting the fibroblast growth factor receptor family in cancer. *Cancer Treat. Rev.* **2016**, *46*, 51–62. [CrossRef]
84. Rankin, E.B.; Giaccia, A.J. The receptor tyrosine kinase AXL in cancer progression. *Cancers* **2016**, *8*, 103. [CrossRef] [PubMed]

85. Das, M.; Wakelee, H. Targeting VEGF in lung cancer. *Expert Opin. Ther. Targets* **2012**, *16*, 395–406. [CrossRef]
86. Kim, I.; Byeon, H.J.; Kim, T.H.; Lee, E.S.; Oh, K.T.; Shin, B.S.; Lee, K.C.; Youn, Y.S. Doxorubicin-loaded porous PLGA microparticles with surface attached TRAIL for the inhalation treatment of metastatic lung cancer. *Biomaterials* **2013**, *34*, 6444–6453. [CrossRef] [PubMed]
87. Nilsson, M.B.; Le, X.; Heymach, J.V. β-Adrenergic Signaling in Lung Cancer: A Potential Role for Beta-Blockers. *J. Neuroimmune Pharmacol.* **2020**, *15*, 27–36. [CrossRef] [PubMed]
88. Thom, I.; Schult-Kronefeld, O.; Burkholder, I.; Goern, M.; Andritzky, B.; Blonski, K.; Kugler, C.; Edler, L.; Bokemeyer, C.; Schumacher, U.; et al. Lectin histochemistry of metastatic adenocarcinomas of the lung. *Lung Cancer* **2007**, *56*, 391–397. [CrossRef]
89. Kalluri, R.; LeBleu, V.S. The biology, function, and biomedical applications of exosomes. *Science* **2020**, *367*, eaau6977. [CrossRef]
90. Xu, R.; Rai, A.; Chen, M.; Suwakulsiri, W.; Greening, D.W.; Simpson, R.J. Extracellular vesicles in cancer—Implications for future improvements in cancer care. *Nat. Rev. Clin. Oncol.* **2018**, *15*, 617–638. [CrossRef]
91. Veerman, R.E.; Akpinar, G.G.; Eldh, M.; Gabrielsson, S. Immune cell-derived extracellular vesicles—Functions and therapeutic applications. *Trends Mol. Med.* **2019**, *25*, 382–394. [CrossRef]
92. Mo, Z.; Cheong, J.Y.A.; Xiang, L.; Le, M.T.; Grimson, A.; Zhang, D.X. Extracellular vesicle-associated organotropic metastasis. *Cell Prolif.* **2021**, *54*, e12948. [CrossRef]
93. Altei, W.F.; Pachane, B.C.; dos Santos, P.K.; Ribeiro, L.N.M.; Sung, B.H.; Weaver, A.M.; Selistre-de-Araujo, H.S. Inhibition of alphavbeta3 integrin impairs adhesion and uptake of tumor-derived small extracellular vesicles. *Cell Commun. Signal.* **2020**, *18*, 158. [CrossRef]
94. Singh, A.; Fedele, C.; Lu, H.; Nevalainen, M.T.; Keen, J.H.; Languino, L.R. Exosome-mediated transfer of αvβ3 integrin from tumorigenic to nontumorigenic cells promotes a migratory phenotype. *Mol. Cancer Res.* **2016**, *14*, 1136–1146. [CrossRef]
95. Hoshino, A.; Costa-Silva, B.; Shen, T.-L.; Rodrigues, G.; Hashimoto, A.; Mark, M.T.; Molina, H.; Kohsaka, S.; di Giannatale, A.; Ceder, S. Tumour exosome integrins determine organotropic metastasis. *Nature* **2015**, *527*, 329–335. [CrossRef]
96. Wiklander, O.P.; Brennan, M.Á.; Lötvall, J.; Breakefield, X.O.; Andaloussi, S.E. Advances in therapeutic applications of extracellular vesicles. *Sci. Transl. Med.* **2019**, *11*, eaav8521. [CrossRef]
97. Murphy, D.E.; de Jong, O.G.; Brouwer, M.; Wood, M.J.; Lavieu, G.; Schiffelers, R.M.; Vader, P. Extracellular vesicle-based therapeutics: Natural versus engineered targeting and trafficking. *Exp. Mol. Med.* **2019**, *51*, 1–12. [CrossRef]
98. Nakase, I.; Noguchi, K.; Aoki, A.; Takatani-Nakase, T.; Fujii, I.; Futaki, S. Arginine-rich cell-penetrating peptide-modified extracellular vesicles for active macropinocytosis induction and efficient intracellular delivery. *Sci. Rep.* **2017**, *7*, 1–11. [CrossRef]
99. Keklikoglou, I.; Cianciaruso, C.; Güç, E.; Squadrito, M.L.; Spring, L.M.; Tazzyman, S.; Lambein, L.; Poissonnier, A.; Ferraro, G.B.; Baer, C. Chemotherapy elicits pro-metastatic extracellular vesicles in breast cancer models. *Nat. Cell Biol.* **2019**, *21*, 190–202. [CrossRef]
100. Liu, Y.; Gu, Y.; Han, Y.; Zhang, Q.; Jiang, Z.; Zhang, X.; Huang, B.; Xu, X.; Zheng, J.; Cao, X. Tumor exosomal RNAs promote lung pre-metastatic niche formation by activating alveolar epithelial TLR3 to recruit neutrophils. *Cancer Cell* **2016**, *30*, 243–256. [CrossRef] [PubMed]
101. Vu, L.T.; Peng, B.; Zhang, D.X.; Ma, V.; Mathey-Andrews, C.A.; Lam, C.K.; Kiomourtzis, T.; Jin, J.; McReynolds, L.; Huang, L. Tumor-secreted extracellular vesicles promote the activation of cancer-associated fibroblasts via the transfer of microRNA-125b. *J. Extracell. Vesicles* **2019**, *8*, 1599680. [CrossRef]
102. Zhou, W.; Fong, M.Y.; Min, Y.; Somlo, G.; Liu, L.; Palomares, M.R.; Yu, Y.; Chow, A.; O'Connor, S.T.F.; Chin, A.R. Cancer-secreted miR-105 destroys vascular endothelial barriers to promote metastasis. *Cancer Cell* **2014**, *25*, 501–515. [CrossRef]
103. Zeng, Z.; Li, Y.; Pan, Y.; Lan, X.; Song, F.; Sun, J.; Zhou, K.; Liu, X.; Ren, X.; Wang, F.; et al. Cancer-derived exosomal miR-25-3p promotes pre-metastatic niche formation by inducing vascular permeability and angiogenesis. *Nat. Commun.* **2018**, *9*, 5395. [CrossRef]
104. Huai, Y.; Hossen, M.N.; Wilhelm, S.; Bhattacharya, R.; Mukherjee, P. Nanoparticle interactions with the tumor microenvironment. *Bioconjugate Chem.* **2019**, *30*, 2247–2263. [CrossRef] [PubMed]
105. Mukhtar, M.; Ali, H.; Ahmed, N.; Munir, R.; Talib, S.; Khan, A.S.; Ambrus, R. Drug delivery to macrophages: A review of nano-therapeutics targeted approach for inflammatory disorders and cancer. *Expert Opin. Drug Deliv.* **2020**, *17*, 1239–1257. [CrossRef] [PubMed]
106. Zhang, M.; He, Y.; Sun, X.; Li, Q.; Wang, W.; Zhao, A.; Di, W. A high M1/M2 ratio of tumor-associated macrophages is associated with extended survival in ovarian cancer patients. *J. Ovarian Res.* **2014**, *7*, 19. [CrossRef] [PubMed]
107. O'Shannessy, D.J.; Somers, E.B.; Wang, L.C.; Wang, H.; Hsu, R. Expression of folate receptors alpha and beta in normal and cancerous gynecologic tissues: Correlation of expression of the beta isoform with macrophage markers. *J. Ovarian Res.* **2015**, *8*, 29. [CrossRef]
108. Hou, Y.C.; Chao, Y.J.; Tung, H.L.; Wang, H.C.; Shan, Y.S. Coexpression of CD44-positive/CD133-positive cancer stem cells and CD204-positive tumor-associated macrophages is a predictor of survival in pancreatic ductal adenocarcinoma. *Cancer* **2014**, *120*, 2766–2777. [CrossRef]
109. Yan, H.; Kamiya, T.; Suabjakyong, P.; Tsuji, N.M. Targeting C-Type Lectin Receptors for Cancer Immunity. *Front. Immunol.* **2015**, *6*, 408. [CrossRef]
110. Patil, T.S.; Deshpande, A.S. Mannosylated nanocarriers mediated site-specific drug delivery for the treatment of cancer and other infectious diseases: A state of the art review. *J. Control. Release* **2020**, *320*, 239–252. [CrossRef]

111. Zang, X.; Zhou, J.; Zhang, X.; Chen, D.; Han, Y.; Chen, X. Dual-targeting Tumor Cells and Tumor Associated Macrophages with Lipid Coated Calcium Zoledronate for Enhanced Lung Cancer Chemoimmunotherapy. *Int. J. Pharm.* **2020**, *594*, 120174. [CrossRef]
112. Gabrilovich, D.I.; Nagaraj, S. Myeloid-derived suppressor cells as regulators of the immune system. *Nat. Rev. Immunol.* **2009**, *9*, 162–174. [CrossRef]
113. Iclozan, C.; Antonia, S.; Chiappori, A.; Chen, D.-T.; Gabrilovich, D. Therapeutic regulation of myeloid-derived suppressor cells and immune response to cancer vaccine in patients with extensive stage small cell lung cancer. *Cancer Immunol. Immunother.* **2013**, *62*, 909–918. [CrossRef]
114. Pan, P.-Y.; Wang, G.X.; Yin, B.; Ozao, J.; Ku, T.; Divino, C.M.; Chen, S.-H. Reversion of immune tolerance in advanced malignancy: Modulation of myeloid-derived suppressor cell development by blockade of stem-cell factor function. *Blood* **2008**, *111*, 219–228. [CrossRef] [PubMed]
115. Ma, J.; Xu, H.; Wang, S. Immunosuppressive role of myeloid-derived suppressor cells and therapeutic targeting in lung cancer. *J. Immunol. Res.* **2018**, *2018*, 6319649. [CrossRef] [PubMed]
116. Zhou, J.; Qu, Z.; Sun, F.; Han, L.; Li, L.; Yan, S.; Stabile, L.P.; Chen, L.-F.; Siegfried, J.M.; Xiao, G. Myeloid STAT3 promotes lung tumorigenesis by transforming tumor immunosurveillance into tumor-promoting inflammation. *Cancer Immunol. Res.* **2017**, *5*, 257–268. [CrossRef]
117. Sawant, A.; Schafer, C.C.; Jin, T.H.; Zmijewski, J.; Hubert, M.T.; Roth, J.; Sun, Z.; Siegal, G.P.; Thannickal, V.J.; Grant, S.C. Enhancement of antitumor immunity in lung cancer by targeting myeloid-derived suppressor cell pathways. *Cancer Res.* **2013**, *73*, 6609–6620. [CrossRef]
118. Srivastava, M.K.; Dubinett, S.; Sharma, S. Targeting MDSCs enhance therapeutic vaccination responses against lung cancer. *Oncoimmunology* **2012**, *1*, 1650–1651. [CrossRef]
119. Ajona, D.; Ortiz-Espinosa, S.; Moreno, H.; Lozano, T.; Pajares, M.J.; Agorreta, J.; Bértolo, C.; Lasarte, J.J.; Vicent, S.; Hoehlig, K. A combined PD-1/C5a blockade synergistically protects against lung cancer growth and metastasis. *Cancer Discov.* **2017**, *7*, 694–703. [CrossRef]
120. Domvri, K.; Petanidis, S.; Anestakis, D.; Porpodis, K.; Bai, C.; Zarogoulidis, P.; Freitag, L.; Hohenforst-Schmidt, W.; Katopodi, T. Dual photothermal MDSCs-targeted immunotherapy inhibits lung immunosuppressive metastasis by enhancing T-cell recruitment. *Nanoscale* **2020**, *12*, 7051–7062. [CrossRef]
121. Yu, G.T.; Rao, L.; Wu, H.; Yang, L.L.; Bu, L.L.; Deng, W.W.; Wu, L.; Nan, X.; Zhang, W.F.; Zhao, X.Z. Myeloid-Derived Suppressor Cell Membrane-Coated Magnetic Nanoparticles for Cancer Theranostics by Inducing Macrophage Polarization and Synergizing Immunogenic Cell Death. *Adv. Funct. Mater.* **2018**, *28*, 1801389. [CrossRef]
122. Erfani, N.; Mehrabadi, S.M.; Ghayumi, M.A.; Haghshenas, M.R.; Mojtahedi, Z.; Ghaderi, A.; Amani, D. Increase of regulatory T cells in metastatic stage and CTLA-4 over expression in lymphocytes of patients with non-small cell lung cancer (NSCLC). *Lung Cancer* **2012**, *77*, 306–311. [CrossRef]
123. Ganesan, A.P.; Johansson, M.; Ruffell, B.; Yagui-Beltran, A.; Lau, J.; Jablons, D.M.; Coussens, L.M. Tumor-infiltrating regulatory T cells inhibit endogenous cytotoxic T cell responses to lung adenocarcinoma. *J. Immunol.* **2013**, *191*, 2009–2017. [CrossRef]
124. Ou, W.; Jiang, L.; Thapa, R.K.; Soe, Z.C.; Poudel, K.; Chang, J.-H.; Ku, S.K.; Choi, H.-G.; Yong, C.S.; Kim, J.O. Combination of NIR therapy and regulatory T cell modulation using layer-by-layer hybrid nanoparticles for effective cancer photoimmunotherapy. *Theranostics* **2018**, *8*, 4574. [CrossRef]
125. Beissert, S.; Schwarz, A.; Schwarz, T. Regulatory T cells. *J. Investig. Dermatol.* **2006**, *126*, 15–24. [CrossRef]
126. Tseng, C.-L.; Wu, S.Y.-H.; Wang, W.-H.; Peng, C.-L.; Lin, F.-H.; Lin, C.-C.; Young, T.-H.; Shieh, M.-J. Targeting efficiency and biodistribution of biotinylated-EGF-conjugated gelatin nanoparticles administered via aerosol delivery in nude mice with lung cancer. *Biomaterials* **2008**, *29*, 3014–3022. [CrossRef]
127. Azarmi, S.; Roa, W.H.; Löbenberg, R. Targeted delivery of nanoparticles for the treatment of lung diseases. *Adv. Drug Deliv. Rev.* **2008**, *60*, 863–875. [CrossRef]
128. Lu, J.; Liong, M.; Zink, J.I.; Tamanoi, F. Mesoporous silica nanoparticles as a delivery system for hydrophobic anticancer drugs. *Small* **2007**, *3*, 1341–1346. [CrossRef] [PubMed]
129. Gasselhuber, A.; Dreher, M.R.; Rattay, F.; Wood, B.J.; Haemmerich, D. Comparison of conventional chemotherapy, stealth liposomes and temperature-sensitive liposomes in a mathematical model. *PLoS ONE* **2012**, *7*, e47453. [CrossRef] [PubMed]
130. Ahmad, J.; Akhter, S.; Rizwanullah, M.; Amin, S.; Rahman, M.; Ahmad, M.Z.; Rizvi, M.A.; Kamal, M.A.; Ahmad, F.J. Nanotechnology-based inhalation treatments for lung cancer: State of the art. *Nanotechnol. Sci. Appl.* **2015**, *8*, 55. [PubMed]
131. Patton, J.S.; Fishburn, C.S.; Weers, J.G. The lungs as a portal of entry for systemic drug delivery. *Proc. Am. Thorac. Soc.* **2004**, *1*, 338–344. [CrossRef]
132. Okamoto, H.; Shiraki, K.; Yasuda, R.; Danjo, K.; Watanabe, Y. Chitosan–interferon-β gene complex powder for inhalation treatment of lung metastasis in mice. *J. Control. Release* **2011**, *150*, 187–195. [CrossRef]
133. Dames, P.; Gleich, B.; Flemmer, A.; Hajek, K.; Seidl, N.; Wiekhorst, F.; Eberbeck, D.; Bittmann, I.; Bergemann, C.; Weyh, T.; et al. Targeted delivery of magnetic aerosol droplets to the lung. *Nat. Nanotechnol.* **2007**, *2*, 495–499. [CrossRef]
134. Ngwa, W.; Kumar, R.; Moreau, M.; Dabney, R.; Herman, A. Nanoparticle drones to target lung cancer with radiosensitizers and cannabinoids. *Front. Oncol.* **2017**, *7*, 208. [CrossRef]
135. Chivere, V.T.; Kondiah, P.P.; Choonara, Y.E.; Pillay, V. Nanotechnology-based biopolymeric oral delivery platforms for advanced cancer treatment. *Cancers* **2020**, *12*, 522. [CrossRef]

136. Sung, J.C.; Pulliam, B.L.; Edwards, D.A. Nanoparticles for drug delivery to the lungs. *Trends Biotechnol.* **2007**, *25*, 563–570. [CrossRef] [PubMed]
137. Gill, S.; Löbenberg, R.; Ku, T.; Azarmi, S.; Roa, W.; Prenner, E.J. Nanoparticles: Characteristics, mechanisms of action, and toxicity in pulmonary drug delivery—A review. *J. Biomed. Nanotechnol.* **2007**, *3*, 107–119. [CrossRef]
138. Lin, W.; Huang, Y.W.; Zhou, X.D.; Ma, Y. Toxicity of cerium oxide nanoparticles in human lung cancer cells. *Int. J. Toxicol.* **2006**, *25*, 451–457. [CrossRef] [PubMed]
139. Aillon, K.L.; Xie, Y.; El-Gendy, N.; Berkland, C.J.; Forrest, M.L. Effects of nanomaterial physicochemical properties on in vivo toxicity. *Adv. Drug Deliv. Rev.* **2009**, *61*, 457–466. [CrossRef] [PubMed]
140. Lin, W.; Huang, Y.-W.; Zhou, X.-D.; Ma, Y. In vitro toxicity of silica nanoparticles in human lung cancer cells. *Toxicol. Appl. Pharmacol.* **2006**, *217*, 252–259. [CrossRef]
141. Rybak-Smith, M. Effect of surface modification on toxicity of nanoparticles. *Encycl. Nanotechnol.* **2012**, *2012*, 645–652.
142. Patra, J.K.; Das, G.; Fraceto, L.F.; Campos, E.V.R.; Rodriguez-Torres, M.D.P.; Acosta-Torres, L.S.; Diaz-Torres, L.A.; Grillo, R.; Swamy, M.K.; Sharma, S.; et al. Nano based drug delivery systems: Recent developments and future prospects. *J. Nanobiotechnol.* **2018**, *16*, 71. [CrossRef]
143. Oyarzun-Ampuero, F.; Kogan, M.J.; Neira-Carrillo, A.; Morales, J.O. Surface-modified nanoparticles to improve drug delivery. *Amino Acids Polyglutamic Acid Polyasparagine* **2014**, *16*, 18.
144. Gautam, M.; Poudel, K.; Yong, C.S.; Kim, J.O. Prussian blue nanoparticles: Synthesis, surface modification, and application in cancer treatment. *Int. J. Pharm.* **2018**, *549*, 31–49. [CrossRef]
145. Yoo, J.-W.; Chambers, E.; Mitragotri, S. Factors that control the circulation time of nanoparticles in blood: Challenges, solutions and future prospects. *Curr. Pharm. Des.* **2010**, *16*, 2298–2307. [CrossRef]
146. Mangal, S.; Gao, W.; Li, T.; Zhou, Q.T. Pulmonary delivery of nanoparticle chemotherapy for the treatment of lung cancers: Challenges and opportunities. *Acta Pharmacol. Sin.* **2017**, *38*, 782–797. [CrossRef] [PubMed]
147. Wen, S.; Liu, H.; Cai, H.; Shen, M.; Shi, X. Targeted and pH-responsive delivery of doxorubicin to cancer cells using multifunctional dendrimer-modified multi-walled carbon nanotubes. *Adv. Healthc. Mater.* **2013**, *2*, 1267–1276. [CrossRef] [PubMed]
148. Meenach, S.A.; Anderson, K.W.; Hilt, J.Z.; McGarry, R.C.; Mansour, H.M. High-performing dry powder inhalers of paclitaxel DPPC/DPPG lung surfactant-mimic multifunctional particles in lung cancer: Physicochemical characterization, in vitro aerosol dispersion, and cellular studies. *AAPS PharmSciTech* **2014**, *15*, 1574–1587. [CrossRef] [PubMed]
149. Li, S.D.; Huang, L. Surface-modified lpd nanoparticles for tumor targeting. *Ann. New York Acad. Sci.* **2006**, *1082*, 1–8. [CrossRef]
150. Gao, J.; Liu, W.; Xia, Y.; Li, W.; Sun, J.; Chen, H.; Li, B.; Zhang, D.; Qian, W.; Meng, Y. The promotion of siRNA delivery to breast cancer overexpressing epidermal growth factor receptor through anti-EGFR antibody conjugation by immunoliposomes. *Biomaterials* **2011**, *32*, 3459–3470. [CrossRef]
151. Grabowski, N.; Hillaireau, H.; Vergnaud, J.; Santiago, L.A.; Kerdine-Romer, S.; Pallardy, M.; Tsapis, N.; Fattal, E. Toxicity of surface-modified PLGA nanoparticles toward lung alveolar epithelial cells. *Int. J. Pharm.* **2013**, *454*, 686–694. [CrossRef]
152. Chung, Y.I.; Kim, J.C.; Kim, Y.H.; Tae, G.; Lee, S.Y.; Kim, K.; Kwon, I.C. The effect of surface functionalization of PLGA nanoparticles by heparin- or chitosan-conjugated Pluronic on tumor targeting. *J. Control. Release* **2010**, *143*, 374–382. [CrossRef]
153. Patil, Y.B.; Toti, U.S.; Khdair, A.; Ma, L.; Panyam, J. Single-step surface functionalization of polymeric nanoparticles for targeted drug delivery. *Biomaterials* **2009**, *30*, 859–866. [CrossRef]
154. Patil, Y.; Sadhukha, T.; Ma, L.; Panyam, J. Nanoparticle-mediated simultaneous and targeted delivery of paclitaxel and tariquidar overcomes tumor drug resistance. *J. Control. Release* **2009**, *136*, 21–29. [CrossRef]
155. Xia, Y.; Chen, Y.; Hua, L.; Zhao, M.; Xu, T.; Wang, C.; Li, Y.; Zhu, B. Functionalized selenium nanoparticles for targeted delivery of doxorubicin to improve non-small-cell lung cancer therapy. *Int. J. Nanomed.* **2018**, *13*, 6929. [CrossRef]
156. Perepelyuk, M.; Maher, C.; Lakshmikuttyamma, A.; Shoyele, S.A. Aptamer-hybrid nanoparticle bioconjugate efficiently delivers miRNA-29b to non-small-cell lung cancer cells and inhibits growth by downregulating essential oncoproteins. *Int. J. Nanomed.* **2016**, *11*, 3533.
157. Li, S.; Fang, C.; Zhang, J.; Liu, B.; Wei, Z.; Fan, X.; Sui, Z.; Tan, Q. Catanionic lipid nanosystems improve pharmacokinetics and anti-lung cancer activity of curcumin. *Nanomedicine* **2016**, *12*, 1567–1579. [CrossRef] [PubMed]
158. Soni, N.; Soni, N.; Pandey, H.; Maheshwari, R.; Kesharwani, P.; Tekade, R.K. Augmented delivery of gemcitabine in lung cancer cells exploring mannose anchored solid lipid nanoparticles. *J. Colloid Interface Sci.* **2016**, *481*, 107–116. [CrossRef] [PubMed]
159. Li, X.-T.; He, M.-L.; Zhou, Z.-Y.; Jiang, Y.; Cheng, L. The antitumor activity of PNA modified vinblastine cationic liposomes on Lewis lung tumor cells: In vitro and in vivo evaluation. *Int. J. Pharm.* **2015**, *487*, 223–233. [CrossRef] [PubMed]
160. Wang, X.-B.; Zhou, H.-Y. Molecularly targeted gemcitabine-loaded nanoparticulate system towards the treatment of EGFR overexpressing lung cancer. *Biomed. Pharmacother.* **2015**, *70*, 123–128. [CrossRef] [PubMed]
161. Li, F.; Mei, H.; Gao, Y.; Xie, X.; Nie, H.; Li, T.; Zhang, H.; Jia, L. Co-delivery of oxygen and erlotinib by aptamer-modified liposomal complexes to reverse hypoxia-induced drug resistance in lung cancer. *Biomaterials* **2017**, *145*, 56–71. [CrossRef] [PubMed]
162. Choi, S.H.; Byeon, H.J.; Choi, J.S.; Thao, L.; Kim, I.; Lee, E.S.; Shin, B.S.; Lee, K.C.; Youn, Y.S. Inhalable self-assembled albumin nanoparticles for treating drug-resistant lung cancer. *J. Control. Release* **2015**, *197*, 199–207. [CrossRef]
163. Singh, R.P.; Sharma, G.; Singh, S.; Patne, S.C.; Pandey, B.L.; Koch, B.; Muthu, M.S. Effects of transferrin conjugated multi walled carbon nanotubes in lung cancer delivery. *Mater. Sci. Eng. C* **2016**, *67*, 313–325. [CrossRef]

164. Almurshedi, A.S.; Radwan, M.; Omar, S.; Alaiya, A.A.; Badran, M.M.; Elsaghire, H.; Saleem, I.Y.; Hutcheon, G.A. A novel pH-sensitive liposome to trigger delivery of afatinib to cancer cells: Impact on lung cancer therapy. *J. Mol. Liq.* **2018**, *259*, 154–166. [CrossRef]
165. Mottaghitalab, F.; Kiani, M.; Farokhi, M.; Kundu, S.C.; Reis, R.L.; Gholami, M.; Bardania, H.; Dinarvand, R.; Geramifar, P.; Beiki, D. Targeted delivery system based on gemcitabine-loaded silk fibroin nanoparticles for lung cancer therapy. *ACS Appl. Mater. Interfaces* **2017**, *9*, 31600–31611. [CrossRef]
166. De Souza Oliveira, R.C.; Corrêa, R.J.; Teixeira, R.S.P.; Queiroz, D.D.; da Silva Souza, R.; Garden, S.J.; de Lucas, N.C.; Pereira, M.D.; Forero, J.S.B.; Romani, E.C. Silica nanoparticles doped with anthraquinone for lung cancer phototherapy. *J. Photochem. Photobiol. B Biol.* **2016**, *165*, 1–9. [CrossRef] [PubMed]
167. Yu, M.K.; Jeong, Y.Y.; Park, J.; Park, S.; Kim, J.W.; Min, J.J.; Kim, K.; Jon, S. Drug-loaded superparamagnetic iron oxide nanoparticles for combined cancer imaging and therapy in vivo. *Angew. Chem. Int. Ed.* **2008**, *47*, 5362–5365. [CrossRef] [PubMed]
168. Peng, G.; Tisch, U.; Adams, O.; Hakim, M.; Shehada, N.; Broza, Y.Y.; Billan, S.; Abdah-Bortnyak, R.; Kuten, A.; Haick, H. Diagnosing lung cancer in exhaled breath using gold nanoparticles. *Nat. Nanotechnol.* **2009**, *4*, 669–673. [CrossRef]
169. Amreddy, N.; Babu, A.; Panneerselvam, J.; Srivastava, A.; Muralidharan, R.; Chen, A.; Zhao, Y.D.; Munshi, A.; Ramesh, R. Chemo-biologic combinatorial drug delivery using folate receptor-targeted dendrimer nanoparticles for lung cancer treatment. *Nanomed. Nanotechnol. Biol. Med.* **2018**, *14*, 373–384. [CrossRef]
170. Ghitman, J.; Biru, E.I.; Stan, R.; Iovu, H. Review of hybrid PLGA nanoparticles: Future of smart drug delivery and theranostics medicine. *Mater. Des.* **2020**, 193. [CrossRef]
171. Li, J.; Zhang, C.; Li, J.; Fan, L.; Jiang, X.; Chen, J.; Pang, Z.; Zhang, Q. Brain delivery of NAP with PEG-PLGA nanoparticles modified with phage display peptides. *Pharm. Res.* **2013**, *30*, 1813–1823. [CrossRef]
172. Karra, N.; Nassar, T.; Ripin, A.N.; Schwob, O.; Borlak, J.; Benita, S. Antibody conjugated PLGA nanoparticles for targeted delivery of paclitaxel palmitate: Efficacy and biofate in a lung cancer mouse model. *Small* **2013**, *9*, 4221–4236. [CrossRef]
173. Patil, M.A.; Upadhyay, A.K.; Hernandez-Lagunas, L.; Good, R.; Carpenter, T.C.; Sucharov, C.C.; Nozik-Grayck, E.; Kompella, U.B. Targeted delivery of YSA-functionalized and non-functionalized polymeric nanoparticles to injured pulmonary vasculature. *Artif. Cells Nanomed. Biotechnol.* **2018**, *46*, S1059–S1066. [CrossRef]
174. Maiz, J.; Verde-Sesto, E.; Asenjo-Sanz, I.; Fouquet, P.; Porcar, L.; Pomposo, J.A.; de Molina, P.M.; Arbe, A.; Colmenero, J. Collective Motions and Mechanical Response of a Bulk of Single-Chain Nano-Particles Synthesized by Click-Chemistry. *Polymers* **2021**, *13*, 50. [CrossRef] [PubMed]
175. De-La-Cuesta, J.; González, E.; Pomposo, J.A. Advances in fluorescent single-chain nanoparticles. *Molecules* **2017**, *22*, 1819. [CrossRef]
176. Verde-Sesto, E.; Arbe, A.; Moreno, A.J.; Cangialosi, D.; Alegría, A.; Colmenero, J.; Pomposo, J.A. Single-chain nanoparticles: Opportunities provided by internal and external confinement. *Mater. Horiz.* **2020**, *7*, 2292–2313. [CrossRef]
177. Benito, A.B.; Aiertza, M.K.; Marradi, M.; Gil-Iceta, L.; Shekhter Zahavi, T.; Szczupak, B.; Jimenez-Gonzalez, M.; Reese, T.; Scanziani, E.; Passoni, L.; et al. Functional Single-Chain Polymer Nanoparticles: Targeting and Imaging Pancreatic Tumors in Vivo. *Biomacromolecules* **2016**, *17*, 3213–3221. [CrossRef]
178. Kroger, A.P.P.; Paulusse, J.M.J. Single-chain polymer nanoparticles in controlled drug delivery and targeted imaging. *J. Control. Release* **2018**, *286*, 326–347. [CrossRef]
179. Kroger, A.P.P.; Komil, M.I.; Hamelmann, N.M.; Juan, A.; Stenzel, M.H.; Paulusse, J.M.J. Glucose Single-Chain Polymer Nanoparticles for Cellular Targeting. *ACS Macro Lett.* **2019**, *8*, 95–101. [CrossRef] [PubMed]
180. Sandoval-Yanez, C.; Castro Rodriguez, C. Dendrimers: Amazing Platforms for Bioactive Molecule Delivery Systems. *Materials* **2020**, *13*, 570. [CrossRef]
181. Taratula, O.; Garbuzenko, O.B.; Kirkpatrick, P.; Pandya, I.; Savla, R.; Pozharov, V.P.; He, H.; Minko, T. Surface-engineered targeted PPI dendrimer for efficient intracellular and intratumoral siRNA delivery. *J. Control. Release* **2009**, *140*, 284–293. [CrossRef] [PubMed]
182. Mishra, V.; Bansal, K.K.; Verma, A.; Yadav, N.; Thakur, S.; Sudhakar, K.; Rosenholm, J.M. Solid Lipid Nanoparticles: Emerging Colloidal Nano Drug Delivery Systems. *Pharmaceutics* **2018**, *10*, 191. [CrossRef]
183. Pooja, D.; Kulhari, H.; Tunki, L.; Chinde, S.; Kuncha, M.; Grover, P.; Rachamalla, S.S.; Sistla, R. Nanomedicines for targeted delivery of etoposide to non-small cell lung cancer using transferrin functionalized nanoparticles. *RSC Adv.* **2015**, *5*, 49122–49131. [CrossRef]
184. Riaz, M.K.; Zhang, X.; Wong, K.H.; Chen, H.; Liu, Q.; Chen, X.; Zhang, G.; Lu, A.; Yang, Z. Pulmonary delivery of transferrin receptors targeting peptide surface-functionalized liposomes augments the chemotherapeutic effect of quercetin in lung cancer therapy. *Int. J. Nanomed.* **2019**, *14*, 2879. [CrossRef]
185. Yang, S.-G.; Chang, J.-E.; Shin, B.; Park, S.; Na, K.; Shim, C.-K. ^{99m}Tc-hematoporphyrin linked albumin nanoparticles for lung cancer targeted photodynamic therapy and imaging. *J. Mater. Chem.* **2010**, *20*, 9042–9046. [CrossRef]
186. Wang, F.; Li, C.; Cheng, J.; Yuan, Z. Recent Advances on Inorganic Nanoparticle-Based Cancer Therapeutic Agents. *Int. J. Env. Res. Public Health* **2016**, *13*, 1182. [CrossRef]
187. Tagliazucchi, M.; Blaber, M.G.; Schatz, G.C.; Weiss, E.A.; Szleifer, I. Optical properties of responsive hybrid au@polymer nanoparticles. *ACS Nano* **2012**, *6*, 8397–8406. [CrossRef] [PubMed]

188. Slaughter, L.S.; Willingham, B.A.; Chang, W.S.; Chester, M.H.; Ogden, N.; Link, S. Toward plasmonic polymers. *Nano Lett.* **2012**, *12*, 3967–3972. [CrossRef]
189. Heo, D.N.; Yang, D.H.; Moon, H.-J.; Lee, J.B.; Bae, M.S.; Lee, S.C.; Lee, W.J.; Sun, I.-C.; Kwon, I.K. Gold nanoparticles surface-functionalized with paclitaxel drug and biotin receptor as theranostic agents for cancer therapy. *Biomaterials* **2012**, *33*, 856–866. [CrossRef] [PubMed]
190. Guo, X.; Zhuang, Q.; Ji, T.; Zhang, Y.; Li, C.; Wang, Y.; Li, H.; Jia, H.; Liu, Y.; Du, L. Multi-functionalized chitosan nanoparticles for enhanced chemotherapy in lung cancer. *Carbohydr. Polym.* **2018**, *195*, 311–320. [CrossRef]
191. Conde, J.; Ambrosone, A.; Sanz, V.; Hernandez, Y.; Marchesano, V.; Tian, F.; Child, H.; Berry, C.C.; Ibarra, M.R.; Baptista, P.V.; et al. Design of multifunctional gold nanoparticles for in vitro and in vivo gene silencing. *ACS Nano* **2012**, *6*, 8316–8324. [CrossRef]
192. Munaweera, I.; Shi, Y.; Koneru, B.; Patel, A.; Dang, M.H.; di Pasqua, A.J.; Balkus, K.J. Nitric oxide- and cisplatin-releasing silica nanoparticles for use against non-small cell lung cancer. *J. Inorg. Biochem.* **2015**, *153*, 23–31. [CrossRef]
193. Vennemann, A.; Alessandrini, F.; Wiemann, M. Differential effects of surface-functionalized zirconium oxide nanoparticles on alveolar macrophages, rat lung, and a mouse allergy model. *Nanomaterials* **2017**, *7*, 280. [CrossRef]
194. Liu, T.; Wang, C.; Gu, X.; Gong, H.; Cheng, L.; Shi, X.; Feng, L.; Sun, B.; Liu, Z. Drug delivery with PEGylated MoS2 nano-sheets for combined photothermal and chemotherapy of cancer. *Adv. Mater.* **2014**, *26*, 3433–3440. [CrossRef]
195. Zhang, W.; Yang, J.; Wu, D. Surface-functionalized MoS2 nanosheets sensor for direct electrochemical detection of PIK3CA gene related to lung cancer. *J. Electrochem. Soc.* **2020**, *167*, 027501. [CrossRef]
196. Saadat, M.; Manshadi, M.K.D.; Mohammadi, M.; Zare, M.J.; Zarei, M.; Kamali, R.; Sanati-Nezhad, A. Magnetic particle targeting for diagnosis and therapy of lung cancers. *J. Control. Release* **2020**, *328*, 776–791. [CrossRef] [PubMed]
197. Mukherjee, S.; Liang, L.; Veiseh, O. Recent Advancements of Magnetic Nanomaterials in Cancer Therapy. *Pharmaceutics* **2020**, *12*, 147. [CrossRef] [PubMed]
198. Bloemen, M.; van Stappen, T.; Willot, P.; Lammertyn, J.; Koeckelberghs, G.; Geukens, N.; Gils, A.; Verbiest, T. Heterobifunctional PEG ligands for bioconjugation reactions on iron oxide nanoparticles. *PLoS ONE* **2014**, *9*, e109475. [CrossRef] [PubMed]
199. Maleki, H.; Simchi, A.; Imani, M.; Costa, B. Size-controlled synthesis of superparamagnetic iron oxide nanoparticles and their surface coating by gold for biomedical applications. *J. Magn. Magn. Mater.* **2012**, *324*, 3997–4005. [CrossRef]
200. Huang, G.; Chen, H.; Dong, Y.; Luo, X.; Yu, H.; Moore, Z.; Bey, E.A.; Boothman, D.A.; Gao, J. Superparamagnetic iron oxide nanoparticles: Amplifying ROS stress to improve anticancer drug efficacy. *Theranostics* **2013**, *3*, 116. [CrossRef]
201. Hauser, A.K.; Mitov, M.I.; Daley, E.F.; McGarry, R.C.; Anderson, K.W.; Hilt, J.Z. Targeted iron oxide nanoparticles for the enhancement of radiation therapy. *Biomaterials* **2016**, *105*, 127–135. [CrossRef]
202. Hauser, A.K.; Mathias, R.; Anderson, K.W.; Hilt, J.Z. The effects of synthesis method on the physical and chemical properties of dextran coated iron oxide nanoparticles. *Mater. Chem. Phys.* **2015**, *160*, 177–186. [CrossRef]
203. Sailor, M.J.; Park, J.H. Hybrid nanoparticles for detection and treatment of cancer. *Adv. Mater.* **2012**, *24*, 3779–3802. [CrossRef]
204. Sacko, K.; Thangavel, K.; Sunday, A. Shoyele. Codelivery of genistein and miRNA-29b to A549 cells using aptamer-hybrid nanoparticle bioconjugates. *Nanomaterials* **2019**, *9*, 1052. [CrossRef]
205. Perepelyuk, M.; Thangavel, C.; Liu, Y.; Den, R.B.; Lu, B.; Snook, A.E.; Shoyele, S.A. Biodistribution and pharmacokinetics study of siRNA-loaded anti-NTSR1-mAb-functionalized novel hybrid nanoparticles in a metastatic orthotopic murine lung cancer model. *Mol. Ther. Nucleic Acids* **2016**, *5*, e282. [CrossRef] [PubMed]
206. Sadhukha, T.; Wiedmann, T.S.; Panyam, J. Inhalable magnetic nanoparticles for targeted hyperthermia in lung cancer therapy. *Biomaterials* **2013**, *34*, 5163–5171. [CrossRef] [PubMed]
207. Mandal, B.; Bhattacharjee, H.; Mittal, N.; Sah, H.; Balabathula, P.; Thoma, L.A.; Wood, G.C. Core–shell-type lipid–polymer hybrid nanoparticles as a drug delivery platform. *Nanomed. Nanotechnol. Biol. Med.* **2013**, *9*, 474–491. [CrossRef] [PubMed]
208. Mandal, B.; Mittal, N.K.; Balabathula, P.; Thoma, L.A.; Wood, G.C. Development and in vitro evaluation of core-shell type lipid-polymer hybrid nanoparticles for the delivery of erlotinib in non-small cell lung cancer. *Eur. J. Pharm. Sci.* **2016**, *81*, 162–171. [CrossRef] [PubMed]
209. Song, Z.; Shi, Y.; Han, Q.; Dai, G. Endothelial growth factor receptor-targeted and reactive oxygen species-responsive lung cancer therapy by docetaxel and resveratrol encapsulated lipid-polymer hybrid nanoparticles. *Biomed. Pharm.* **2018**, *105*, 18–26. [CrossRef] [PubMed]
210. Guo, Y.; Wang, L.; Lv, P.; Zhang, P. Transferrin-conjugated doxorubicin-loaded lipid-coated nanoparticles for the targeting and therapy of lung cancer. *Oncol. Lett.* **2015**, *9*, 1065–1072. [CrossRef] [PubMed]
211. Ramasamy, T.; Tran, T.H.; Choi, J.Y.; Cho, H.J.; Kim, J.H.; Yong, C.S.; Choi, H.G.; Kim, J.O. Layer-by-layer coated lipid-polymer hybrid nanoparticles designed for use in anticancer drug delivery. *Carbohydr. Polym.* **2014**, *102*, 653–661. [CrossRef] [PubMed]
212. Li, S.; Wang, L.; Li, N.; Liu, Y.; Su, H. Combination lung cancer chemotherapy: Design of a pH-sensitive transferrin-PEG-Hz-lipid conjugate for the co-delivery of docetaxel and baicalin. *Biomed. Pharm.* **2017**, *95*, 548–555. [CrossRef]
213. Itani, R.; Al Faraj, A. siRNA Conjugated Nanoparticles-A Next Generation Strategy to Treat Lung Cancer. *Int. J. Mol. Sci.* **2019**, *20*, 6088. [CrossRef]
214. Yang, X.Z.; Dou, S.; Wang, Y.C.; Long, H.Y.; Xiong, M.H.; Mao, C.Q.; Yao, Y.D.; Wang, J. Single-step assembly of cationic lipid-polymer hybrid nanoparticles for systemic delivery of siRNA. *ACS Nano* **2012**, *6*, 4955–4965. [CrossRef] [PubMed]
215. Li, J.; Yang, Y.; Huang, L. Calcium phosphate nanoparticles with an asymmetric lipid bilayer coating for siRNA delivery to the tumor. *J. Control. Release* **2012**, *158*, 108–114. [CrossRef] [PubMed]

216. Foldbjerg, R.; Dang, D.A.; Autrup, H. Cytotoxicity and genotoxicity of silver nanoparticles in the human lung cancer cell line, A549. *Arch. Toxicol.* **2011**, *85*, 743–750. [CrossRef]
217. Wang, C.; Li, X.; Wang, Y.; Liu, Z.; Fu, L.; Hu, L. Enhancement of radiation effect and increase of apoptosis in lung cancer cells by thio-glucose-bound gold nanoparticles at megavoltage radiation energies. *J. Nanopart. Res.* **2013**, *15*, 1–12. [CrossRef]
218. Liu, F.; Sun, J.; Yu, W.; Jiang, Q.; Pan, M.; Xu, Z.; Mo, F.; Liu, X. Quantum dot-pulsed dendritic cell vaccines plus macrophage polarization for amplified cancer immunotherapy. *Biomaterials* **2020**, *242*, 119928. [CrossRef] [PubMed]
219. Zhang, T.; Chen, Y.; Ge, Y.; Hu, Y.; Li, M.; Jin, Y. Inhalation treatment of primary lung cancer using liposomal curcumin dry powder inhalers. *Acta Pharm. Sin. B* **2018**, *8*, 440–448. [CrossRef] [PubMed]
220. Samadi, S.; Moradkhani, M.; Beheshti, H.; Irani, M.; Aliabadi, M. Fabrication of chitosan/poly (lactic acid)/graphene oxide/TiO_2 composite nanofibrous scaffolds for sustained delivery of doxorubicin and treatment of lung cancer. *Int. J. Biol. Macromol.* **2018**, *110*, 416–424. [CrossRef] [PubMed]
221. Kim, S.-W.; Lee, Y.K.; Lee, J.Y.; Hong, J.H.; Khang, D. PEGylated anticancer-carbon nanotubes complex targeting mitochondria of lung cancer cells. *Nanotechnology* **2017**, *28*, 465102. [CrossRef] [PubMed]
222. Mohamad Saimi, N.I.; Salim, N.; Ahmad, N.; Abdulmalek, E.; Abdul Rahman, M.B. Aerosolized Niosome Formulation Containing Gemcitabine and Cisplatin for Lung Cancer Treatment: Optimization, Characterization and In Vitro Evaluation. *Pharmaceutics* **2021**, *13*, 59. [CrossRef] [PubMed]
223. Hamishehkar, H.; Bahadori, M.B.; Vandghanooni, S.; Eskandani, M.; Nakhlband, A.; Eskandani, M. Preparation, characterization and anti-proliferative effects of sclareol-loaded solid lipid nanoparticles on A549 human lung epithelial cancer cells. *J. Drug Deliv. Sci. Technol.* **2018**, *45*, 272–280. [CrossRef]
224. Rossi, S.M.; Ryan, B.K.; Kelly, H.M. Evaluation of the activity of a chemo-ablative, thermoresponsive hydrogel in a murine xenograft model of lung cancer. *Br. J. Cancer* **2020**, *123*, 369–377. [CrossRef]
225. Zhang, Q.; Liu, Q.; Du, M.; Vermorken, A.; Cui, Y.; Zhang, L.; Guo, L.; Ma, L.; Chen, M. Cetuximab and Doxorubicin loaded dextran-coated Fe_3O_4 magnetic nanoparticles as novel targeted nanocarriers for non-small cell lung cancer. *J. Magn. Magn. Mater.* **2019**, *481*, 122–128. [CrossRef]
226. Md, S.; Alhakamy, N.A.; Aldawsari, H.M.; Husain, M.; Kotta, S.; Abdullah, S.T.; Fahmy, U.A.; Alfaleh, M.A.; Asfour, H.Z. Formulation Design, Statistical Optimization, and In Vitro Evaluation of a Naringenin Nanoemulsion to Enhance Apoptotic Activity in A549 Lung Cancer Cells. *Pharmaceuticals* **2020**, *13*, 152. [CrossRef]
227. Wang, M.; Wang, K.; Deng, G.; Liu, X.; Wu, X.; Hu, H.; Zhang, Y.; Gao, W.; Li, Q. Mitochondria-Modulating Porous Se@ SiO_2 Nanoparticles Provide Resistance to Oxidative Injury in Airway Epithelial Cells: Implications for Acute Lung Injury. *Int. J. Nanomed.* **2020**, *15*, 2287. [CrossRef] [PubMed]
228. Wei, D.; Xin, Y.; Rong, Y.; Li, Y.; Zhang, C.; Chen, Q.; Qin, S.; Wang, W.; Hao, Y. A Mesoporous Gd-MOF with Lewis Basic Sites for 5-Fu Delivery and Inhibition of Human Lung Cancer Cells In Vivo and In Vitro. *J. Inorg. Organomet. Polym. Mater.* **2019**, *30*, 1–11. [CrossRef]
229. Hoffner, B.; Leighl, N.B.; Davies, M. Toxicity management with combination chemotherapy and programmed death 1/programmed death ligand 1 inhibitor therapy in advanced lung cancer. *Cancer Treat. Rev.* **2020**, *85*, 101979. [CrossRef] [PubMed]
230. Lv, S.; Tang, Z.; Li, M.; Lin, J.; Song, W.; Liu, H.; Huang, Y.; Zhang, Y.; Chen, X. Co-delivery of doxorubicin and paclitaxel by PEG-polypeptide nanovehicle for the treatment of non-small cell lung cancer. *Biomaterials* **2014**, *35*, 6118–6129. [CrossRef] [PubMed]
231. Jabbari, S.; Ghamkhari, A.; Javadzadeh, Y.; Salehi, R.; Davaran, S. Doxorubicin and chrysin combination chemotherapy with novel pH-responsive poly [(lactide-co-glycolic acid)-block-methacrylic acid] nanoparticle. *J. Drug Deliv. Sci. Technol.* **2018**, *46*, 129–137. [CrossRef]
232. Chatterjee, D.K.; Diagaradjane, P.; Krishnan, S. Nanoparticle-mediated hyperthermia in cancer therapy. *Ther. Deliv.* **2011**, *2*, 1001–1014. [CrossRef]
233. Huang, J.Y.; Chen, M.H.; Kuo, W.T.; Sun, Y.J.; Lin, F.H. The characterization and evaluation of cisplatin-loaded magnetite–hydroxyapatite nanoparticles (mHAp/CDDP) as dual treatment of hyperthermia and chemotherapy for lung cancer therapy. *Ceram. Int.* **2015**, *41*, 2399–2410. [CrossRef]
234. Desai, N.; Trieu, V.; Yao, Z.; Louie, L.; Ci, S.; Yang, A.; Tao, C.; De, T.; Beals, B.; Dykes, D.; et al. Increased antitumor activity, intratumor paclitaxel concentrations, and endothelial cell transport of cremophor-free, albumin-bound paclitaxel, ABI-007, compared with cremophor-based paclitaxel. *Clin. Cancer Res.* **2006**, *12*, 1317–1324. [CrossRef] [PubMed]
235. Kim, D.W.; Kim, S.Y.; Kim, H.K.; Kim, S.W.; Shin, S.W.; Kim, J.S.; Park, K.; Lee, M.Y.; Heo, D.S. Multicenter phase II trial of Genexol-PM, a novel Cremophor-free, polymeric micelle formulation of paclitaxel, with cisplatin in patients with advanced non-small-cell lung cancer. *Ann. Oncol.* **2007**, *18*, 2009–2014. [CrossRef]
236. Lee, K.S.; Chung, H.C.; Im, S.A.; Park, Y.H.; Kim, C.S.; Kim, S.B.; Rha, S.Y.; Lee, M.Y.; Ro, J. Multicenter phase II trial of Genexol-PM, a Cremophor-free, polymeric micelle formulation of paclitaxel, in patients with metastatic breast cancer. *Breast Cancer Res. Treat.* **2008**, *108*, 241–250. [CrossRef] [PubMed]
237. Ventola, C.L. Progress in Nanomedicine: Approved and Investigational Nanodrugs. *Pharm. Ther.* **2017**, *42*, 742–755.
238. Havel, H.; Finch, G.; Strode, P.; Wolfgang, M.; Zale, S.; Bobe, I.; Youssoufian, H.; Peterson, M.; Liu, M. Nanomedicines: From Bench to Bedside and Beyond. *AAPS J.* **2016**, *18*, 1373–1378. [CrossRef] [PubMed]
239. Bobo, D.; Robinson, K.J.; Islam, J.; Thurecht, K.J.; Corrie, S.R. Nanoparticle-Based Medicines: A Review of FDA-Approved Materials and Clinical Trials to Date. *Pharm. Res.* **2016**, *33*, 2373–2387. [CrossRef]

240. Caster, J.M.; Patel, A.N.; Zhang, T.; Wang, A. Investigational nanomedicines in 2016: A review of nanotherapeutics currently undergoing clinical trials. *Wiley Interdiscip. Rev. Nanomed. Nanobiotechnol.* **2017**, *9*. [CrossRef]
241. Sainz, V.; Conniot, J.; Matos, A.I.; Peres, C.; Zupancic, E.; Moura, L.; Silva, L.C.; Florindo, H.F.; Gaspar, R.S. Regulatory aspects on nanomedicines. *Biochem. Biophys. Res. Commun.* **2015**, *468*, 504–510. [CrossRef]
242. U.S. Food & Drug Administration. Novel Drug Approvals for 2017. FDA 2018. Available online: https://www.fda.gov/drugs/new-drugs-fda-cders-new-molecular-entities-and-new-therapeutic-biological-products/novel-drug-approvals-2018 (accessed on 31 March 2021).
243. U.S. Food & Drug Administration. *Novel Drug Approvals for 2016*; U.S. Food & Drug Administration: Washington, DC, USA, 2016.
244. Tse, T.; Fain, K.M.; Zarin, D.A. How to avoid common problems when using ClinicalTrials.gov in research: 10 issues to consider. *BMJ* **2018**, *361*, k1452. [CrossRef]

processes

MDPI

Review

Effect of Alumina Additives on Mechanical and Fresh Properties of Self-Compacting Concrete: A Review

Hoofar Shokravi [1,*], Seyed Esmaeil Mohammadyan-Yasouj [2], Seyed Saeid Rahimian Koloor [3], Michal Petrů [4] and Mahshid Heidarrezaei [5]

[1] School of Civil Engineering, Faculty of Engineering, Universiti Teknologi Malaysia, Skudai 81310, Johor, Malaysia

[2] Department of Civil Engineering, Najafabad Branch, Islamic Azad University, Najafabad 8514143131, Iran; semy2016@pci.iaun.ac.ir

[3] Institute for Nanomaterials, Advanced Technologies, and Innovation (CXI), Technical University of Liberec (TUL), Studentska 2, 461 17 Liberec, Czech Republic; s.s.r.koloor@gmail.com

[4] Technical University of Liberec (TUL), Studentska 2, 461 17 Liberec, Czech Republic; michal.petru@tul.cz

[5] School of Chemical & Energy Engineering, Faculty of Engineering, Universiti Teknologi Malaysia, Skudai 81310, Johor, Malaysia; mahshidheidarrezaei2@graduate.utm.my

* Correspondence: shokravihoofar@utm.my

Citation: Shokravi, H.; Mohammadyan-Yasouj, S.E.; Koloor, S.S.R.; Petrů, M.; Heidarrezaei, M. Effect of Alumina Additives on Mechanical and Fresh Properties of Self-Compacting Concrete: A Review. *Processes* **2021**, *9*, 554. https://doi.org/10.3390/pr9030554

Academic Editor: Shaghayegh Hamzehlou

Received: 18 February 2021
Accepted: 15 March 2021
Published: 22 March 2021

Abstract: Self-compacting concrete (SCC) has been increasingly used in the construction sector due to its favorable characteristics in improving various durability and rheology aspects of concrete such as deformability and segregation resistance. Recently, the studies on the application of nano-alumina (NA) produced from factory wastes have been significantly considered to enhancing the performance, and mechanical strength, of SCC. Many experimental works show that NA can be used in SCC with appropriate proportion to enjoy the benefits of improved microstructure, fresh and hardened properties, durability, and resistance to elevated temperature. However, a limited detailed review is available to particularly study using NA to improve the performance of SCC, so far. Hence, the present study is conducted to fill the existing gap of knowledge. In this study, the effect of using NA in improving rheological, mechanical parameters, and elevated temperature resistance of SCC is reviewed. This research summarized the studies in this area, which have been different from the previous researches, and provided a discussion on limitations, practical implications, and suggestions for future studies.

Keywords: self-compacting concrete; self-consolidating concrete; waste alumina; nano alumina; nanoparticles

1. Introduction

Concrete is a construction material that is widely used in buildings [1,2], bridges [3–5], and other civil structures [6,7]. The application of nanoparticles in concrete has received great attention recently because of the ultrafine size of their particles [8]. A limited number of nanoparticles have demonstrated utility for improving the durability and mechanical properties of concrete. Nazari and Riahi [9] reported that the addition of nano-alumina (NA) particles into concrete mixtures can enhance strength gaining, water permeability, and the pore structure characteristics of concrete. Khoshakhlagh et al. [10] indicated that the inclusion of Fe_2O_3 nanoparticles in cementitious materials improves the compressive strength. Moreover, Fe_2O_3 nanoparticles act as nanofillers to recover the pore structure enhancing the water permeability of concrete.

Self-compacting concrete (SCC) is concrete with enhanced fresh properties that allow pouring without external compaction [11,12]. SCC was first reported in Japan in 1988 [13]. SCC contains the same components as conventional concrete but with different proportions and fresh characteristics [14]. The main fresh characteristic of SCC is high workability that enables the concrete to fill formwork to achieve full compaction without vibration [15]. The compressive strength of SCC compared to the ordinary concrete with the same water to cement ratio is considerably higher [16]. The higher compressive strength of SCC is attributed to its dense microstructure as compared to conventional concrete [17]. Apart from high workability, SCC must possess a high filling ability, passing ability, and resistance to static and dynamic segregation [18]. Nazari and Riahi [19] indicated that the inclusion of SiO_2 nanoparticles enhances the flexural strength of SCC and accelerates cement hydration. The inclusion of TiO_2 nanoparticles also can improve the formation of C-S-H gel in SCC resulting in faster hydration and improved growth of the mechanical and durability properties of concrete [20–22].

NA is a chemical compound containing aluminum and oxygen [23,24]. The addition of NA to concrete can significantly affect the fresh properties of concrete due to their high ratio of surface area to volume [25]. NA has high chemical reactivity and behaves as pozzolanic reaction promoters owing to its high ratio of the surface area [26]. NA can improve the mechanical parameters of cementitious composites exposed to elevated temperatures [27]. It was reported that the inclusion of NA in SCC can accelerate the formation of hydrated products, and pore structure while reducing the workability of fresh concrete, water absorption of hardened specimens [28]. Table 1 shows the advantages and disadvantages of SCC which has been exploited from literature. As can be seen from the presented pros and cons analysis, the advantages are significantly higher than the inconveniences provoked. In Figure 1 (a and b) the application of roller alumina in tile manufacturing factories and the factory wastes of the alumina rollers are shown, respectively.

(a) (b)

Figure 1. The application of roller alumina in tile manufacturing factories; (**a**) ceramic roller kiln, (**b**) factory wastes.

Since the introduction of SCC, a large number of studies are conducted on enhancing the engineering properties of the SCC as well as the prediction of the loading capacity of SCC based on the associated mixture design. Despite the large volume of literature on the production of SCC and the relevant admixtures such as nano-particles, little attention has been given to predicting the complex interaction between incorporated admixtures and mechanical characteristics of the SCC. To the authors' knowledge, there is no review paper on the prediction model for the compressive strength of SCC. Some papers have discussed the prediction of the mechanical and rhetorical behavior of SCC but their main focus is only on mixed design. This study attempts to fill part of this void in the literature by offering a discussion on the application of NA in SCC.

Table 1. The advantages and disadvantages of self-compacting concrete (SCC).

Advantages	Ref.	Disadvantages	Ref.
Speeded up construction	[29]	Prolonged demolding time	[30]
Improved the construction quality	[31]	Increased risk and associated uncertainty	[32]
Safer work conditions	[33]	Lowered elevated temperature resistance	[34]
The increased service life of formworks due to the elimination of vibration	[31]	Higher formwork pressure means higher formwork costs.	[35]
Improved quality of the final product	[33]	Not fully known fire behavior	[36]
Reduced manpower	[33]	Maintaining ready-mixed is not easy under the construction site	[37]
Improved ecological footprint	[38]	Not appropriate for every application	[39]
Improved economic	[38]	Unsuitable choice for horizontal castings	[40]
Enhanced filling spaces in dense reinforcement or inaccessible voids	[38]	Higher associated costs for ready-mixed	[41]
Improved freeze-thaw resistance	[42]	Using conventional drum mixers are not suitable for the distribution	[43]
Noise-free working atmosphere	[31]	Not standardized mix design	[44]

2. Nanoparticles

Nanotechnology has become a popular and necessary part of science and technology in recent years by addressing nanoparticles in atomic or molecular size [45]. Nanoparticles are defined as materials where at least one dimension of a particle is less than 100 nm. Partial replacement of nano-materials with cement in the mix design can enhance the physical and chemical characteristics of fresh and hardened concrete [46]. The addition of nanoparticles can improve the microstructural properties, filler effect, compactness, and durability, as well as accelerating cement hydration [47]. Nanoparticles can also act as pozzolanic materials and produce the additional formation of calcium–silicate–hydrates (C–S–H) gel and, by taking place of pozzolanic reactions. The C–S–H gel Formation can improve stiffness, flexural, tensile, and shear strength of cement-concrete [48]. Uniform dispersion of nanoparticles in concrete is the key issue in obtaining the expected mechanical and chemical characteristics [49,50]. The applied nanoparticles in the concrete mix to partially replace cement are spherical shapes cementitious materials.

The addition of nanoparticles can improve the microstructural properties, compactness, and durability of hardened concrete [19,48]. The effect of nanoparticles as partial cement replacement in the concrete mix has been studied by many researchers in recent years. Nano-SiO_2 is the most widely investigated nanoparticle [19,51–53] while the effect of adding other partials such as nano-TiO_2 [27,54,55], nano-Al_2O_3 [9,56,57], nano-ZnO_2 [58,59], nano-Fe_2O_3 [10,60], nano-CuO [61], nano-SnO_2, nano-ZrO_2 [62], nano-TiO_2 [63], carbon nanotubes [64], carbon nano-fibers [45], polycarboxylates [65], nano-Cr_2O_3 [48,66], nano-clay [45] and nano-$CaCO_3$ [62,67] in properties of fresh and hardened concrete has been investigated so far. For example, Tawfik et al. [68] evaluated the effect of nano-waste ceramic and nano-silica on the mechanical properties of hardened concrete. The obtained results showed improvement in the performance of concrete but also resolved the footprint caused by this waste. Jalal et al. [69] studied the effect of TiO_2 nanoparticle inclusion in tensile strength, thermal, rheological, transport, and microstructural properties of SCC. The chemical effect of TiO_2 as partial replacement of cement accelerates the formation of C–S–H gel and hydration resulting in increased split tensile strength of concrete specimens. Moreover adding TiO_2 nanoparticles can enhance the pore structure of concrete by shifting the distributed pores into a less harmful configuration. Joshaghani et al. [20] evaluated the fresh, mechanical, and durability properties of nano-TiO_2, nano-Al_2O_3 and nano-Fe_2O_3, on SCC two different contents of 3% and 5%. It was observed that addition of 3% nanoparticles can slightly improve workability properties of the mixes by increasing the water demand. Calcium ferric hydrate (C-F-H) gel formation enhanced the compressive strength and durability properties in nano-Fe_2O_3, nano-Al_2O_3 and nano-TiO_2. It was reported that nanoparticles addition controlled the formation of C-S-H gel, lowering permeability to penetration of malicious ions of chloride.

Fresh properties of concrete containing nanoparticles are one of the most investigated subjects. Workability, flowability, and consistency of concrete are greatly affected by the addition of nanoparticles [70]. Generally, the flowability of SCC mixes is reduced by the addition of nanomaterials [71]. This reduction is mainly attributed to the ability of nanoparticles to absorb more water molecules due to their large area surface [20]. Mechanical properties of hardened concrete including flexural, tensile, shear, and compressive strength and their change due to incorporation of the microparticle in concrete are another most important study in recent years [72]. It is stated that nanoparticles act as nuclei to form hydration products filling micropores [73]. The formation of a dense C-S-H gel is facilitated by altering cement hydration that leads to an increase in compressive strength [74]. Adding excessive amounts of nanoparticles may adversely affect the compressive strength due to restricting the $Ca(OH)_2$ crystals growth. Several research studies have investigated the influence of nanoparticles on the durability of hardened concrete. It was stated that the water absorption of SCC mixes with nanoparticles is different from the control specimens due to the formation of hydrated products [75]. On the other hand, the addition of nanoparticles may affect the capillary permeability of concrete and the specimens containing nanoparticles can better resist chloride penetration [76,77].

3. Production Processes of Nano Alumina

There are different methods for the extraction and production of alumina nanoparticles. Alumina is mainly extracted from two main resources of clays and coal fly ash (CFA) [78]. The row materials undergo several chemical processes for extracting their alumina contents. The production phase of nano alumina is performed by arc plasma, precipitation, hydrothermal, and sol-gel methods. Functionalization is the last step in NA production aiming at improving surface characteristics [79]. The functionalization process prevents agglomeration between alumina nano-particles that is mainly caused by the high surface energy and activity of nanoparticles [80,81].

3.1. Extraction of Alumina Nanoparticles

Clay is a natural mineral that is widely used to produce nano alumina due to its quite abundant availability and low cost. Kaoline is clay made from kaolinite $Al_2O_3.2SiO_22H_2O$ that contains high alumina content ranged between 25 to 40%. Kaolin is a product of weathering of all granitic rocks which is characterized by its fire resistance, good plasticity, and other unique chemical and physical properties [82]. Kaolin is a chemically inert material within a wide pH range and it is not listed as a hazardous material [83]. The alumina extraction process from clays is performed using acids such as nitric acid, hydrochloric, or sulfuric acid to dissolve the alumina followed by clay roasting. Heavy metal ores of clay are extracted using acid leachants [78]. The chemical reaction for removing heavy metals from clays can be shown below.

$$Al_2O_3H_2O + 3H_2SO_4 \rightarrow Al_2(SO_4)_3 + 4H_2O \quad (1)$$

CFA is another major source of alumina nanoparticles. CFA is rich in alumina and alumina contents in CFA is found around 50%. The process of alumina extraction from CFA includes three important steps namely sintering [84], hydro-chemical [85], and acid processes [86]. The process of extracting alumina particles from CFA is shown in Figure 2.

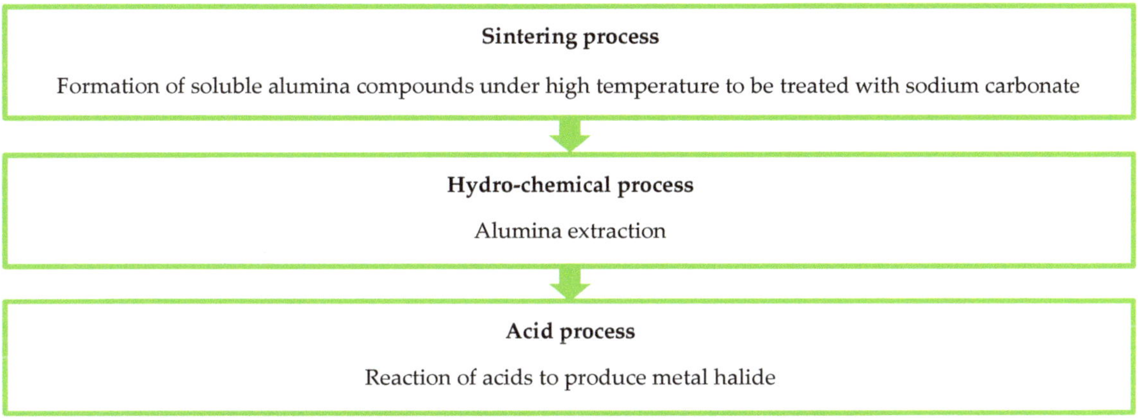

Figure 2. The process of alumina extraction from coal fly ash (CFA).

3.2. Production of Alumina Nanoparticles

Different methods are developed to produce alumina nano-particles that include arc plasma, hydrothermal, sol-gel, and precipitation. The arc plasma process is defined as a thermal treatment of solid feedstock utilizing an arc between anode and cathode generally made of graphitic carbon for generating plasma. Using plasma for the production of alumina has lower residue as well as lower production and maintenance costs compared to traditional gas and oil burners [87]. Saravanakumar et al. [88] converted aluminum dross into ultrafine alumina powder using a plasma arc melting process. It was reported that the amount of the conversion of Al dross to alumina powder substantially correlates with plasma power. Madhu Kumar et al. [89] and Kumar et al. [90] used a D.C. arc plasma reactor for the preparation of alumina nano-particles. Fu et al. [91] adopted microwave oxygen plasma to prepare alumina nanoparticles sized 21–24 nm. Stanislaus et al. [92] investigated the hydrothermal process for the production of alumina nanoparticles in the presence and absence of various additives. Noguchi et al. [6] used hydrothermal reaction in supercritical water using a continuous flow reactor to prepare alumina -crystalline nano particles.

Sol-gel and coprecipitation, two common methods to extract nano alumina with different sizes and morphologies. spherical particles of alumina can be prepared by using the sol-gel method while spherical and hexagonal alumina particles can be formed by the co-precipitation method [93,94]. Mirjalili et al. [95] synthesized alumina nanoparticles by a sol-gel method. It was shown that the addition of surfactant and incorporated stirring time are parameters that affect the shape and size of the formed particle. Belekar et al. [96] obtained alumina granules with an average size of 30 nm by a modified sol-gel method. The process included hydrolysis of Al $(NO_3)_3$ in aqueous media. Esmaeilirad et al. [97] prepared alumina by heterogeneous deposition-precipitation, sol-gel, and the co-precipitation methods. The alumina prepared by the sol-gel method using La-Cu/AlSE showed the best performance. Feng et al. [98] used aluminum powders as the aluminum source and acetic acid as precipitants to prepare alumina powders in a precipitation process.

4. Sustainability of Nano Alumina

The use of nanoparticles is not yet a cost-effective option for the construction industry and particularly as a concrete additive. Hence, meaningful approaches are needed to be carried out to overcome that limitation. The valorization of industrial waste is a sustainable way to wisely utilize renewable resources. Aluminum dross is a toxic industrial waste generated from aluminum refining industries that contain aluminum oxynitride (AlON), aluminum oxide (Al_2O_3), aluminum metal (Al), and impurities such as potassium chloride and sodium chloride [88]. Based on reports nearly 95% of aluminum dross is landfilled without treatment that is hazardous to the environment in China [99]. Using conventional disposal or landfilling practices of aluminum wastes without proper treatment and recycling strategy can adversely contribute to human health due to the toxic nature of materials [100]. Hence aluminum dross should be converted into inert or less toxic products [88].

EI-Katatny et al. [101] used caustic soda for leaching an aluminum factory waste under atmospheric and high-pressure aluminum to form alumina. David and Kopac [102] proposed an alumina extraction method from aluminum dross using a chemical route. Dash et al. [103] recycled aluminum dross using acid dissolution and salt treatment for recovering residual aluminum. Das et al. [104] adopted acid treatment for the recycling of aluminum dross to obtain alumina generated by Indian aluminum industries. Sarker et al. [105] used an acid dissolution process to extract alumina from Bangladesh foundry industries. Singh et al. [106] optimized a chemical process for the production of alumina through recycling aluminum dross. How et al. [107] recovered alumina from the wastes of an aluminum production factory in Malaysia. The recovering process of alumina consisted of acid leaching, alkaline precipitation, and calcinations steps. Most of the aforementioned processes incorporated alkaline salts for the treatment of aluminum dross that are hazardous to groundwater and agricultural soil and it is important to remove these chemicals before discharging them to the environment.

Plasma processing of materials is a fast and environmentally-friendly method to treat and high volume reduction of various types of wastes. Szente et al. [87] conducted a comparative study on using plasma systems and traditional oil and gas burners for recovering aluminum from dross. It was reported that the plasma process provides cheaper operation costs (at least 23%) with lower residues than oil/gas burners. Yang et al. [108] treated aluminum dross using a radio frequency-based plasma to recover high-purity fine aluminum oxide (with the size of 8 μm). Saravanakumar et al. [88] reported using arc plasma to convert dross of aluminum into alumina powder in an eco-friendly manner. The obtained results indicated that the application of arc plasma can be efficiently used to treat aluminum dross to recover alumina powder.

5. Self-Compacting Concrete (SCC)

SCC mixes always contain a large number of powder materials, viscosity-modifying admixtures, and superplasticizer [109–111]. Higher cement content in concrete has some negative effects such as a rise in material cost, increased thermal stress, and shrinkage [112]. The requirement for cement replacements in SCC is usually met by the use of filler materials such as fly ash (FA), pulverized fuel ash (PFA), marble powder (MP), basalt powder (BP), granulated ground blast-furnace slag (GGBS), limestone powder, etc. [112]. Uysal and Sumer [113] investigated the effect of different mineral admixtures on the properties of SCC such as durability, workability, and reducing cement content. The replacement of Portland cement with FA, GGBS, BP, and MP increases the fluidity and can improve mechanical properties, and durability of the SCC against sulfate attack. Fathi et al. [114] also testified that fibers reduce the slump and compressive strength of SCC but increase its flexural tensile strength. Talking about mechanical properties, Ahmad et al. [115] compared the mechanical properties of normal concrete (NC), SCC, and glass fiber reinforced SCC. It was observed that addition of fiber glass to SCC decreased the workability of the concrete but it significantly enhanced flexural of ruptures in test specimens. The change in compressive strength by addition of glass fibers was small enough to be ignorable.

SCC is extensively used in various types of structures such as commercial buildings and industrial structures that are subjected to high temperatures or accidental fires. Hence, gaining proper information on the effects of high temperatures on the performance of SCC is necessary. The effect of high-temperature on the behavior of SCC was studied by many researchers. Anand et al. [116] reviewed the effect of the elevated temperature on the chemical and mechanical properties of concrete. Distinct behavior was found in mechanical properties of normal, high strength, and SCC when they are exposed to high temperatures. It was revealed that parameters such as the compressive, tensile, and flexural strength of concrete, water-cement ratio, cement type, the density of concrete, aggregate type, reinforcement percentage, and reinforcement cover are some of the important factors that affect the concrete performance at elevated temperature.

Annerel et al. [117] revealed that the thermal influence of raised temperature on the mechanical behavior of SCC is much significant compared to normal concrete. Pineaud et al. [118] studied the mechanical properties of high-performance SCC under raised temperatures ranged from 20 to 600 ° C. The results of experiments on 11 different mix designs showed that increasing the temperature reduces their E-value and compressive strength significantly. Andiç-Çakır and Hızal [119] explored the properties of SCC under raised temperature ranged from 300 to 900 °C. It was shown that the aggregate type and w/c ratio are the most influential parameters in the modulus of elasticity and compressive strength of the SCC while the aggregate type is the main influential parameter in tensile strength. Table 2 shows some of these changes within a various temperature range of exposure. Further studies on the influence of raised temperature on the mechanical properties of concrete can be found somewhere else [120–122]. There are limited studies on the behavior of SCC with nanoparticles and subjected to high-temperature [123,124].

Table 2. The changes in physical and chemical parameters of concrete due to exposure to elevated temperature (data are exploited from [120]).

Investigated Parameter	Temperature Range	Effect of Temperature Rise
Compressive strength	100–800 °C.	• Decreases in a linear rate
Porosity and pore size	100–800 °C. Above 1000 °C,	• Increase of porosity and pore sizes • Porosities are smaller and better structured
Elastic modulus.	100–800 °C.	• Decreases in a linear rate
Splitting tensile strength	100–800 °C.	• Decreases in a linear rate
Stress-strain relationship	100–800 °C	• Flatter stress-strain curves, downwards and rightwards shift of the peak stress
Residual flexural strength	100–800 °C	• Decreases in a linear rate
Water evaporation	At 105 °C At 400 °C	• Free water and physically absorbed water are completely lost. Chemically bonded water start to lose • Capillary water is lost completely

Table 2. *Cont.*

Investigated Parameter	Temperature Range	Effect of Temperature Rise
Hydration	Up to 300 °C	• Hydration of un-hydrated cement is improved
Microstructure	Up to 200 °C 200 °C–400 °C	• No micro-cracks • The intensity of micro-cracks increases

6. Nano Alumina (NA) Applications in SCC

Nanotechnology is a research area that has revolutionized the mechanical and chemical properties of materials [125,126]. Nanotechnology is a promising research field with applications to improve the quality of the product and the performance of concretes [127]. The nanoparticle is applied in SCC aiming to reduce segregation and to modify fresh properties and mechanical strength., and. NA is a kind of ultra-fine chemical compound of aluminum and oxygen with a large surface area, high density, high melting point, high hardness, and good chemical stability with particle sizes in the range of 1~100 nm [128]. The advantages of using NA in concrete are presented in Table 3.

Table 3. Advantages of using nano-alumina (NA) in concrete.

Advantages	Reference
Reduced porosity of the microstructure as the voids were filled by NS.	[129]
Decreased in water absorption	[130]
Improved frost resistance of concrete	[131]
Controlled the setting time of the cement through a faster hydration process will be.	[132]
Reduced amount of un-hydrated cement in the mix	[129]
Increased modulus of elasticity of cement mortar.	[133]
Reduced segregation and flocculation.	[124]
Refined voids in the hydration gel as a nanofiller.	[124]
Reduced coefficients of permeability by 1–3 orders of magnitude.	[133]

Sua-iam et al. [134] studied the effect of using recycled NA and fly ash in SCC. It was shown that using recycled NA and fly ash could considerably enhance the compressive strength and workability of SCC. Mohseni et al. studied the effects of NA and rice husk ash (RHA) in polypropylene fiber (PPF) reinforced concrete. The combined inclusion of, NA, PPF, and RHA reduced the water absorption and drying shrinkage of mortars and increased flexural strength. Farzadnia et al. [135] investigated the chemical composition, microstructure and mechanical properties of NA-based high strength mortars subjected to elevated temperatures ranged from 100 °C to 1000 °C. It was indicated that the addition of NA improved 16% of the compressive strength of samples. Behfarnia and Salemi [136] studied the influence of NS and NA on frost resistance and mechanical properties of concrete. Higher frost-resistance was achieved for concrete with NA while concrete containing NS had higher compressive strength. Zhan et al. [137] analyzed the effect of NA in the hydration of cement. Accelerated cement hydration and enhanced compressive strength at all ages were recorded. Owing to accelerated cement hydration, the strongest growth at 28 days was less obvious. Mohammadyan-Yasouj et al. [27] investigated the compressive strength and modulus of elasticity of each SCC mix design under temperatures of 27 °C, 100 °C, 200 °C, 300 °C, 450 °C, and 600 °C for specimens cured in 7 and 28-days. It was observed that addition of NA into the mix enhances the compressive strength of SCC for samples cured at 28-day in temperature under 100 °C. E-value of the samples cured on 28-day exhibited increasing trend. Figure 3 shows the comparison compressive strength of self-compacting concrete (SCC) at target temperatures for specimens cured at 28-days.

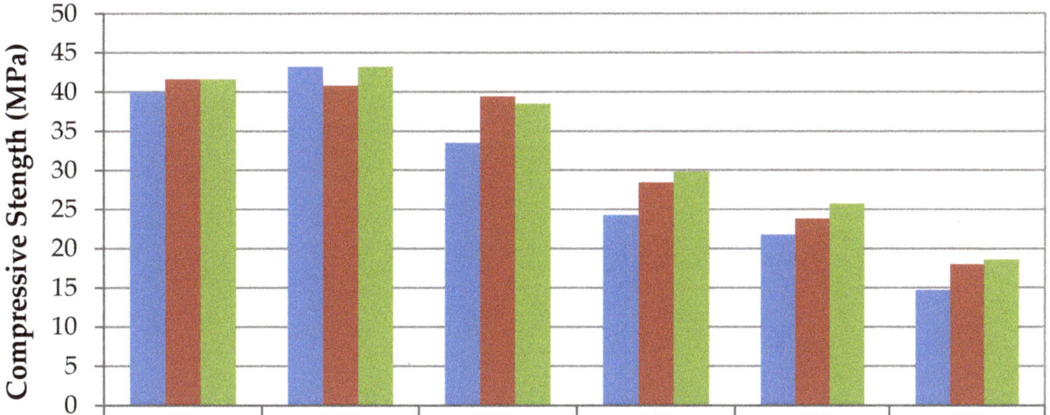

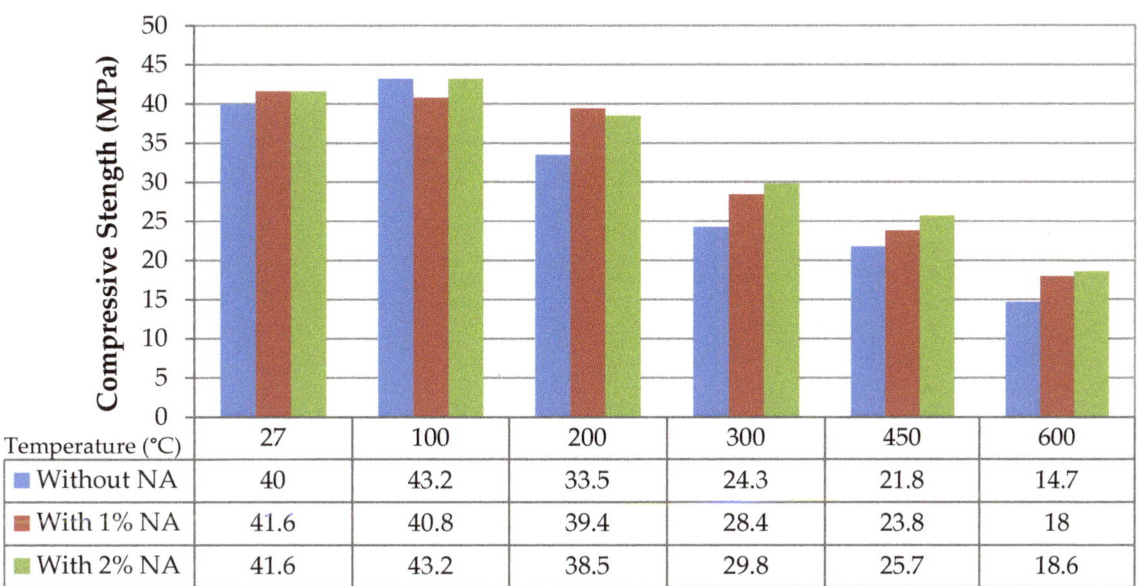

Temperature (°C)	27	100	200	300	450	600
Without NA	40	43.2	33.5	24.3	21.8	14.7
With 1% NA	41.6	40.8	39.4	28.4	23.8	18
With 2% NA	41.6	43.2	38.5	29.8	25.7	18.6

Figure 3. Comparison of 28-day compressive strength of self-compacting concrete (SCC) at target temperatures [27].

Szymanowski and Sadowski [138] explored the applicability of NA mortars for overlaying concrete floors. It was shown that the mortar with Al_2O_3 nano-powder used to make the overlay was less porous than the reference mortar. Li et al. [139] investigated the effect of NA on the elastic modulus of cement composite. Based on the experiment, the elastic modulus of mortars was increased by incorporating NA into the matrix. Nazari and Riahi [9] investigated the effect of curing medium on the physical, mechanical, and thermal properties of NA concrete. It was observed that curing of the specimens in saturated limewater resulted in accelerated setting time. Barbhuiya et al. [140] reported the effects of NA inclusion on the early-age microstructural properties of cement paste. It was revealed a denser microstructure with larger crystals of portlandite in the cement matrix. Mohseni et al. [77] conducted a comparative study on the single and the combined effects of adding NS), nano alumina (NA), and NT on the durability and mechanical properties of SCC. The ternary combination of NA, NS, and NT had the best permeability to chloride ingress and electrical resistivity at 3%. Heikal et al. [141] studied the resistance of cement pastes with NA against fire. By enhancing the hydration reaction of cement phases higher compressive strength, bulk density, gel/space ratio was reported for the prepared cement paste. Khaliq and Khan [121] investigated the material properties of calcium aluminate cement concrete (CACC) under elevated temperatures. It was revealed that the presence of alumina as a binding agent significantly enhances the mechanical performance and axial strain in CACC. Niewiadomski et al. [42] presented a comparative study to assess the effect of NA, NS, and NT nanoparticle additives. It was indicated that the addition of up to 4 wt.% of NA and NT showed no improvement in concrete strength. Table 4 summarizes the literature on the application of NA in SCC.

Table 4. Summary of the literature on the application of nano-alumina (NA) in SCC.

Ref.	Research Topic	Test Age (day)	Sample Size	W/C (%)	Alumina				Remarks
					Type	Size (mm)	Amount of Alumina	Investigated Parameters	
Suaiam and Makul [134,142]	Rheological and mechanical properties SCC with high volumes of alumina	28 and 90 days	Φ150 mm × 300 mm	0.38	Alumina	Below ≈5 mm	0, 25, 50, 75, and 100% of the total fine aggregate	• Velocity measurements • Compressive strength, • V-funnel, • Blocking flow assessment • J-ring flow • Ultrasonic pulse • Slump flow,	• SCC mixtures containing waste alumina had 75% higher compressive strength.
Mohseni et al. [143]	Cement mortars containing rice husk ash, Polypropylene fiber (PPF), and nano-alumina (NA)	28 and 90 days	50 × 50 × 50 mm^3 and 50 × 50 × 200 mm^3	0.49	NA	20 nm	0, 1, 2 and 3%,	• Compressive strength, • Flexural strength, • Water absorption and • Drying shrinkage	• NA improved the compressive strength of mortar. • The addition of RHA, NA, and PPF reduced both drying shrinkage and water absorption and increased flexural strength.
Gowda et al. [129]	Durability and microstructural properties of mortars with NA	28 days	70.5× 70.5× 70.5 mm^3	0.79	NA	-	1, 3 and 5%	• Water absorption • Electrical resistivity • Scanning electron microscope (SEM)	• Increasing of the NA increased water absorption • Electrical resistivity was almost the same for 3% and 5% NA. • The addition of NA gave a denser microstructure in the mortar.

Table 4. *Cont.*

Ref.	Research Topic	Test Age (day)	Sample Size	W/C (%)	Alumina				Remarks
					Type	Size (mm)	Amount of Alumina	Investigated Parameters	
Farzadnia et al. [135]	Elevated temperature's effect on NA -based mortars	1, 7, and 28 days	50 × 50 × 50 mm^3 and Φ20 mm × 50 mm	0.35	NA	the average size of 13 nm	0, 1, 2 and 3%	• Differential scanning calorimetry (DSC) • X-ray powder diffractometry (XRD) • SEM tests • Residual compressive strength, • Elastic modulus, and • Measurement of gas permeability coefficient	• NA enhanced the compressive strength of mortars by 16%. • NA improved residual properties of mortars at temperatures from 0 °C to 400 °C. • The addition of 1% NA reduced gas permeability of mortars at temperatures from 0 °C to 600 °C. • The inclusion of NA reduced the intensity of CH formation at room temperature and 400 °C. • The NA inclusion transformed the CH crystallization from well to ill.
Behfarnia and Salemi [136]	Frost resistance of NA concrete	7, 28 and 120 days	70 × 70 × 70 mm^3.	0.48	NA	15 nm	1% and more than 2%	• Compressive strength • Loss of mass measurement • Change in length • Water absorption test	• NA increased the compressive strength of concrete. • Frost resistance of concrete mixes improved by the addition of NA.

Table 4. *Cont.*

Ref.	Research Topic	Test Age (day)	Sample Size	W/C (%)	Alumina				Remarks
					Type	Size (mm)	Amount of Alumina	Investigated Parameters	
Zhan et al. [137]	Effect of NA on early hydration and mechanical properties of cement pastes	1, 3, 7, and 28 days	cubes of size 20 mm × 20 mm × 20 mm cubes of size 6 mm × 6 mm × 13 mm	0.456	NA	30 nm	0%, 1%, 2% and 4%	• Compressive strength • Isothermal conduction calorimetry • Thermogravimetric analysis • Backscattered electron imaging and energy dispersive spectroscopy • Mercury intrusion porosimetry (MIP)	• NA speeded the aluminate and silicate phases reactions in ordinary cement. • NA improved compressive strength at all ages. • NA modified pastes at 12 h.
Szymanowski and Sadowski [138]	NA-based cement mortars for concrete floors	28 days	$71 \times 71 \times 71$ mm^3, $11 \times 11 \times 11$ mm^3, and $40 \times 40 \times 80$ mm^3	0.3	NA	Below 50 nm	0.5, 1 and 1.5%	• Compressive and flexural strength. • Subsurface tensile strength • Abrasion resistance • Hardness.	• The addition of 0.5% of NA increased subsurface tensile strength and reduced abrasion resistance • The addition of 0.5% of NA resulted in a less porous mortar than the reference.

Table 4. *Cont.*

Ref.	Research Topic	Test Age (day)	Sample Size	W/C (%)	Alumina				Remarks
					Type	Size (mm)	Amount of Alumina	Investigated Parameters	
Li et al. [39]	Preparation and mechanical properties of the NA based cement composite	3 days, 7 days, 28 days	Φ20 × 40 mm	0.4	NA	Below 150 nm	5% and 7%	• Elastic modulus • Compressive strength	• In 5% of NA, the E-value of composites improved by 143%. • The compressive strength of 5% NA composites increased by 30% at 7 days.
Nazari and Riahi [9]	Different curing media for NA in concrete	7, 28 and 90 days	Cubes of 100 mm edge for compressive strength tests, cubes with 200 mm × 50 mm × 50 mm	0.40	NA	15 nm	0.5%, 1.5% and 2%	• Compressive strength water absorption • XRD analysis Split tensile strength Flexural strength • Thermogravimetric analyzer (TGA)	• Al_2O_3 nanoparticles have significantly higher strength. • The optimum level of NA content was achieved by 1.0%. • The pore structure of SCC containing NA is improved and the content of all mesopores and macropores is increased.

Table 4. *Cont.*

Ref.	Research Topic	Test Age (day)	Sample Size	W/C (%)	Alumina				Remarks
					Type	Size (mm)	Amount of Alumina	Investigated Parameters	
Barbhuiya et al. [140]	Early-age microstructural properties of adding NA to cement paste	1, 3, 7, and 28 days	50 × 50 × 50 mm^3	0.40	NA	27–43 nm	2% and 4%	• Compressive strength water absorption • XRD analysis • FTIR analysis • Electron microscopy Scanning	• No noticeable change was observed in the early-age compressive strength by NA addition. • The addition of NA induced no phase change.
Mohseni et al. [77]	Effects of NA on mechanical, rheological, and durability properties of SCC	3, 7, 28 and 90 days	50 × 50 × 50 mm^3 and Φ100 × 50 mm	0.4	NA	15 nm	1%, 3%, and 5%	• Water absorption • Electrical resistivity • Rapid chloride permeability tests	• NA improved the flexural and compressive strengths of specimens with 10% and 20% RHA. • The most effective amount of NA was 3% by weight of the binder. • Water absorption decreased with the increase of NA dosage up to 3%. • The formation of denser microstructure with NA addition.

Table 4. *Cont.*

Ref.	Research Topic	Test Age (day)	Sample Size	W/C (%)	Alumina				Remarks
					Type	Size (mm)	Amount of Alumina	Investigated Parameters	
Heikal et al. [141]	Physico-mechanical, microstructure characteristics and fire resistance of cement pastes with NA -	3, 7, 28, and 90 days.	-		NA	15 ± 3 nm	1, 2, 4 and 6 %	• Setting times, • Bulk density • Gel/space ratio • Compressive strength • XRD • TEM • Differential thermal analysis (DTA)/TGA • Free lime contents (FL, %)	• NA addition shortens the setting times. • 1% NA enhanced the compressive strength of the pastes up to 27.22%. • NA speeded up the hydration. • 1% NA enhanced the fire resistance.
Khaliq and Khan [121]	High-temperature material properties of calcium aluminate cement concrete	3, 14, and 28 days	Φ100 mm × 200 mm	0.5	calcium aluminate cement concrete	-	-	• SEM • Compressive and tensile strength • Elastic modulus • Compressive toughness • Visual investigations	• Better tensile and compressive strength was achieved in temperature up to 800 °C. • stress-strain response and E value were improved. • Compressive toughness was improved in temperature up to 800 °C.

Table 4. *Cont.*

Ref.	Research Topic	Test Age (day)	Sample Size	W/C (%)	Alumina				Remarks
					Type	Size (mm)	Amount of Alumina	Investigated Parameters	
Niewiadomski et al. [42]	Self-compacting concrete modified with NA	28 and 90 days	100 × 100 × 100 mm^3, 40 × 40 × 160 mm^3, and Φ25 mm × 20 mm	0.42	NA	Below 50 nm	Cement replacement 0.5 wt.%, 1.0 wt.%, 2.0 wt.% and 3.0 wt.% of the.	• Compressive strength • Flexural strength • hardness • Elastic modulus	• The fluidity of concrete deteriorated with an increased amount of NA. • The samples with the addition of NA had a larger size of air pores than the reference sample. • Concretes with NA showed no improvement in compressive and flexural strengths.

7. Conclusions

This review is aimed to investigate the effect of NA on rheology and mechanical parameters of SCC subjected to elevated temperature. Then the potential of predicting the mechanical parameters of SCC is further studied. It was concluded from the literature review that the rheology parameters of the fresh SCC are significantly affected by the addition of NA due to their high surface-area-to-volume ratio. The addition of NA reduces slump flow diameter while the variation of the workability is contributed with the replacement ratio. NA has lower water and chloride ions permeability that is due to the accelerated hydration process. Moreover, enhanced durability and compressive strength are achieved by adding a small amount of NA to SCC. Therefore, the use of NA enhances the performance of SCC. The addition of NA to SCC results in a denser microstructure as compared to normal SCC. The authors believe that future research in the field of NA should further concentrate on enhancing the elevated temperature resistance and thermal behavior of SCC to enable more sustainable and durable construction and to extend the SCC use to more application fields.

Author Contributions: Conceptualization, H.S., and S.E.M.-Y.; methodology, H.S., S.E.M.-Y. and M.H.; software, H.S.; validation, H.S., and S.E.M.-Y.; formal analysis, H.S., S.E.M.-Y. and M.H.; investigation, H.S., and S.E.M.-Y.; resources, H.S., S.E.M.-Y., M.H., S.S.R.K., and M.P.; data curation, H.S.; writing—original draft preparation, H.S., S.E.M.-Y. and M.H.; writing—review and editing, H.S., S.E.M.-Y., M.H., M.P. and S.S.R.K.; visualization, H.S., S.E.M.-Y., M.H. S.S.R.K., and M.P.; supervision, H.S.; project administration, H.S., S.E.M.-Y. M.H., S.S.R.K., and M.P. funding acquisition H.S., S.E.M.-Y., M.H., S.S.R.K., and M.P. All authors have read and agreed to the published version of the manuscript.

Funding: The research was supported by the Islamic Azad University, Najafabad Branch, and the Ministry of Education, Youth, and Sports of the Czech Republic and the European Union (European Structural and Investment Funds Operational Program Research, Development, and Education) in the framework of the project "Modular platform for autonomous chassis of specialized electric vehicles for freight and equipment transportation", Reg. No. CZ.02.1.01/0.0/0.0/16_025/0007293.

Acknowledgments: The authors would like to acknowledge the invaluable help of N. Heidari during the preparation of this manuscript.

Conflicts of Interest: The authors declare no conflict of interest.

References

1. Shokravi, H.; Shokravi, H.; Bakhary, N.; Heidarrezaei, M.; Rahimian Koloor, S.S.; Petrů, M. Application of the subspace-based methods in health monitoring of civil structures: A Systematic review and meta-analysis. *Appl. Sci.* **2020**, *10*, 3607. [CrossRef]
2. Shokravi, H.; Shokravi, H.; Shokravi, N.; Bakhary, N.; Koloor, S.S.R.; Petrů, M. Health monitoring of civil infrastructures by subspace system identification method: An overview. *Appl. Sci.* **2020**, *10*, 2786. [CrossRef]
3. Shokravi, H.; Shokravi, H.; Shokravi, N.; Bakhary, N.; Heidarrezaei, M.; Koloor, S.R.K.; Petru, M. A review on vehicle classification methods and the potential of using smart-vehicle-assisted techniques. *Sensors* **2020**, *20*, 3274. [CrossRef] [PubMed]
4. Shokravi, H.; Shokravi, H.; Shokravi, N.; Bakhary, N.; Heidarrezaei, M.; Koloor, S.R.K.; Petru, M. Vehicle-assisted techniques for health monitoring of bridges. *Sensors* **2020**, *20*, 3460. [CrossRef] [PubMed]
5. Shokravi, H.; Shokravi, H.; Shokravi, N.; Bakhary, N.; Koloor, S.S.R.; Petrů, M. A comparative study of the data-driven stochastic subspace methods for health monitoring of structures: A bridge case study. *Appl. Sci.* **2020**, *10*, 3132. [CrossRef]
6. Shokravi, H.; Mohammadyan-Yasouj, S.E.; Rafiei, P.; Rahimian Koloor, S.S.; Petrů, M. Temperature impact on engineered cementitious composite containing basalt fibers. *J. Mater. Res. Technol.* **2021**. (Under review).
7. Mohammadyan-Yasouj, S.E.; Ahangar, H.A.; Oskoei, N.A.; Shokravi, H.; Koloor, S.S.R.; Petrů, M. Thermal performance of alginate concrete reinforced with basalt fiber. *Crystals* **2020**, *10*, 779. [CrossRef]
8. Praveenkumar, T.R.; Vijayalakshmi, M.M.; Meddah, M.S. Strengths and durability performances of blended cement concrete with TiO_2 nanoparticles and rice husk ash. *Constr. Build. Mater.* **2019**, *217*, 343–351. [CrossRef]
9. Nazari, A.; Riahi, S. Improvement compressive strength of concrete in different curing media by Al_2O_3 nanoparticles. *Mater. Sci. Eng. A* **2011**, *528*, 1183–1191. [CrossRef]
10. Khoshakhlagh, A.; Nazari, A.; Khalaj, G. Effects of Fe_2O_3 nanoparticles on water permeability and strength assessments of high strength self-compacting concrete. *J. Mater. Sci. Technol.* **2012**, *28*, 73–82. [CrossRef]
11. Karatas, M.; Benli, A.; Arslan, F. The effects of kaolin and calcined kaolin on the durability and mechanical properties of self-compacting mortars subjected to high temperatures. *Constr. Build. Mater.* **2020**, *265*, 120300. [CrossRef]

12. Maia, L.; Neves, D. Developing a commercial self-compacting concrete without limestone filler and with volcanic aggregate materials. *Procedia Struct. Integr.* **2017**, *5*, 147–154. [CrossRef]
13. Ibrahim, M.H.W.; Hamzah, A.F.; Jamaluddin, N.; Ramadhansyah, P.J.; Fadzil, A.M. Split tensile strength on self-compacting concrete containing coal bottom ash. *Procedia Soc. Behav. Sci.* **2015**, *195*, 2280–2289.
14. Saha, P.; Debnath, P.; Thomas, P. Prediction of fresh and hardened properties of self-compacting concrete using support vector regression approach. *Neural Comput. Appl.* **2019**, *32*, 1–16. [CrossRef]
15. Ramkumar, K.B.; Rajkumar, P.R.K.; Ahmmad, S.N.; Jegan, M. A review on performance of self-compacting concrete–use of mineral admixtures and steel fibres with artificial neural network application. *Constr. Build. Mater.* **2020**, *261*, 120215. [CrossRef]
16. Salami, B.A.; Rahman, S.M.; Oyehan, T.A.; Maslehuddin, M.; Al Dulaijan, S.U. Ensemble machine learning model for corrosion initiation time estimation of embedded steel reinforced self-compacting concrete. *Measurement* **2020**, *165*, 108141. [CrossRef]
17. Saafan, M.A.A.; Bait AL-Shab, T. Behavior of self-compacting concrete in simulated hot weather. *ERJ. Eng. Res. J.* **2020**, *43*, 223–230. [CrossRef]
18. Benaicha, M.; Roguiez, X.; Jalbaud, O.; Burtschell, Y.; Alaoui, A.H. New approach to determine the plastic viscosity of self-compacting concrete. *Front. Struct. Civ. Eng.* **2016**, *10*, 198–208. [CrossRef]
19. Nazari, A.; Riahi, S. The effects of SiO_2 nanoparticles on physical and mechanical properties of high strength compacting concrete. *Compos. Part B Eng.* **2011**, *42*, 570–578. [CrossRef]
20. Joshaghani, A.; Balapour, M.; Mashhadian, M.; Ozbakkaloglu, T. Effects of nano-TiO_2, nano-Al_2O_3, and nano-Fe_2O_3 on rheology, mechanical and durability properties of self-consolidating concrete (SCC): An experimental study. *Constr. Build. Mater.* **2020**, *245*. [CrossRef]
21. Nazari, A.; Riahi, S. The effects of TiO_2 nanoparticles on physical, thermal and mechanical properties of concrete using ground granulated blast furnace slag as binder. *Mater. Sci. Eng. A* **2011**, *528*, 2085–2092. [CrossRef]
22. Nazari, A.; Riahi, S. The effect of TiO_2 nanoparticles on water permeability and thermal and mechanical properties of high strength self-compacting concrete. *Mater. Sci. Eng. A* **2010**, *528*, 756–763. [CrossRef]
23. Faez, A.; Sayari, A.; Manie, S. Mechanical and rheological properties of self-compacting concrete containing Al_2O_3 nanoparticles and silica fume. *Iran. J. Sci. Technol. Trans. Civ. Eng.* **2020**, *44*, 1–11. [CrossRef]
24. Kaplan, G.; Demircan, R.K.; Cakmak, C.; Gultekin, A.B. Microwave curing effect on internal sulfate damage in nano alumina reinforced white cement. *IOP Conf. Ser. Mater. Sci. Eng.* **2019**, *471*, 032038. [CrossRef]
25. Meddah, M.S.; Praveenkumar, T.R.; Vijayalakshmi, M.M.; Manigandan, S.; Arunachalam, R. Mechanical and microstructural characterization of rice husk ash and Al_2O_3 nanoparticles modified cement concrete. *Constr. Build. Mater.* **2020**, *255*, 119358. [CrossRef]
26. Salemi, N.; Behfarnia, K. Effect of nano-particles on durability of fiber-reinforced concrete pavement. *Constr. Build. Mater.* **2013**, *48*, 934–941. [CrossRef]
27. Mohammadyan-Yasouj, S.E.; Heidari, N.; Shokravi, H. Influence of waste alumina powder on self-compacting concrete resistance under elevated temperature. *J. Build. Eng.* **2021**. [CrossRef]
28. De Schutter, G.; Bartos, P.J.M.; Domone, P.; Gibbs, J. *Self-Compacting Concrete*; Whittles Publishing: Caithness, Scotland, 2008.
29. Tande, S.N.; Mohite, P.B. Applications of self compacting concrete. In Proceedings of the International Conference Concrete Structure, Singapore, 12 March 2007.
30. Aslani, F.; Liu, Y.; Wang, Y. The effect of NiTi shape memory alloy, polypropylene and steel fibres on the fresh and mechanical properties of self-compacting concrete. *Constr. Build. Mater.* **2019**, *215*, 644–659. [CrossRef]
31. Basu, P.; Agrawal, V.; Gupta, R.C. Permeation characteristics of self-compacting concrete containing sandstone slurry. *Mater. Today Proc.* **2020**, *26*, 1590–1593. [CrossRef]
32. Simonsson, P.; Emborg, M. Industrialized construction: Benefits using SCC in cast in-situ construction. *Nord. Concr. Res.* **2009**, *39*, 33–52.
33. Georgiadis, A.S.; Sideris, K.K.; Anagnostopoulos, N.S. Properties of SCC produced with limestone filler or viscosity modifying admixture. *J. Mater. Civ. Eng.* **2010**, *22*, 352–360. [CrossRef]
34. Bilir, T.; Kockal, N.U.; Khatib, J.M. Properties of SCC at elevated temperature. In *Self-Compacting Concrete: Materials, Properties and Applications*; Elsevier: Amsterdam, The Netherlands, 2020; pp. 195–218.
35. Kwon, S.H.; Shah, S.P.; Phung, Q.T.; Kim, J.H.; Lee, Y. Intrinsic model to predict formwork pressure. *ACI Mater. J.* **2010**, *107*, 20.
36. Reinhardt, H.W.; Stegmaier, M. Self-consolidating concrete in fire. *ACI Mater. J.* **2006**, *103*, 130.
37. Kim, J.H.; Noemi, N.; Shah, S.P. Effect of powder materials on the rheology and formwork pressure of self-consolidating concrete. *Cem. Concr. Compos.* **2012**, *34*, 746–753. [CrossRef]
38. Omrane, M.; Kenai, S.; Kadri, E.-H.; Aït-Mokhtar, A. Performance and durability of self compacting concrete using recycled concrete aggregates and natural pozzolan. *J. Clean. Prod.* **2017**, *165*, 415–430. [CrossRef]
39. Koehler, E.P. Aggregates in Self-Consolidating Concrete. Ph.D. Thesis, University of Texas, Austin, TX, USA, 2007.
40. Skarendahl, A. Changing concrete construction through use of self-compacting concrete. In Proceedings of the First International Symposium on Design, Performance and Use of Self-Consolidating Concrete, Changsha, China, 26–28 May 2005; RILEM Publications SARL: Hunan, Kina, 2005; pp. 17–24.
41. D'Souza, B.; Yamamiya, H. Applications of smart dynamic concrete. In Proceedings of the International Symposium on Design, Performance and Use of Self-Consolidating Concrete, Kyoto, Japan, 19–21 August 2013; pp. 18–21.

42. Niewiadomski, P.; Hoła, J.; Ćwirzeń, A. Study on properties of self-compacting concrete modified with nanoparticles. *Arch. Civ. Mech. Eng.* **2018**, *18*, 877–886. [CrossRef]
43. Hemalatha, T.; Sundar, K.R.R.; Murthy, A.R.; Iyer, N.R. Influence of mixing protocol on fresh and hardened properties of self-compacting concrete. *Constr. Build. Mater.* **2015**, *98*, 119–127. [CrossRef]
44. Pai, B.H.V.; Nandy, M.; Krishnamoorthy, A.; Sarkar, P.K.; George, P. Comparative study of self compacting concrete mixes containing fly ash and rice husk ash. *Am. J. Eng. Res.* **2014**, *3*, 150–154.
45. Dolatabad, Y.A.; Kamgar, R.; Nezad, I.G. Rheological and mechanical properties, acid resistance and water penetrability of lightweight self-compacting concrete containing nano-SiO_2, nano-TiO_2 and nano-Al_2O_3. *Iran. J. Sci. Technol. Trans. Civ. Eng.* **2019**, *44*, 1–16.
46. Sumesh, M.; Alengaram, U.J.; Jumaat, M.Z.; Mo, K.H.; Alnahhal, M.F. Incorporation of nano-materials in cement composite and geopolymer based paste and mortar—A review. *Constr. Build. Mater.* **2017**, *148*, 62–84. [CrossRef]
47. Arefi, M.R.; Rezaei-Zarchi, S. Synthesis of zinc oxide nanoparticles and their effect on the compressive strength and setting time of self-compacted concrete paste as cementitious composites. *Int. J. Mol. Sci.* **2012**, *13*, 4340–4350. [CrossRef] [PubMed]
48. Nazari, A.; Riahi, S. The effects of Cr_2O_3 nanoparticles on strength assessments and water permeability of concrete in different curing media. *Mater. Sci. Eng. A* **2011**, *528*, 1173–1182. [CrossRef]
49. Lin, K.L.; Chang, W.C.; Lin, D.F.; Luo, H.L.; Tsai, M.C. Effects of nano-SiO_2 and different ash particle sizes on sludge ash–cement mortar. *J. Environ. Manag.* **2008**, *88*, 708–714. [CrossRef] [PubMed]
50. Qing, Y.; Zenan, Z.; Deyu, K.; Rongshen, C. Influence of nano-SiO_2 addition on properties of hardened cement paste as compared with silica fume. *Constr. Build. Mater.* **2007**, *21*, 539–545. [CrossRef]
51. Nazari, A.; Riahi, S. Microstructural, thermal, physical and mechanical behavior of the self compacting concrete containing SiO_2 nanoparticles. *Mater. Sci. Eng. A* **2010**, *527*, 7663–7672. [CrossRef]
52. Abhilash, P.P.; Nayak, D.K.; Sangoju, B.; Kumar, R.; Kumar, V. Effect of nano-silica in concrete; a review. *Constr. Build. Mater.* **2021**, *278*, 122347.
53. Kooshafar, M.; Madani, H. An investigation on the influence of nano silica morphology on the characteristics of cement composites. *J. Build. Eng.* **2020**, *30*, 101293. [CrossRef]
54. Orakzai, M.A. Hybrid effect of nano-alumina and nano-titanium dioxide on mechanical properties of concrete. *Case Stud. Constr. Mater.* **2021**, *14*, e00483.
55. Wang, X.; Dong, S.; Ashour, A.; Zhang, W.; Han, B. Effect and mechanisms of nanomaterials on interface between aggregates and cement mortars. *Constr. Build. Mater.* **2020**, *240*, 117942. [CrossRef]
56. Shao, Q.; Zheng, K.; Zhou, X.; Zhou, J.; Zeng, X. Enhancement of nano-alumina on long-term strength of Portland cement and the relation to its influences on compositional and microstructural aspects. *Cem. Concr. Compos.* **2019**, *98*, 39–48. [CrossRef]
57. Zhang, A.; Yang, W.; Ge, Y.; Du, Y.; Liu, P. Effects of nano-SiO_2 and nano-Al_2O_3 on mechanical and durability properties of cement-based materials: A comparative study. *J. Build. Eng.* **2020**, *34*, 101936. [CrossRef]
58. Nazari, A.; Riahi, S. The effects of zinc dioxide nanoparticles on flexural strength of self-compacting concrete. *Compos. Part B Eng.* **2011**, *42*, 167–175. [CrossRef]
59. Yang, J.; Mohseni, E.; Behforouz, B.; Khotbehsara, M.M. An experimental investigation into the effects of Cr_2O_3 and ZnO_2 nanoparticles on the mechanical properties and durability of self-compacting mortar. *Int. J. Mater. Res.* **2015**, *106*, 886–892. [CrossRef]
60. Heikal, M.; Zaki, M.E.A.; Ibrahim, S.M. Characterization, hydration, durability of nano-Fe_2O_3-composite cements subjected to sulphates and chlorides media. *Constr. Build. Mater.* **2021**, *269*, 121310. [CrossRef]
61. Madandoust, R.; Mohseni, E.; Mousavi, S.Y.; Namnevis, M. An experimental investigation on the durability of self-compacting mortar containing nano-SiO_2, nano-Fe_2O_3 and nano-CuO. *Constr. Build. Mater.* **2015**, *86*, 44–50. [CrossRef]
62. Khotbehsara, M.M.; Miyandehi, B.M.; Naseri, F.; Ozbakkaloglu, T.; Jafari, F.; Mohseni, E. Effect of SnO_2, ZrO_2, and $CaCO_3$ nanoparticles on water transport and durability properties of self-compacting mortar containing fly ash: Experimental observations and ANFIS predictions. *Constr. Build. Mater.* **2018**, *158*, 823–834. [CrossRef]
63. Massa, M.A.; Covarrubias, C.; Bittner, M.; Fuentevilla, I.A.; Capetillo, P.; Von Marttens, A.; Carvajal, J.C. Synthesis of new antibacterial composite coating for titanium based on highly ordered nanoporous silica and silver nanoparticles. *Mater. Sci. Eng. C* **2014**, *45*, 146–153. [CrossRef] [PubMed]
64. Morsy, M.S.; Alsayed, S.H.; Aqel, M. Hybrid effect of carbon nanotube and nano-clay on physico-mechanical properties of cement mortar. *Constr. Build. Mater.* **2011**, *25*, 145–149. [CrossRef]
65. Felekoğlu, B.; Sarıkahya, H. Effect of chemical structure of polycarboxylate-based superplasticizers on workability retention of self-compacting concrete. *Constr. Build. Mater.* **2008**, *22*, 1972–1980. [CrossRef]
66. Nazari, A.; Riahi, S. Effects of Cr_2O_3 nanoparticles on properties of SCC with GGBFS binder. *Mag. Concr. Res.* **2012**, *64*, 433–444. [CrossRef]
67. Bankir, M.B.; Ozturk, M.; Sevim, U.K.; Depci, T. Effect of n-$CaCO_3$ on fresh, hardened properties and acid resistance of granulated blast furnace slag added mortar. *J. Build. Eng.* **2020**, *29*, 101209. [CrossRef]
68. Tawfik, T.A.; Aly Metwally, K.; EL-Beshlawy, S.A.; Al Saffar, D.M.; Tayeh, B.A.; Soltan Hassan, H. Exploitation of the nanowaste ceramic incorporated with nano silica to improve concrete properties. *J. King Saud Univ. Eng. Sci.* **2020**. [CrossRef]

69. Jalal, M.; Ramezanianpour, A.A.; Pool, M.K. Split tensile strength of binary blended self compacting concrete containing low volume fly ash and TiO_2 nanoparticles. *Compos. Part B Eng.* **2013**, *55*, 324–337. [CrossRef]
70. Quercia, G.; Spiesz, P.; Hüsken, G.; Brouwers, H.J.H. SCC modification by use of amorphous nano-silica. *Cem. Concr. Compos.* **2014**, *45*, 69–81. [CrossRef]
71. Beigi, M.H.; Berenjian, J.; Omran, O.L.; Nik, A.S.; Nikbin, I.M. An experimental survey on combined effects of fibers and nanosilica on the mechanical, rheological, and durability properties of self-compacting concrete. *Mater. Des.* **2013**, *50*, 1019–1029. [CrossRef]
72. Vassaux, S.; Gaudefroy, V.; Boulangé, L.; Pévère, A.; Mouillet, V.; Barragan-Montero, V. Towards a better understanding of wetting regimes at the interface asphalt/aggregate during warm-mix process of asphalt mixtures. *Constr. Build. Mater.* **2017**, *133*, 182–195. [CrossRef]
73. Feng, D.; Xie, N.; Gong, C.; Leng, Z.; Xiao, H.; Li, H.; Shi, X. Portland cement paste modified by TiO_2 nanoparticles: A microstructure perspective. *Ind. Eng. Chem. Res.* **2013**, *52*, 11575–11582. [CrossRef]
74. Chen, J.; Kou, S.; Poon, C. Hydration and properties of nano-TiO_2 blended cement composites. *Cem. Concr. Compos.* **2012**, *34*, 642–649. [CrossRef]
75. Oltulu, M.; Şahin, R. Single and combined effects of nano-SiO_2, nano-Al_2O_3 and nano-Fe_2O_3 powders on compressive strength and capillary permeability of cement mortar containing silica fume. *Mater. Sci. Eng. A* **2011**, *528*, 7012–7019. [CrossRef]
76. Mohseni, E.; Naseri, F.; Amjadi, R.; Khotbehsara, M.M.; Ranjbar, M.M. Microstructure and durability properties of cement mortars containing nano-TiO_2 and rice husk ash. *Constr. Build. Mater.* **2016**, *114*, 656–664. [CrossRef]
77. Mohseni, E.; Miyandehi, B.M.; Yang, J.; Yazdi, M.A. Single and combined effects of nano-SiO_2, nano-Al_2O_3 and nano-TiO_2 on the mechanical, rheological and durability properties of self-compacting mortar containing fly ash. *Constr. Build. Mater.* **2015**, *84*, 331–340. [CrossRef]
78. Said, S.; Mikhail, S.; Riad, M. Recent processes for the production of alumina nano-particles. *Mater. Sci. Energy Technol.* **2020**, *3*, 344–363. [CrossRef]
79. De Dios, A.S.; Díaz-García, M.E. Multifunctional nanoparticles: Analytical prospects. *Anal. Chim. Acta* **2010**, *666*, 1–22. [CrossRef]
80. Neouze, M.-A.; Schubert, U. Surface modification and functionalization of metal and metal oxide nanoparticles by organic ligands. *Monatshefte Chem. Chem. Mon.* **2008**, *139*, 183–195. [CrossRef]
81. Xue, Y.; Yu, W.; Mei, J.; Jiang, W.; Lv, X. A clean process for alumina extraction and ferrosilicon alloy preparation from coal fly ash via vacuum thermal reduction. *J. Clean. Prod.* **2019**, *240*, 118262. [CrossRef]
82. Bertolino, L.C.; Rossi, A.M.; Scorzelli, R.B.; Torem, M.L. Influence of iron on kaolin whiteness: An electron paramagnetic resonance study. *Appl. Clay Sci.* **2010**, *49*, 170–175. [CrossRef]
83. Murray, H.H. Structure and composition of the clay minerals and their physical and chemical properties. *Dev. Clay Sci.* **2006**, *2*, 7–31.
84. Lin, H.Y.; Wan, L.; Yang, Y.F. Aluminium hydroxide ultrafine powder extracted from fly ash. *Adv. Mater. Res.* **2012**, *512*, 1548–1553. [CrossRef]
85. Li, H.; Hui, J.; Wang, C.; Bao, W.; Sun, Z. Extraction of alumina from coal fly ash by mixed-alkaline hydrothermal method. *Hydrometallurgy* **2014**, *147*, 183–187. [CrossRef]
86. Ding, J.; Ma, S.; Shen, S.; Xie, Z.; Zheng, S.; Zhang, Y. Research and industrialization progress of recovering alumina from fly ash: A concise review. *Waste Manag.* **2017**, *60*, 375–387. [CrossRef]
87. Szente, R.N.; Schroeter, R.A.; Garcia, M.G.; Bender, O.W. Recovering aluminum via plasma processing. *JOM* **1997**, *49*, 52–55. [CrossRef]
88. Saravanakumar, R.; Ramachandran, K.; Laly, L.G.; Ananthapadmanabhan, P.V.; Yugeswaran, S. Plasma assisted synthesis of γ-alumina from waste aluminium dross. *Waste Manag.* **2018**, *77*, 565–575. [CrossRef]
89. Kumar, P.M.; Balasubramanian, C.; Sali, N.D.; Bhoraskar, S.V.; Rohatgi, V.K.; Badrinarayanan, S. Nanophase alumina synthesis in thermal arc plasma and characterization: Correlation to gas-phase studies. *Mater. Sci. Eng. B* **1999**, *63*, 215–227. [CrossRef]
90. Kumar, P.M.; Borse, P.; Rohatgi, V.K.; Bhoraskar, S.V.; Singh, P.; Sastry, M. Synthesis and structural characterization of nanocrystalline aluminium oxide. *Mater. Chem. Phys.* **1994**, *36*, 354–358. [CrossRef]
91. Fu, L.; Johnson, D.L.; Zheng, J.G.; Dravid, V.P. Microwave plasma synthesis of nanostructured γ-Al_2O_3 powders. *J. Am. Ceram. Soc.* **2003**, *86*, 1635–1637. [CrossRef]
92. Stanislaus, A.; Al-Dolama, K.; Absi-Halabi, M. Preparation of a large pore alumina-based HDM catalyst by hydrothermal treatment and studies on pore enlargement mechanism. *J. Mol. Catal. A Chem.* **2002**, *181*, 33–39. [CrossRef]
93. Akia, M.; Alavi, S.M.; Rezaei, M.; Yan, Z.-F. Optimizing the sol–gel parameters on the synthesis of mesostructure nanocrystalline γ-Al_2O_3. *Microporous Mesoporous Mater.* **2009**, *122*, 72–78. [CrossRef]
94. Weitkamp, J.; Sing, K.S.W.; Schüth, F. *Handbook of Porous Solids*; Wiley: Hoboken, NJ, USA, 2002.
95. Mirjalili, F.; Hasmaliza, M.; Abdullah, L.C. Size-controlled synthesis of nano α-alumina particles through the sol–gel method. *Ceram. Int.* **2010**, *36*, 1253–1257. [CrossRef]
96. Belekar, R.M.; Dhoble, S.J. Activated alumina granules with nanoscale porosity for water defluoridation. *Nano Struct. Nano Objects* **2018**, *16*, 322–328. [CrossRef]
97. Esmaeilirad, M.; Zabihi, M.; Shayegan, J.; Khorasheh, F. Oxidation of toluene in humid air by metal oxides supported on γ-alumina. *J. Hazard. Mater.* **2017**, *333*, 293–307. [CrossRef]

98. Feng, G.; Jiang, F.; Jiang, W.; Liu, J.; Zhang, Q.; Hu, Q.; Miao, L.; Wu, Q. Effect of oxygen donor alcohol on nonaqueous precipitation synthesis of alumina powders. *Ceram. Int.* **2019**, *45*, 354–360. [CrossRef]
99. Hong, J.-P.; Jun, W.; Chen, H.-Y.; Sun, B.-D.; Li, J.-J.; Chong, C. Process of aluminum dross recycling and life cycle assessment for Al-Si alloys and brown fused alumina. *Trans. Nonferrous Met. Soc. China* **2010**, *20*, 2155–2161. [CrossRef]
100. Calder, G.V.; Stark, T.D. Aluminum reactions and problems in municipal solid waste landfills. *Pract. Period. Hazardous Toxic Radioact. Waste Manag.* **2010**, *14*, 258–265. [CrossRef]
101. El Katatny, E.A.; Halawy, S.A.; Mohamed, M.A.; Zaki, M.I. Recovery of high surface area alumina from aluminium dross tailings. *J. Chem. Technol. Biotechnol. Int. Res. Process. Environ. Clean Technol.* **2000**, *75*, 394–402. [CrossRef]
102. David, E.; Kopac, J. Aluminum recovery as a product with high added value using aluminum hazardous waste. *J. Hazard. Mater.* **2013**, *261*, 316–324. [CrossRef]
103. Dash, B.; Das, B.R.; Tripathy, B.C.; Bhattacharya, I.N.; Das, S.C. Acid dissolution of alumina from waste aluminium dross. *Hydrometallurgy* **2008**, *92*, 48–53. [CrossRef]
104. Das, J.; Patra, B.S.; Baliarsingh, N.; Parida, K.M. Adsorption of phosphate by layered double hydroxides in aqueous solutions. *Appl. Clay Sci.* **2006**, *32*, 252–260. [CrossRef]
105. Sarker, M.S.R.; Alam, M.Z.; Qadir, M.R.; Gafur, M.A.; Moniruzzaman, M. Extraction and characterization of alumina nanopowders from aluminum dross by acid dissolution process. *Int. J. Miner. Metall. Mater.* **2015**, *22*, 429–436. [CrossRef]
106. Singh, U.; Ansari, M.S.; Puttewar, S.P.; Agnihotri, A. Studies on process for conversion of waste aluminium dross into value added products. *Russ. J. Non Ferrous Met.* **2016**, *57*, 296–300. [CrossRef]
107. How, L.F.; Islam, A.; Jaafar, M.S.; Taufiq-Yap, Y.H. Extraction and characterization of Γ-alumina from waste aluminium dross. *Waste Biomass Valoriz.* **2017**, *8*, 321–327. [CrossRef]
108. Yang, S.-F.; Wang, T.-M.; Shie, Z.-Y.J.; Jiang, S.-J.; Hwang, C.-S.; Tzeng, C.-C. Fine Al_2O_3 powder produced by radio-frequency plasma from aluminum dross. *IEEE Trans. Plasma Sci.* **2014**, *42*, 3751–3755. [CrossRef]
109. Ashish, D.K.; Verma, S.K. Robustness of self-compacting concrete containing waste foundry sand and metakaolin: A sustainable approach. *J. Hazard. Mater.* **2020**, *401*, 123329. [CrossRef]
110. Vali, R.H.; Wani, M.M. The effect of mixed nano-additives on performance and emission characteristics of a diesel engine fuelled with diesel-ethanol blend. *Mater. Today Proc.* **2021**. [CrossRef]
111. Uysal, M. Self-compacting concrete incorporating filler additives: Performance at high temperatures. *Constr. Build. Mater.* **2012**, *26*, 701–706. [CrossRef]
112. Zhu, W.; Gibbs, J.C. Use of different limestone and chalk powders in self-compacting concrete. *Cem. Concr. Res.* **2005**, *35*, 1457–1462. [CrossRef]
113. Uysal, M.; Sumer, M. Performance of self-compacting concrete containing different mineral admixtures. *Constr. Build. Mater.* **2011**, *25*, 4112–4120. [CrossRef]
114. Fathi, H.; Lameie, T.; Maleki, M.; Yazdani, R. Simultaneous effects of fiber and glass on the mechanical properties of self-compacting concrete. *Constr. Build. Mater.* **2017**, *133*, 443–449. [CrossRef]
115. Ahmad, S.; Umar, A.; Masood, A. Properties of normal concrete, self-compacting concrete and glass fibre-reinforced self-compacting concrete: An experimental study. *Procedia Eng.* **2017**, *173*, 807–813. [CrossRef]
116. Anand, N.; Arulraj, G.P. The effect of elevated temperature on concrete materials-A literature review. *Int. J. Civ. Struct. Eng.* **2011**, *1*, 928–938.
117. Annerel, E.; Taerwe, L.; Vandevelde, P. Assessment of temperature increase and residual strength of SCC after fire exposure. In Proceedings of the 5th International RILEM Symposium on Self-Compacting Concrete, Ghent, Belgium, 5–7 September 2007; pp. 715–720.
118. Pineaud, A.; Pimienta, P.; Rémond, S.; Carré, H. Mechanical properties of high performance self-compacting concretes at room and high temperature. *Constr. Build. Mater.* **2016**, *112*, 747–755. [CrossRef]
119. Andiç-Çakır, Ö.; Hızal, S. Influence of elevated temperatures on the mechanical properties and microstructure of self consolidating lightweight aggregate concrete. *Constr. Build. Mater.* **2012**, *34*, 575–583. [CrossRef]
120. Ma, Q.; Guo, R.; Zhao, Z.; Lin, Z.; He, K. Mechanical properties of concrete at high temperature—A review. *Constr. Build. Mater.* **2015**, *93*, 371–383. [CrossRef]
121. Khaliq, W.; Khan, H.A. High temperature material properties of calcium aluminate cement concrete. *Constr. Build. Mater.* **2015**, *94*, 475–487. [CrossRef]
122. Lea, F.C. The resistance to fire of concrete and reinforced concrete. *J. Soc. Chem. Ind.* **1922**, *41*, 395R–396R. [CrossRef]
123. Qadir, S.S. Strength and behavior of self compacting concrete with crushed ceramic tiles as partial replacement for coarse aggregate and subjected to elevated temperature. *Int. J. Eng. Technol. Manag. Appl. Sci.* **2015**, *3*, 278–286.
124. Norhasri, M.S.M.; Hamidah, M.S.; Fadzil, A.M. Applications of using nano material in concrete: A review. *Constr. Build. Mater.* **2017**, *133*, 91–97. [CrossRef]
125. Ashoor, M.; Khorshidi, A.; Sarkhosh, L. Appraisal of new density coefficient on integrated-nanoparticles concrete in nuclear protection. *Kerntechnik* **2020**, *85*, 9–14. [CrossRef]
126. Mousavi, S.S.; Mousavi Ajarostaghi, S.S.; Bhojaraju, C. A critical review of the effect of concrete composition on rebar–concrete interface (RCI) bond strength: A case study of nanoparticles. *SN Appl. Sci.* **2020**, *2*, 1–23. [CrossRef]

127. Niewiadomski, P.; Stefaniuk, D. Creep assessment of the cement matrix of self-compacting concrete modified with the addition of nanoparticles using the indentation method. *Appl. Sci.* **2020**, *10*, 2442. [CrossRef]
128. Xu, S.Y.; Zhang, Y.M.; Pan, L.S.; Tan, W. Research on characterization and chemical composition of baoluo kaolin in hainan province. *Adv. Mater. Res. Trans Tech Publ.* **2011**, *233*, 2258–2262. [CrossRef]
129. Gowda, R.; Narendra, H.; Nagabushan, B.M.; Rangappa, D.; Prabhakara, R. Investigation of nano-alumina on the effect of durability and micro-structural properties of the cement mortar. *Mater. Today Proc.* **2017**, *4*, 12191–12197. [CrossRef]
130. Rasin, F.A.; Abbas, L.K.; Kadhim, M.J. Study the effects of nano-materials addition on some mechanical properties of cement mortar. *Eng. Technol. J.* **2017**, *35*, 348–355.
131. Liu, F.; Zhang, T.; Luo, T.; Zhou, M.; Ma, W.; Zhang, K. The effects of Nano-SiO_2 and Nano-TiO_2 addition on the durability and deterioration of concrete subject to freezing and thawing cycles. *Materials* **2019**, *12*, 3608. [CrossRef] [PubMed]
132. Al-Luhybi, A.S.; Altalabani, D. The Influence of Nano-Silica on the Properties and Microstructure of Lightweight Concrete: A Review. *IOP Conf. Ser. Mater. Sci. Eng.* **2021**, *1094*, 012075. [CrossRef]
133. Mukhopadhyay, S.C.; Ihara, I. Sensors and technologies for structural health monitoring: A review. *New Dev. Sens. Technol. Struct. Heal. Monit.* **2011**, *96*, 1–14.
134. Sua-iam, G.; Makul, N. Rheological and mechanical properties of cement–fly ash self-consolidating concrete incorporating high volumes of alumina-based material as fine aggregate. *Constr. Build. Mater.* **2015**, *95*, 736–747. [CrossRef]
135. Farzadnia, N.; Abang Ali, A.A.; Demirboga, R. Characterization of high strength mortars with nano alumina at elevated temperatures. *Cem. Concr. Res.* **2013**, *54*, 43–54. [CrossRef]
136. Behfarnia, K.; Salemi, N. The effects of nano-silica and nano-alumina on frost resistance of normal concrete. *Constr. Build. Mater.* **2013**, *48*, 580–584. [CrossRef]
137. Zhan, B.J.; Xuan, D.X.; Poon, C.S. The effect of nanoalumina on early hydration and mechanical properties of cement pastes. *Constr. Build. Mater.* **2019**, *202*, 169–176. [CrossRef]
138. Szymanowski, J.; Sadowski, L. The development of nanoalumina-based cement mortars for overlay applications in concrete floors. *Materials* **2019**, *12*, 3465. [CrossRef] [PubMed]
139. Li, Z.; Wang, H.; He, S.; Lu, Y.; Wang, M. Investigations on the preparation and mechanical properties of the nano-alumina reinforced cement composite. *Mater. Lett.* **2006**, *60*, 356–359. [CrossRef]
140. Barbhuiya, S.; Mukherjee, S.; Nikraz, H. Effects of nano-Al_2O_3 on early-age microstructural properties of cement paste. *Constr. Build. Mater.* **2014**, *52*, 189–193. [CrossRef]
141. Heikal, M.; Ismail, M.N.; Ibrahim, N.S. Physico-mechanical, microstructure characteristics and fire resistance of cement pastes containing Al_2O_3 nano-particles. *Constr. Build. Mater.* **2015**, *91*, 232–242. [CrossRef]
142. Sua-iam, G.; Makul, N. Use of recycled alumina as fine aggregate replacement in self-compacting concrete. *Constr. Build. Mater.* **2013**, *47*, 701–710. [CrossRef]
143. Mohseni, E.; Khotbehsara, M.M.; Naseri, F.; Monazami, M.; Sarker, P. Polypropylene fiber reinforced cement mortars containing rice husk ash and nano-alumina. *Constr. Build. Mater.* **2016**, *111*, 429–439. [CrossRef]

processes

MDPI

Review

Processes and Properties of Ionic Liquid-Modified Nanofiller/Polymer Nanocomposites—A Succinct Review

Ahmad Adlie Shamsuri [1,*], Siti Nurul Ain Md. Jamil [2,3,*] and Khalina Abdan [1,*]

1 Laboratory of Biocomposite Technology, Institute of Tropical Forestry and Forest Products, Universiti Putra Malaysia, Serdang UPM 43400, Selangor, Malaysia
2 Department of Chemistry, Faculty of Science, Universiti Putra Malaysia, Serdang UPM 43400, Selangor, Malaysia
3 Centre of Foundation Studies for Agricultural Science, Universiti Putra Malaysia, Serdang UPM 43400, Selangor, Malaysia
* Correspondence: adlie@upm.edu.my (A.A.S.); ctnurulain@upm.edu.my (S.N.A.M.J.); khalina@upm.edu.my (K.A.)

Abstract: Ionic liquids can typically be synthesized via protonation, alkylation, metathesis, or neutralization reactions. The many types of ionic liquids have increased their attractiveness to researchers for employment in various areas, including in polymer composites. Recently, ionic liquids have been employed to modify nanofillers for the fabrication of polymer nanocomposites with improved physicochemical properties. In this succinct review, four types of imidazolium-based ionic liquids that are employed as modifiers—specifically alkylimidazolium halide, alkylimidazolium hexafluorophosphate, alkylimidazolium tetrafluoroborate, and alkylimidazolium bistriflimide—are reviewed. Additionally, three types of ionic liquid-modified nanofiller/polymer nanocomposites—namely ionic liquid-nanofiller/thermoplastic nanocomposites, ionic liquid-nanofiller/elastomer nanocomposites, and ionic liquid-nanofiller/thermoset nanocomposites—are described as well. The effect of imidazolium-based ionic liquids on the thermo-mechanico-chemical properties of the polymer nanocomposites is also succinctly reviewed. This review can serve as an initial guide for polymer composite researchers in modifying nanofillers by means of ionic liquids for improving the performance of polymer nanocomposites.

Keywords: ionic liquid; nanofiller; polymer nanocomposite; thermal; mechanical; chemical

Citation: Shamsuri, A.A.; Jamil, S.N.A.M.; Abdan, K. Processes and Properties of Ionic Liquid-Modified Nanofiller/Polymer Nanocomposites—A Succinct Review. *Processes* **2021**, *9*, 480. https://doi.org/10.3390/pr9030480

Academic Editor: Shaghayegh Hamzehlou

Received: 27 January 2021
Accepted: 7 February 2021
Published: 8 March 2021

1. Introduction

Recently, the employment of ionic liquids in green chemical technology has grown because of the environmental responsiveness and growth of the sustainable chemical industry. Ionic liquids are liquid salts that generally possess a lower melting point than the boiling point of water. They also possess a very low vapor pressure, resulting in them being non-volatile. The recyclability of ionic liquids has led them to be regarded as an environmentally benign solvent [1]. Furthermore, ionic liquids possess intriguing solvent properties, such as non-flammability, high polarity, good electrical conductivity, high thermal stability, and good solubility with many organic solvents, including water. They also have the ability to dissolve most organic and some inorganic materials, as well as biopolymers. Additionally, ionic liquids can be custom made, with properties corresponding to the requirements of the application. They can also be used as cationic surfactants if they have a long alkyl chain ($\geq C_{10}$) [2].

Polymer nanocomposites are polymers that incorporate nanometer-scale fillers (usually called nanofillers) [3]. The modification of nanofillers using ionic liquids is a promising process in the fabrication of multifunctional polymer nanocomposites. In various types of ionic liquids, imidazolium-based ionic liquids with distinct oppose anions are regularly employed for modification of nanofillers. This was due to their effectiveness in a wide variety of chemical structures. Table 1 shows examples of imidazolium-based ionic liquids employed for modification of nanofillers. It can be observed that some ionic liquids possess similar alkylimidazolium cations with different counter anions; therefore, in this concise review, they have been classified into four types, namely alkylimidazolium halide, alkylimidazolium hexafluorophosphate, alkylimidazolium tetrafluoroborate, and alkylimidazolium bistriflimide. This classification is made since the ionic liquids must undergo distinct preparation processes with different counter anions.

Table 1. Examples of imidazolium-based ionic liquids employed for modification of nanofillers.

Imidazolium-Based Ionic Liquid	Abbreviation	References
1-Allyl-3-methylimidazolium chloride	[Amim][Cl]	[4–7]
1-(3-Aminopropyl)-2-methyl-3-butylimidazole bromide	[Apmbim][Br]	[8]
1-(3-Aminopropyl)-3-methylimidazolium bromide	[Apmim][Br]	[8,9]
1-Butyl-3-methylimidazolium chloride	[Bmim][Cl]	[10]
1-Benzyl-3-methylimidazolium chloride	[Bzmim][Cl]	[11–16]
1-Decyl-3-methylimidazolium bromide	[Dmim][Br]	[17]
1-Decyl-3-methylimidazolium chloride	[Dmim][Cl]	[18]
1-(Ethoxycarbonyl)methyl-3-methylimidazolium bromide	[Ecmmim][Br]	[19]
1-Ethyl-3-methylimidazolium bromide	[Emim][Br]	[20]
1-Ethyl-3-methylimidazolium chloride	[Emim][Cl]	[21]
1-Hexadecyl-3-methylimidazolium bromide	[Hdmim][Br]	[22]
1-Hexadecyl-3-methylimidazolium chloride	[Hdmim][Cl]	[10,23]
1-Hexyl-3-methylimidazolium bromide	[Hmim][Br]	[20]
1-Hexyl-3-methylimidazolium chloride	[Hmim][Cl]	[10]
1-Methyl-3-carboxymethylimidazolium chloride	[Mcmim][Cl]	[24]
1-Methyl-3-dodecylimidazolium bromide	[Mddim][Br]	[25,26]
1-Methyl-3-octadecylimidazolium bromide	[Modim][Br]	[25]
1-Methyl-3-octylimidazolium bromide	[Moim][Br]	[25]
1-Methyl-3-octylimidazolium chloride	[Moim][Cl]	[4]
1-Methylimidazolium chloride	[Mim][Cl]	[27]
1-Allyl-3-methylimidazolium hexafluorophosphate	[Amim][PF_6]	[28]
1-Butyl-3-methylimidazolium hexafluorophosphate	[Bmim][PF_6]	[28–34]
1-Hexadecyl-3-methylimidazolium hexafluorophosphate	[Hdmim][PF_6]	[22]
1-Hexyl-3-methylimidazolium hexafluorophosphate	[Hmim][PF_6]	[34]
1-(3-Butoxy-2-hydroxy-propyl)-3-methylimidazolium tetrafluoroborate	[Bhpmim][BF_4]	[35]
1-Butyl-3-methylimidazolium tetrafluoroborate	[Bmim][BF_4]	[36–39]
1-Ethyl-3-methylimidazolium tetrafluoroborate	[Emim][BF_4]	[28,40]
1-Hexyl-3-methylimidazolium tetrafluoroborate	[Hmim][BF_4]	[36]
1-Vinyl-3-ethylimidazolium tetrafluoroborate	[Veim][BF_4]	[41]
1-Butyl-3-methylimidazolium bis(trifluoromethylsulfonyl)imide	[Bmim][NTf_2]	[42–51]
1-Carboxyethyl-3-methylimidazolium bis(trifluoromethylsulfonyl)imide	[Cemim][NTf_2]	[52]
1-Ethyl-2,3-dimethylimidazolium bis(trifluoromethylsulfonyl)imide	[Edmim][NTf_2]	[53]
1-Ethyl-3-methylimidazolium bis(trifluoromethylsulfonyl)imide	[Emim][NTf_2]	[54,55]
1-Methyl-3-dodecylimidazolium bis(trifluoromethylsulfonyl)imide	[Mddim][NTf_2]	[26]

Meanwhile, Table 2 exhibits examples of polymer matrices and nanofillers used for fabrication of polymer nanocomposites. Thermoplastics, such as EVM [40], ImPEEK [8], Pebax [9], PEI [32], PEO [38], PI [39], PLLA [19], PMMA [26,34], PVA [7], and PVDF [22,28,54] have been used as polymer matrices for fabricating polymer nanocomposites. On the other hand, elastomers, for example, BIIR [29], AEM [4], EPDM [17], NR [5], NRL [20,51], CR [42–44,46–50], PDMS [55], and SBR [11–16,18,45,53], have frequently been used as polymer matrices. In addition, polymer nanocomposites can be fabricated by using thermosets like DGEAC [35], DGEBA [33,37], and PUF [21,31]. Furthermore, PEDOT [36] and PTh [25] are conductive polymers that can also be used for fabrication of polymer nanocomposites. Additionally, nanomaterials, such as MWCNTs [4,32,34,40,55], MWCNT-COOH [8,30,52], GO [6,27], GN [19,39], GNP [26], nano-SiO_2 [5,17], nano-TiO_2 [24], ZnO NPs [38], CB NPs [41], and MMT [10,23] have been used as nanofillers for polymer nanocomposites because of their high aspect ratio, large surface area, high stiffness, and low density [56].

Table 2. Examples of polymer matrices and nanofillers used for fabrication of polymer nanocomposites.

Polymer Matrix	Abbreviation	Nanofiller	Abbreviation
Ethylene–vinyl acetate	EVM	Multi-walled carbon nanotubes	MWCNTs
Imidazolium-based poly(ether ether ketone)	ImPEEK	Carboxylated multi-walled carbon nanotubes	MWCNT-COOH
Polyether block amide	Pebax	Graphene oxide	GO
Polyetherimide	PEI	Graphene nanosheets	GN
Poly(ethylene oxide)	PEO	Graphene nanoplatelets	GNP
Polyimide	PI	Nanosilica	nano-SiO_2
Poly(L-lactic acid)	PLLA	Titanium dioxide nanoparticles	nano-TiO_2
Poly(methyl methacrylate)	PMMA	Zinc oxide nanoparticles	ZnO NPs
Polyvinyl alcohol	PVA	Carbon black nanoparticles	CB NPs
Poly(vinylidene fluoride)	PVDF	Montmorillonite	MMT
Bromobutyl rubber	BIIR		
Ethylene acrylic elastomer	AEM		
Ethylene propylene–diene	EPDM		
Natural rubber	NR		
Natural rubber latex	NRL		
Polychloroprene rubber	CR		
Polydimethylsiloxane	PDMS		
Styrene butadiene rubber	SBR		
Diglycidyl ester of aliphatic cyclo	DGEAC		
Diglycidyl ether of bisphenol A	DGEBA		
Polyurethane foam	PUF		
Poly(3,4-ethylene dioxythiophene)	PEDOT		
Polythiophene	PTh		

In the past ten years, numerous nanofiller modification processes have been proposed with the aim of improving the physicochemical properties of polymer nanocomposites. The employment of ionic liquids as modifiers could give a benefit due to their unique chemical structure, which possesses both bulky cations and anions that are poorly coordinated. These ions are capable of interacting with organic polymers and inorganic nanofillers. The presence of intermolecular interactions could have a modification effect on polymer nanocomposites and subsequently enhance the interface linkage between the polymer matrix and the nanofiller. As far as the authors know, no succinct review has been created focusing on the study of processes and thermo-mechanico-chemical properties of ionic liquid-modified nanofiller/polymer nanocomposites. That is the intention of creating the classified review presented in this paper. Moreover, this succinct review is general, and even though not specific, it is nevertheless relevant to other related studies.

2. Types of Imidazolium-Based Ionic Liquids

2.1. *Alkylimidazolium Halide Ionic Liquids*

Imidazolium-based ionic liquids with halide counter anions (denoted as alkylimidazolium halide ionic liquids) can be synthesized via two reaction processes, specifically protonation and alkylation reactions [57]. Alkylimidazolium halide ionic liquids that are synthesized via protonation reaction can be prepared by reacting *N*-alkylimidazole with hydrohalic acid in a polar solvent, such as ethanol, at slightly above room temperature while stirring [27]. Meanwhile, the alkylation reaction can be conducted by reacting *N*-alkylimidazole with alkyl halide under reflux conditions at an elevated temperature while stirring [58]. Figure 1 shows the schematic of the protonation reaction of *N*-alkylimidazole with hydrohalic acid to prepare alkylimidazolium halide ionic liquid. Hydrohalic acids, such as hydrochloric acid and hydrobromic acid, are frequently used for protonation reaction, since they have the capacity to protonate (or hydronate) the nitrogen atom situated at the third position of the imidazole ring. Figure 2 indicates the schematic of the alkylation reaction of *N*-alkylimidazole with alkyl halide to prepare alkylimidazolium halide ionic liquid. Alkyl halides, such as allyl-, butyl-, decyl-, and ethyl halides, are commonly used for alkylation reactions in the production of ionic liquids with different alkyl chain lengths. Additionally, these ionic liquids can be applied as precursors in the preparation of other ionic liquids [22,26].

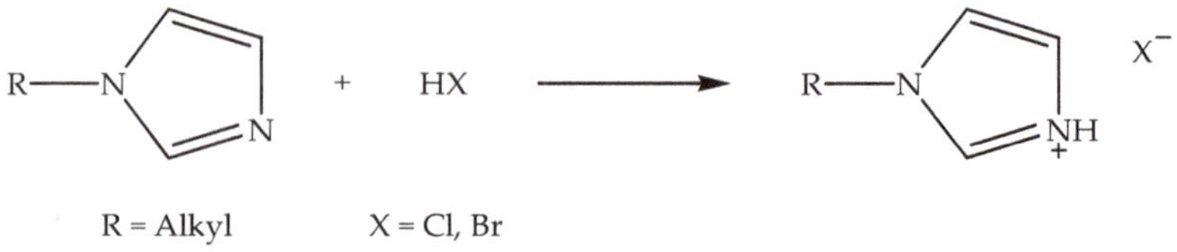

Figure 1. Schematic of the protonation reaction of *N*-alkylimidazole with hydrohalic acid to prepare alkylimidazolium halide ionic liquid.

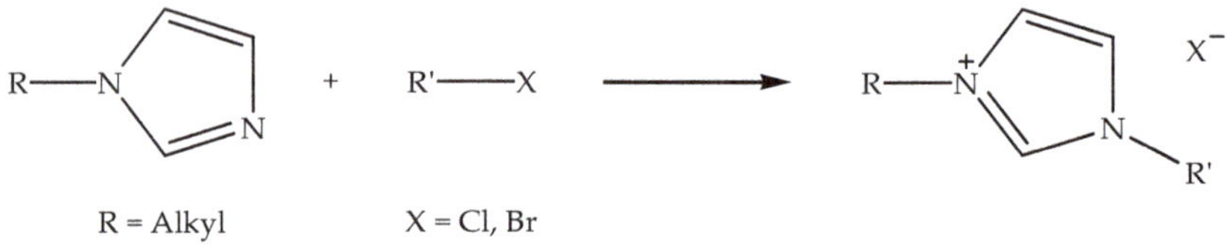

Figure 2. Schematic of the alkylation reaction of *N*-alkylimidazole with alkyl halide to prepare alkylimidazolium halide ionic liquid.

Additionally, alkylimidazolium halide ionic liquids are generally used for dissolution of biopolymers as ionic liquids with carboxylate anions, such as alkylimidazolium acetate and alkylimidazolium propionate [58]. Table 3 demonstrates the types of imidazolium-based ionic liquids, nanofillers, and modification processes of nanofillers. It can be observed that different types of nanofillers, for example, MWCNTs, nano-SiO_2, GO, MWCNT-COOH, MMT, GN, and GNP, can be modified with alkylimidazolium halide ionic liquids using different modification processes. The most common processes are ultrasonication and sonication, with both processes having almost the same concept, but different apparatuses, e.g., the ultrasonication process can be done using an ultrasonic bath [4,25,54]. In contrast, the sonication process can be performed by using an ultrasonic homogenizer probe sonicator [7,26,55]. Moreover, the processes do not require high temperatures and a long time, like refluxation [8]. On top of that, modification of nanofillers with alkylimidazolium halide ionic liquids can also be accomplished using low-cost processes, such as stirring and grinding. However, the modification of GO via stirring requires a higher temperature than MMT [9].

Table 3. Types of imidazolium-based ionic liquids, nanofillers, and modification processes of nanofillers.

Ionic Liquid	Nanofiller	Modification Process	Modification Temperature (°C)	Time (Hour)	References
[Amim][Cl]	MWCNTs	Ultrasonication	R	U	[4]
[Amim][Cl]	nano-SiO_2	Ultrasonication	R	1	[5]
[Amim][Cl]	GO	Ultrasonication	R	0.5	[6]
[Apmbim][Br]	MWCNT-COOH	Refluxation	60	24	[8]
[Apmim][Br]	GO	Stirring	80	24	[9]
[Bmim][Cl]	MMT	Stirring	R	U	[10]
[Bzmim][Cl]	MWCNTs	Sonication	R	0.5	[11]
[Dmim][Br]	nano-SiO_2	Ultrasonication	R	0.25	[17]
[Dmim][Cl]	MWCNTs	Sonication	R	2	[18]
[Ecmmim][Br]	GN	Sonication	R	4	[19]
[Emim][Br]	MWCNTs	Ultrasonication	R	2	[20]
[Hdmim][Br]	MWCNTs	Sonication	R	0.5	[22]
[Hdmim][Cl]	MMT	Stirring	R	24	[23]
[Mddim][Br]	GNP	Grinding	R	0.17	[26]
[Mim][Cl]	GO	Stirring	35	12	[27]
[Amim][PF_6]	MWCNTs	Grinding	R	8	[28]
[Bmim][PF_6]	GO	Ultrasonication	R	0.5	[29]
[Bmim][PF_6]	MWCNT-COOH	Grinding	R	0.25	[30]
[Bmim][PF_6]	nano-SiO_2	Stirring	R	U	[31]
[Bmim][PF_6]	MWCNTs	Grinding	R	0.5	[32]
[Bmim][PF_6]	MWCNTs	Ultrasonication	R	1	[33]
[Hdmim][PF_6]	MWCNTs	Sonication	R	0.5	[22]
[Hmim][PF_6]	MWCNTs	Milling	R	0.17	[34]
[Bhpmim][BF_4]	GO	Irradiation	20	0.02	[35]
[Bmim][BF_4]	MWCNTs	Grinding	R	0.3	[37]
[Bmim][BF_4]	ZnO NPs	Sonication	R	U	[38]
[Bmim][BF_4]	GN	Stirring	80	24	[39]
[Emim][BF_4]	MWCNTs	Grinding	R	8	[28]
[Emim][BF_4]	MWCNTs	Grinding	R	0.3	[40]
[Hmim][BF_4]	GO	Stirring	R	U	[36]
[Veim][BF_4]	CB NPs	Grinding	R	U	[41]
[Bmim][NTf_2]	MWCNTs	Grinding	R	U	[42]
[Bmim][NTf_2]	MWCNTs	Sonication	R	0.3	[51]
[Cemim][NTf_2]	MWCNT-COOH	Ultrasonication	R	0.25	[52]
[Edmim][NTf_2]	MWCNTs	Sonication	R	0.5	[53]
[Emim][NTf_2]	MWCNTs	Ultrasonication	R	0.5	[54]
[Emim][NTf_2]	MWCNTs	Grinding	R	U	[55]
[Mddim][NTf_2]	GNP	Grinding	R	0.17	[26]

R = room, and U = unspecified.

2.2. Alkylimidazolium Hexafluorophosphate Ionic Liquids

Alkylimidazolium hexafluorophosphate ionic liquids can be synthesized through metathesis reaction by using alkylimidazolium halide ionic liquids as precursors [59]. The reaction process can be carried out by exchanging halide anion with a hexafluorophosphate anion by using a metal hexafluorophosphate salt. The ion exchange is usually conducted in a polar solvent, such as ethanol, at room temperature while stirring [58]. Figure 3 shows the schematic of the metathesis reaction of alkylimidazolium chloride with potassium hexafluorophosphate to prepare alkylimidazolium hexafluorophosphate ionic liquid. The metathesis reaction is a general process for exchanging counter anions. When the reaction is carried out in ethanol, the potassium chloride salt precipitates from the reaction solvent, since it is insoluble in ethanol, and the alkylimidazolium hexafluorophosphate ionic liquid remains in the solution. Moreover, the metathesis reaction is often applied for the production of hydrophobic and low-temperature ionic liquids, for instance, 1-butyl-3-methylimidazolium- and 1-hexyl-3-methylimidazolium hexafluorophosphate [33,34].

Nevertheless, the physicochemical properties of the ionic liquids are different from the ionic liquids prepared via alkylation reaction, as a result of alkylimidazolium hexafluorophosphate possessing large-sized anions.

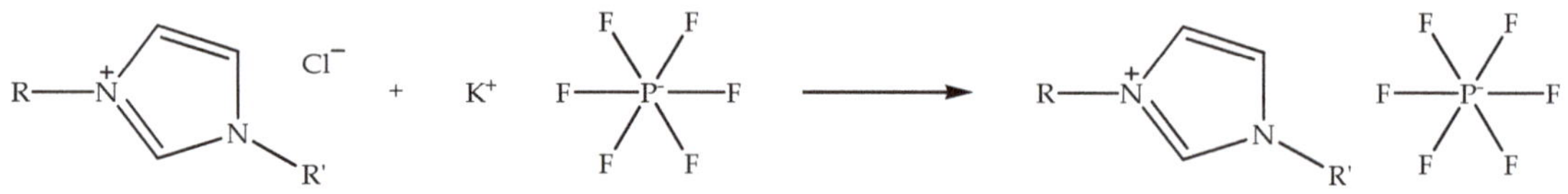

Figure 3. Schematic of the metathesis reaction of alkylimidazolium chloride with potassium hexafluorophosphate to prepare alkylimidazolium hexafluorophosphate ionic liquid.

Additionally, alkylimidazolium hexafluorophosphate ionic liquids can be applied in the chemical demulsification process. A previous review revealed that the ionic liquids could effectively demulsify emulsions compared to alkylimidazolium halide ionic liquids [60]. It can be seen in Table 3 that the nanofillers MWCNTs and MWCNT-COOH are frequently modified using alkylimidazolium hexafluorophosphate ionic liquids, and the grinding process is regularly performed to modify them [28,30,32]. Other processes, such as ultrasonication, sonication, and milling, can also be used for the modification of MWCNTs at room temperature [22,33,34]. The processes induce the ionic liquids to be coated on the surface of the MWCNTs, forming bucky gels. Moreover, the hydrophobic nature of alkylimidazolium hexafluorophosphate ionic liquids makes them appropriate for application as modifiers for hydrophobic nanofillers, such as MWCNTs, GO, and nano-SiO_2 [8,31]. This property can also form interactions between the nanofillers and hydrophobic polymer matrices by acting as an interface linker of polymer nanocomposites in order to tune their properties.

2.3. *Alkylimidazolium Tetrafluoroborate Ionic Liquids*

Alkylimidazolium tetrafluoroborate ionic liquids can not only be synthesized via metathesis reaction, but also through acid-base reaction by neutralizing alkylimidazolium hydroxide with fluoroboric acids such as tetrafluoroboric acid. Figure 4 shows the schematic of the neutralization of alkylimidazolium hydroxide with tetrafluoroboric acid to prepare alkylimidazolium tetrafluoroborate ionic liquid. The reaction is typically carried out in a polar solvent like ethanol while stirring at room temperature [58]. Unlike the alkylimidazolium hexafluorophosphate ionic liquids, alkylimidazolium tetrafluoroborate can present two different properties, specifically hydrophilic and hydrophobic properties. The presence of both properties is influenced by the length of the alkylimidazolium chain; for example, 1-ethyl-3-methylimidazolium tetrafluoroborate is hydrophilic in nature, but 1-hexyl-3-methylimidazolium tetrafluoroborate possesses the hydrophobic property. In addition, the melting point of the ionic liquids is also affected by their alkyl chain length, whereby the longer alkylimidazolium chain ionic liquids have a higher melting point, and the shorter alkylimidazolium chain ionic liquids exhibit the opposite characteristic. Therefore, the hydrophobicity and melting point of alkylimidazolium tetrafluoroborate ionic liquids can be tweaked by controlling the length of their alkyl chain.

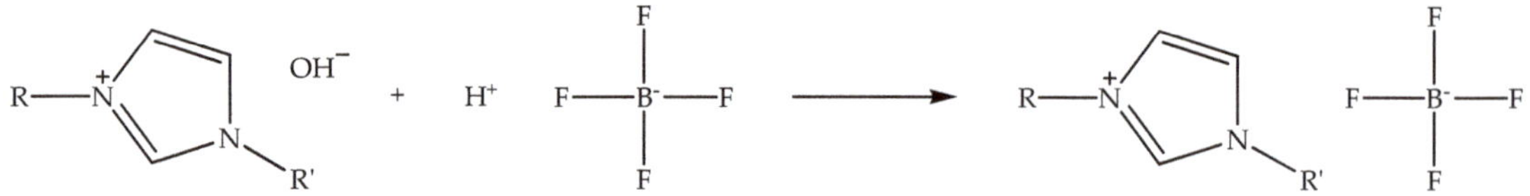

Figure 4. Schematic of the neutralization of alkylimidazolium hydroxide with tetrafluoroboric acid to prepare alkylimidazolium tetrafluoroborate ionic liquid.

Furthermore, alkylimidazolium tetrafluoroborate ionic liquids are typically exploited as plasticizing agents for synthetic biodegradable polymers [19]. In Table 3, it can be observed that the ionic liquids are able to modify various types of nanofillers, including GO, MWCNTs, ZnO NPs, GN, and CB NPs. The grinding process is commonly applied for the modification of nanofillers, as this does not consume solvent [28,37], as is required in sonication and stirring processes [38,39]. It can also be seen that the irradiation process by means of microwaves is the fastest process for modifying GO at 20 °C [35]. Nevertheless, the modification of GN through the stirring process is a long process in comparison to the other processes, requiring 24 h at 80 °C [39]. The processes for modifying different types of nanofillers using ionic liquids are generally optimized to improve the dispersion of nanofillers in the polymer matrices and enhancing the interface compatibility between them. The improved dispersion and compatibility of nanofillers results in high-performance polymer nanocomposites.

2.4. Alkylimidazolium Bistriflimide Ionic Liquids

Alkylimidazolium bis(trifluoromethylsulfonyl)imide ionic liquids (denoted as alkylimidazolium bistriflimide ionic liquids) can also be synthesized through metathesis reaction by exchanging halide anion of alkylimidazolium halide with a bistriflimide anion by using a metal bis(trifluoromethylsulfonyl)imide salt. The exchange process is generally performed in an aqueous solvent, such as distilled water, at an elevated temperature under nitrogen atmosphere [26]. Figure 5 displays the schematic of the metathesis reaction of alkylimidazolium bromide with lithium bis(trifluoromethylsulfonyl)imide to prepare alkylimidazolium bistriflimide ionic liquid. When the reaction is conducted in distilled water, the alkylimidazolium bistriflimide liquid separates from the aqueous solution, since it possesses the hydrophobic property, and the unreacted reactants remain in solution. Unlike the alkylimidazolium tetrafluoroborate ionic liquids, alkylimidazolium bistriflimide only exists as hydrophobic ionic liquids with a wide range of structures [17]. In addition, the ionic liquids are more air- and water-stable than alkylimidazolium halide ionic liquids. Additionally, the bistriflimide anion is not easily decomposed in the presence of water, as compared to tetrafluoroborate and hexafluorophosphate anions. Moreover, alkylimidazolium bistriflimide ionic liquids have a low melting point and are less viscous [48].

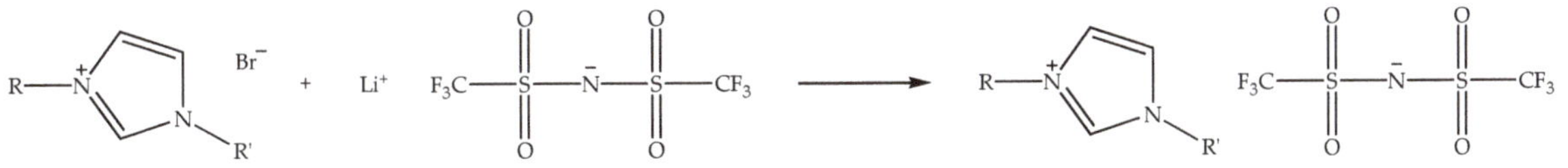

Figure 5. Schematic of the metathesis reaction of alkylimidazolium bromide with lithium bis(trifluoromethylsulfonyl)imide to prepare alkylimidazolium bistriflimide ionic liquid.

Furthermore, alkylimidazolium bistriflimide ionic liquids can be used as a compatibilizer for polymer blends. The preceding study determined that the ionic liquids were able to more efficiently compatibilize bioplastic/biopolymer blends in comparison to the alkylimidazolium halide ionic liquids [1]. It can be observed in Table 3 that the carbon-based nanofillers, such as MWCNTs, MWCNT-COOH, and GNP, are typically modified using alkylimidazolium bistriflimide ionic liquids. This is due to the ionic liquids having the capacity to interact with the surface of the carbon materials via cation-π, π-π, or van der Waals interactions [26,51,54]. On top of that, the modification process can be done either through grinding, sonication, or ultrasonication processes in solvent or solvent-free mediums [42,52,53]. Additionally, the processes are usually carried out at room temperature in less than an hour. Hence, the modification of nanofillers using alkylimidazolium bistriflimide ionic liquids possesses significant advantages, such as low materials depletion, simple modification process, and low energy consumption.

3. Types of Ionic Liquid-Modified Nanofiller/Polymer Nanocomposites

3.1. Ionic Liquid-Nanofiller/Thermoplastic Nanocomposites

Thermoplastics are widely used as polymer matrices in the fabrication of polymer nanocomposites. This is due to the fact that they can be processed either in the molten state or in the solution state. Table 4 displays the types of polymer matrices, nanofillers, ionic liquids, and fabrication processes of polymer nanocomposites. It can be observed that PVDF thermoplastics are more often incorporated with ionic liquid-modified nanofillers than other thermoplastics. This can be attributed to the fact that ionic liquid-nanofiller/PVDF nanocomposites can potentially be applied in the production of dielectric [41], energy storage [23], piezoresistive [30], and hybrid membrane [24] materials. The mixing process for PVDF with ionic liquid-modified nanofillers can be carried out through melt blending at 190 °C [23,41] and solution blending in different solvents [10,22,24]. Meanwhile, other thermoplastics, such as EVM, ImPEEK, Pebax, PEI, PEO, PI, PLLA, PMMA, and PVA, can also be mixed with ionic liquid-modified nanofillers via the same processes at different temperatures or with different solvents. In addition, sonication or ultrasonication processes can also be applied during solution blending [7,9,19,26,32]. The final process is practically important for shaping the polymer nanocomposites prior to thermo-mechanico-chemical characterization. The process can be done through hot pressing or compression molding and solution or solvent casting processes.

Table 4. Types of polymer matrices, nanofillers, ionic liquids, and fabrication processes of polymer nanocomposites.

Polymer Matrix	Nanofiller	Ionic Liquid	Mixing Process	Final Process	References
EVM	MWCNTs	[Emim][BF_4]	Melt blending	Hot pressing	[40]
ImPEEK	MWCNT-COOH	[Apmbim][Br]	Solution blending	Solution casting	[8]
Pebax	GO	[Apmbim][Br]	Sonication	Solution casting	[9]
PEI	MWCNTs	[Bmim][PF_6]	Sonication	Solution casting	[32]
PEO	ZnO NPs	[Bmim][BF_4]	Solution blending	Solution casting	[38]
PI	GN	[Bmim][BF_4]	Solution blending	Solution casting	[39]
PLLA	GN	[Ecmmim][Br]	Ultrasonication	Hot pressing	[19]
PMMA	GNP	[Mddim][NTf_2]	Sonication	Compression molding	[26]
PMMA	MWCNTs	[Hmim][PF_6]	Melt blending	Compression molding	[34]
PVA	GO	[Amim][Cl]	Sonication	Solvent casting	[7]
PVDF	CB NPs	[Veim][BF_4]	Melt blending	Hot pressing	[41]
PVDF	MMT	[Hdmim][Cl]	Melt blending	Compression molding	[23]
PVDF	MMT	[Hmim][Cl]	Solution blending	Solution casting	[10]
PVDF	MWCNTs	[Hdmim][PF_6]	Solution blending	Compression molding	[22]
PVDF	MWCNT-COOH	[Bmim][PF_6]	Melt blending	Compression molding	[30]
PVDF	nano-TiO_2	[Mcmim][Cl]	Solution blending	Hot pressing	[24]
AEM	MWCNTs	[Amim][Cl]	Hot mixing	Heat curing	[4]
BIIR	GO	[Bmim][PF_6]	Mill mixing	Heat curing	[29]
CR	MWCNTs	[Bmim][NTf_2]	Mill mixing	Heat curing	[42]
CR	MWCNT-COOH	[Cemim][NTf_2]	Mill mixing	Heat curing	[52]
EPDM	nano-SiO_2	[Dmim][Br]	Mill mixing	Heat curing	[17]
NR	nano-SiO_2	[Amim][Cl]	Mill mixing	Heat curing	[5]
NRL	MWCNTs	[Emim][Br]	Mill mixing	Heat curing	[20]
PDMS	MWCNTs	[Emim][NTf_2]	Speed mixing	Heat curing	[55]
SBR	GO	[Amim][Cl]	Mill mixing	Heat curing	[6]
SBR	MWCNTs	[Bzmim][Cl]	Mill mixing	Heat curing	[11]
DGEAC	GO	[Bhpmim][BF_4]	Solvent-free blending	Microwave curing	[35]
DGEBA	MWCNTs	[Bmim][PF_6]	Solution blending	Heat curing	[33]
PUF	GO	[Mim][Cl]	Mill mixing	Compression molding	[27]
PUF	nano-SiO_2	[Bmim][PF_6]	Solvent-free blending	Cast molding	[31]
PUF	nano-SiO_2	[Emim][Cl]	Solvent-free blending	Cast molding	[21]

3.2. Ionic Liquid-Nanofiller/Elastomer Nanocomposites

Ionic liquid-modified nanofiller/polymer nanocomposites can also be fabricated by using elastomers as polymer matrices. Ionic liquid-nanofiller/elastomer nanocomposites are more environmentally friendly than conventional elastomer composites due to the cleaner processes involving the use of ionic liquids [20]. It can be seen in Table 4 that elastomers such as CR and SBR are commonly incorporated with ionic liquid-modified nanofillers. The nanocomposites made from the elastomers can potentially be used in elastic conductors [42], electrochemistry [52], automobile components [6], and electromagnetic interference shielding [11]. Meanwhile, the mixing process for elastomers and ionic liquid-modified nanofillers can be performed by means of mill mixing, hot mixing, and speed mixing processes [4,16,55]. On the other hand, the elastomers require curatives (curing agents) [4,17,42,52] for crosslinking to occur in the nanocomposites. Elastomers that are admixed with nanofiller-ionic liquid can be mixed with curatives prior to the final process. The heat curing process is typically conducted at an elevated temperature under moderate compression with an optimal curing time.

3.3. Ionic Liquid-Nanofiller/Thermoset Nanocomposites

Thermosets can also be used as polymer matrices for the fabrication of polymer nanocomposites. However, only a limited number of ionic liquid-nanofiller/thermoset nanocomposites have been found. In Table 4, it can be observed that the PUF thermoset has more frequently been incorporated with ionic liquid-modified nanofillers than other thermosets. This is probably due to the thermoset being able to be applied for the production of polymer nanocomposite foams with controlled porosity [27], improved physicomechanical properties [31], and enhanced flame retardancy [21]. The mixing process between the PUF and ionic liquid-modified nanofillers can be accomplished via solvent-free blending [21,31] and mill mixing [27] at room temperature. Meanwhile, other thermosets, such as DGEAC and DGEBA, can be mixed with ionic liquid-modified nanofillers as well through solvent-free blending or solution blending [33,35], followed by the addition of curing agents. The final process is significant for the crosslinking reaction of the polymer nanocomposites prior to physicochemical characterization. The process can be completed via microwave curing, heat curing, compression molding, and cast molding, either at room temperature or at an elevated temperature.

4. Effect of Imidazolium-Based Ionic Liquids on the Thermo-Mechanico-Chemical Properties

4.1. Effect of Alkylimidazolium Halide Ionic Liquids

Table 5 shows the thermo-mechanico-chemical properties of imidazolium-based ionic liquid-modified nanofiller/polymer nanocomposites. Nano-SiO_2 modified with [Amim][Cl] ionic liquid has been employed for the fabrication of [Amim][Cl]-modified nano-SiO_2/NR nanocomposites [5]. The thermal, mechanical, and chemical properties of the nanocomposites and their components were characterized by means of differential scanning calorimeter, universal testing machine, and FTIR spectrometer. The thermal properties, such as the glass transition temperature, of the [Amim][Cl]-nano-SiO_2 increased by up to 14% in comparison to the pristine [Amim][Cl]. This was due to the motion of [Amim][Cl] being restricted by the strong interaction between the [Amim][Cl] and nano-SiO_2. On the other hand, the mechanical properties, such as the tensile strength, tensile modulus, and elongation at break, of the modified nano-SiO_2/NR nanocomposites increased by up to 102%, 28%, and 24%, respectively, compared to the unmodified nano-SiO_2/NR nanocomposite. This was caused by the combinatory effects giving the homogenous dispersion of [Amim][Cl]-nano-SiO_2 and increasing the interactions between the nano-SiO_2 and [Amim][Cl]. Additionally, the chemical properties, such as the absorption bands of the C−H stretching vibrations, of the [Amim][Cl]-nano-SiO_2 shifted to higher wavenumber regions in comparison to the C−H stretching vibrations of the pristine [Amim][Cl]. This indicated that the detachment of chloride anions from the $C{-}H\cdots Cl^-$ occurred because of the formation of the hydrogen bonding between the chloride anions and hydroxyl groups on the surface of nano-SiO_2 [5]. Furthermore, interaction between the modified nano-SiO_2 and NR matrix was formed as well. Therefore, it can be deduced that the modification of nano-SiO_2 with [Amim][Cl] ionic liquid provides NR nanocomposites with high tensile strength and modulus properties, as well as good interaction.

Table 5. Thermo-mechanico-chemical properties of imidazolium-based ionic liquid-modified nanofiller/polymer nanocomposites.

Ionic Liquid	Polymer Nanocomposite	Thermo-Mechanico-Chemical Properties *								References
		T_g	T_c	T_m	T_d	TS	TM	EB	Ch	
[Amim][Cl]	nano-SiO_2/NR	▲	○	○	○	▲	▲	▲	▲	[5]
[Bmim][Cl]	MMT/PVDF	○	▲	○	▲	▲	▲	▲	▲	[10]
[Bzmim][Cl]	MWCNT/SBR	▼	○	○	○	▲	▲	▼	▲	[13]
[Bmim][PF_6]	MWCNT/PEI	▲	○	○	▲	▲	▲	▼	▲	[32]
[Bmim][PF_6]	nano-SiO_2/PUF	▲	○	○	▲	▲	○	▼	▲	[31]
[Hdmim][PF_6]	MWCNT/PVDF	○	▲	▲	○	▼	○	▼	▲	[22]
[Bhpmim][BF_4]	GO/DGEAC	▲	○	○	▲	▲	▲	○	▲	[35]
[Bmim][BF_4]	GN/PI	▲	○	○	▲	▲	▲	▼	▲	[39]
[Emim][BF_4]	MWCNT/EVM	▲	○	▲	■	▲	○	▲	▲	[40]
[Bmim][NTf_2]	MWCNT/CR	▲	○	○	▲	○	▲	▲	▲	[42]
[Bmim][NTf_2]	MWCNT/NRL	▲	○	○	▼	▼	▲	▼	▲	[51]
[Emim][NTf_2]	MWCNT/PVDF	▼	○	▼	○	▲	▼	▲	▲	[54]

T_g = glass transition temperature, T_c = crystallization temperature, T_m = melting temperature, T_d = decomposition temperature, TS = tensile strength, TM = tensile modulus, EB = elongation at break, Ch = chemical property. * The symbol '▲' corresponds to an increase in the properties, and '▼' to a decrease in the properties, while '■' and '○' describe unchanged and not available, respectively.

Meanwhile, MMT modified with [Bmim][Cl] ionic liquid has been employed for the fabrication of [Bmim][Cl]-modified MMT/PVDF nanocomposites [10]. The thermal, mechanical, and chemical properties of the nanocomposites and their components were characterized by using differential scanning calorimeter, thermogravimetric analyzer, universal testing machine, and FTIR spectrometer. The thermal properties, such as the crystallization temperature and maximum decomposition temperature, of the modified MMT/PVDF nanocomposites slightly increased by up to 2.9% and 0.6%, respectively, compared to the unmodified MMT/PVDF nanocomposite. This was attributed to the presence of interaction between the CH_2-CF_2 of PVDF and the imidazolium ring of [Bmim][Cl]. In addition, the mechanical properties, such as the tensile strength, tensile modulus, and elongation at break, of the modified MMT/PVDF nanocomposites increased by up to 45%, 16%, and 125%, respectively, compared to the unmodified MMT/PVDF nanocomposite. This was caused by the formation of a structure with molecular orientation composed of an extended chain crystal and folded chain crystal in the modified MMT/PVDF nanocomposites. On top of that, the chemical properties, such as the absorption band of the CF_2-CH_2 stretching vibration, of the modified MMT/PVDF nanocomposite shifted to a lower wavenumber region in comparison to the CF_2-CH_2 stretching vibration of the neat PVDF. In contrast, the absorption band of the CF_2-CH_2 bending vibration of the modified MMT/PVDF nanocomposite shifted to a higher wavenumber region compared to the CF_2-CH_2 bending vibration of the neat PVDF. This was due to the formation of the β-phase due to the coulombic interaction between the imidazolium cations in the MMT interface and the negatively polarized CF_2 groups in PVDF [10]. Thus, it can be inferred that the modification of MMT with [Bmim][Cl] ionic liquid gives PVDF nanocomposites high tensile strength and elongation properties, as well as good interaction.

Additionally, MWCNTs modified with [Bzmim][Cl] ionic liquid have been employed for the fabrication of [Bzmim][Cl]-modified MWCNT/SBR nanocomposites [13]. The thermal, mechanical, and chemical properties of the nanocomposites and their components were characterized by means of dynamic mechanical analyzer, universal testing machine, and Raman spectrometer. The thermal properties, such as the glass transition temperature, of the modified MWCNT/SBR nanocomposites decreased, which was attributed to the plasticizing effect of ionic liquid on the SBR matrix. This was also induced by the formation of the strong interaction between the modified MWCNT and SBR. On the other hand, the mechanical properties, such as the tensile strength and tensile modulus, of the modified MWCNT/SBR nanocomposites significantly increased by up to 317% and 348%, respectively, compared to the neat SBR. This was due to the [Bzmim][Cl] assisting the dispersion of MWCNTs, which enhanced stress transfer from the matrix to the nanofiller. Moreover, the good adhesion at the interfacial of MWCNTs/SBR was also able to improve the mechanical properties, due to the formation of the physical adsorption and the chemical bonding between the modified nanofiller and polymer matrix. Nevertheless, the elongation at break of the modified MWCNT/SBR nanocomposites decreased because of the existence of physical crosslinks in SBR nanocomposites, which acted as blocks for deformation. In addition, the chemical properties, such as the G band, of the modified MWCNT/SBR nanocomposite shifted to a lower wavenumber region in comparison to the G band of the pristine MWCNTs. This was due to the realignment of [Bzmim][Cl]-MWCNTs themselves in the SBR matrix, which prompted an improved arrangement of MWCNTs in the nanocomposites [13]. Hence, it can be concluded that the modification of MWCNTs with [Bzmim][Cl] ionic liquid grants SBR nanocomposites high tensile strength and modulus properties, as well as good adhesion.

4.2. *Effect of Alkylimidazolium Hexafluorophosphate Ionic Liquids*

MWCNTs modified with [Bmim][PF_6] ionic liquid have been employed for the fabrication of [Bmim][PF_6]-modified MWCNT/PEI nanocomposites [32]. The thermal, mechanical, and chemical properties of the nanocomposites and their components were characterized by using differential scanning calorimeter, thermogravimetric analyzer, universal testing machine, and Raman spectrometer. The thermal properties, such as the glass transition temperature, of the modified MWCNT/PEI nanocomposites increased by up to 14% in comparison to the PEI containing ionic liquid (Table 5). This was caused by the restricting effect of MWCNTs as the movement of the PEI molecular chains is reduced. Additionally, the increment of the temperature was also due to the well-dispersed MWCNTs in the nanocomposites modified by [Bmim][PF_6]. Additionally, the initial decomposition temperature of the modified MWCNT/PEI nanocomposites increased by up to 33% compared to the PEI containing ionic liquid. This was attributed to the existence of strong interfacial interaction between the [Bmim][PF_6]-MWCNT and PEI, which slowed down the decomposition rate of the nanocomposites. Moreover, the molecular chains of PEI were able to wrap the well-dispersed MWCNTs and provide restricted segments that resulted in high thermal stability of the nanocomposites. On top of that, the mechanical properties, such as the tensile strength and tensile modulus, of the modified MWCNT/PEI nanocomposites increased by up to 41% and 32%, respectively, compared to the neat PEI. This was owing to the strong interfacial interactions between the modified MWCNT and PEI matrix, which gave the nanocomposites stiffer and stronger properties. However, the elongation at break of the modified MWCNT/PEI nanocomposites decreased because of the formation of the MWCNTs networks in the PEI matrix. Furthermore, the chemical properties, such as the D band, of the modified MWCNT shifted to a higher wavenumber region in comparison to the D band of the pristine MWCNTs. This was due to the adhesion of the [Bmim][PF_6] molecules via cation-π or π-π interactions on the MWCNT surface [32]. Therefore, it can be deduced that the modification of MWCNTs with [Bmim][PF_6] ionic liquid provides PEI nanocomposites with high thermal stability and high tensile strength and modulus properties, as well as good interaction.

Meanwhile, nano-SiO_2 modified with [Bmim][PF_6] ionic liquid has been employed for the fabrication of [Bmim][PF_6]-modified nano-SiO_2/PUF nanocomposites [31]. The thermal, mechanical, and chemical properties of the nanocomposites and their components were characterized by means of dynamic mechanical analyzer, thermogravimetric analyzer, testing machine, and FTIR spectrometer. The thermal properties, such as the glass transition temperature, of the modified nano-SiO_2/PUF nanocomposites increased by up to

20% in comparison to the unmodified nano-SiO_2/PUF nanocomposite. This was due to an improvement in the rigidity of the nanocomposites, which was induced by high crosslink density with the incorporation of [Bmim][PF_6]-nano-SiO_2. In addition, the initial decomposition temperature of the modified nano-SiO_2/PUF nanocomposites increased by up to 10% compared to the unmodified nano-SiO_2/PUF nanocomposite. This was attributed to the good dispersion of the modified nano-SiO_2 and homogeneous microstructure of the nanocomposites, which significantly enhanced the thermal stability of the nanocomposites. On the other hand, the mechanical properties, such as the flexural strength of the modified nano-SiO_2/PUF nanocomposites, increased by up to 15% in comparison to the unmodified nano-SiO_2/PUF nanocomposite. This was because of the compatibilizing effect of [Bmim][PF_6] on the nanocomposites. Nevertheless, the elongation at break of the modified nano-SiO_2/PUF nanocomposites decreased, which was caused by the presence of nano-SiO_2 aggregates that probably acted as defects for the PUF matrix, reducing the extension of the nanocomposites. Moreover, the chemical properties, such as the absorption band of the C=O stretching vibration of the carbonyl groups of the modified nano-SiO_2/PUF nanocomposite shifted to a higher wavenumber region in comparison to the C=O stretching vibration of the neat PUF. This was due to the hydrogen bonding between the NH and C=O of the matrix disturbed by the incorporation of the modified nano-SiO_2, which improved the interfacial adhesion between the PUF and nano-SiO_2 [31]. Thus, it can be inferred that the modification of nano-SiO_2 with [Bmim][PF_6] ionic liquid gives PUF nanocomposites high thermal and tensile strength properties, as well as good adhesion.

Additionally, MWCNTs modified with [Hdmim][PF_6] ionic liquid have been employed for the fabrication of [Hdmim][PF_6]-modified MWCNT/PVDF nanocomposites [22]. The thermal, mechanical, and chemical properties of the nanocomposites and their components were characterized by using differential scanning calorimeter, universal testing machine, and FTIR spectrometer, respectively. The thermal properties, such as the crystallization temperature of the modified MWCNT/PVDF nanocomposites moderately increased by up to 4.3% in comparison to the unmodified MWCNT/PVDF nanocomposite. This was due to the efficiency of modified MWCNT to act as a nucleation agent for the PVDF compared to the pristine MWCNTs, which accelerated the crystallization of PVDF. In addition, the melting temperature of the modified MWCNT/PVDF nanocomposites moderately increased by up to 3.3% in comparison to the unmodified MWCNT/PVDF nanocomposite. This was attributed to the disruption of [Hdmim][PF_6], which complemented the effect of promoting on crystallization of PVDF. Nevertheless, the mechanical properties, such as the tensile strength and elongation at break of the modified MWCNT/PVDF nanocomposites slightly decreased possibly because of the presence of micro-stress, which required less force for deformation, and consequently reduced the strength and elongation. However, the elongation at break of the modified MWCNT/PVDF nanocomposites is significantly higher compared to neat PVDF, which indicated that the plasticizing effect of ionic liquid on the polymer matrix. On the other hand, the chemical properties, such as the absorption band of the $-CF_2$ stretching vibration, of the modified MWCNT/PVDF nanocomposites shifted to a higher wavenumber region in comparison to the $-CF_2$ stretching vibration of the unmodified MWCNT/PVDF nanocomposite. This was due to the role of [Hdmim][PF_6], which acted as a linker for improving the compatibility between the MWCNTs surfaces and molecular chains of PVDF [22]. Hence, it can be concluded that the modification of MWCNTs with [Hdmim][PF_6] ionic liquid grants PVDF nanocomposites moderate crystallization and melting temperatures, as well as good compatibility.

4.3. *Effect of Alkylimidazolium Tetrafluoroborate Ionic Liquids*

GO modified with [Bhpmim][BF_4] ionic liquid has been employed for the fabrication of [Bhpmim][BF_4]-modified GO/DGEAC nanocomposites [35]. The thermal, mechanical, and chemical properties of the nanocomposites and their components were characterized by means of dynamic mechanical thermal analyzer, thermogravimetric analyzer, mechanical testing machine, and Raman spectrometer. The thermal properties, such as the glass transition temperature, of the modified GO/DGEAC nanocomposites slightly increased by up to 2.4% in comparison to the unmodified GO/DGEAC nanocomposite (Table 5). This was due to the introduction of [Bhpmim][BF_4] efficiently enhancing the interfacial bonding between the GO nanofiller and DGEAC matrix, which efficiently strengthened the nanocomposite system. In addition, the maximum decomposition temperature of the [Bhpmim][BF_4]-GO increased by approximately up to 10% compared to the pristine [Bhpmim][BF_4]. This could probably be attributed to the slow thermal elimination of the ester bond. On the other hand, the mechanical properties, such as the tensile strength and tensile modulus, of the modified GO/DGEAC nanocomposites moderately increased by up to 4.3% and 4.0%, respectively, compared to the unmodified GO/DGEAC nanocomposite. This was caused by the capability of the modified GO to disperse well in the DGEAC matrix and improve the interfacial bonding with the matrix. Additionally, the reinforcing ability of modified GO on the nanocomposites was well exerted compared to the pristine GO. Furthermore, the chemical properties, such as the D band, of the modified GO/DGEAC nanocomposites shifted to higher wavenumber regions in comparison to the D band of the unmodified GO/DGEAC nanocomposite. This was due to the exceptional interfacial bonding between the modified GO and DGEAC matrix [35]. Therefore, it can be deduced that the modification of GO with [Bhpmim][BF_4] ionic liquid provides DGEAC nanocomposites with moderate thermal and mechanical properties, as well as good interfacial bonding.

Meanwhile, GN modified with [Bmim][BF_4] ionic liquid has been employed for the fabrication of [Bmim][BF_4]-modified GN/PI nanocomposites [39]. The thermal, mechanical, and chemical properties of the nanocomposites and their components were characterized by using differential scanning calorimeter, thermogravimetric analyzer, universal tensile tester, and Raman spectrometer. The thermal properties, such as the glass transition temperature, of the modified GN/PI nanocomposites moderately increased by up to 4.1% in comparison to the unmodified GN/PI nanocomposite. This was caused by the uniform dispersion of modified GN in the PI phase and strong adhesion to the polymer matrix, which restricted the movement of the molecular chains of PI during the glass phase transition, requiring more heat. Additionally, the initial decomposition temperature of the modified GN/PI nanocomposites increased by up to 9.2% compared to the unmodified GN/PI nanocomposite. This was because of the presence of π-π, cation-π, and van der Waals interactions between the [Bmim][BF_4] and GN, which enhanced the dispersion of modified GN in the PI matrix. Consequently, the enhanced interfacial compatibility and stability of the modified GN with the matrix and the relaxation of the molecular chains of the PI required more fracture energy during the heating process. On top of that, the mechanical properties, such as the tensile strength and tensile modulus, of the modified GN/PI nanocomposites increased by up to 35% and 38%, respectively, compared to the unmodified GN/PI nanocomposite. This was due to the outstanding dispersion and excessive degree of orientation of GN modified by [Bmim][BF_4] in the polymer matrix, as well as the effective load transfer from PI to modified GN. Nonetheless, the elongation at break of the modified GN/PI nanocomposites decreased owing to the modified GN acted as a physical crosslinking agent, which restrained the flexibility of the PI molecular chains, subsequently increasing the brittleness of the nanocomposites, and preventing deformation. On the other hand, the chemical properties, such as the G band, of the modified GN shifted to a lower wavenumber region in comparison to the G band of the pristine GN. This was attributed to the modification of GN by [Bmim][BF_4] affecting the lattice of GN [39]. Thus, it can be inferred that the modification of GN with [Bmim][BF_4] ionic liquid gives PI nanocomposites moderate thermal properties and high tensile strength and modulus properties, as well as good interaction.

Additionally, MWCNTs modified with [Emim][BF_4] ionic liquid have been employed for the fabrication of [Emim][BF_4]-modified MWCNT/EVM nanocomposites [40]. The thermal, mechanical, and chemical properties of the nanocomposites and their components were characterized by means of differential scanning calorimeter, thermogravimetric analyzer, tensile testing machine, and Raman spectrometer. The thermal properties, such as the glass transition temperature and melting temperature, of the modified MWCNT/EVM nanocomposites moderately increased by up to 2.2% and 6.2%, respectively, compared to the unmodified MWCNT/EVM nanocomposite. This was attributed to the special interactions among MWCNTs, [Emim][BF_4], and EVM, which influenced the glass transition and melting temperatures of the nanocomposites. Nevertheless, the decomposition temperature remained almost unchanged for both of the nanocomposites. On the other hand, the mechanical properties, such as the tensile strength and elongation at break, of the modified MWCNT/EVM nanocomposites increased by up to 20% and 42%, respectively, compared to the unmodified MWCNT/EVM nanocomposite. This was because of the reinforcing ability of the modified MWCNT and the plasticizing effect of [Emim][BF_4] on the nanocomposites, which increased the strength, stretchability, and flexibility of the modified MWCNT/EVM nanocomposites. Furthermore, the chemical properties, such as the D band, of the modified MWCNT shifted to a higher wavenumber region in comparison to the D band of the pristine MWCNTs. This was caused by the formation of a strong interaction between the imidazolium ring of [Emim][BF_4] and the surface of MWCNTs through cation-π or π-π interactions [40]. Hence, it can be concluded that the modification of MWCNTs with [Emim][BF_4] ionic liquid grants EVM nanocomposites moderate thermal properties and high tensile strength and elongation properties, as well as good interaction.

4.4. *Effect of Alkylimidazolium Bistriflimide Ionic Liquids*

MWCNTs modified with [Bmim][NTf_2] ionic liquid have been employed for the fabrication of [Bmim][NTf_2]-modified MWCNT/CR nanocomposites [42]. The thermal, mechanical, and chemical properties of the nanocomposites and their components were characterized by using differential scanning calorimeter, thermogravimetric analyzer, universal testing machine, and Raman spectrometer. The thermal properties, such as the glass transition temperature of the modified MWCNT/CR nanocomposites, slightly increased by up to 2.6% compared to the unmodified MWCNT/CR nanocomposite (Table 5). This is probably due to the presence of specific interactions between the ionic liquid and nanofiller. Moreover, the maximum decomposition temperature of the [Bmim][NTf_2]-MWCNT increased by up to 5.2% compared to the pristine [Bmim][NTf_2]. This was attributed to the adhesion of [Bmim][NTf_2] to MWCNTs via cation-π, π-π, or van der Waals interactions, which enhanced the thermal stability of [Bmim][NTf_2]. On top of that, the mechanical properties, such as the tensile modulus and hardness, of the modified MWCNT/CR nanocomposites increased by up to 48% and 2.7%, respectively, compared to the unmodified MWCNT/CR nanocomposite. This was caused by the presence of [Bmim][NTf_2], which acted as a coupling agent for the nanocomposites. In addition, the chemical properties, such as the G band, of the modified MWCNT/CR nanocomposite shifted to a higher wavenumber region in comparison to the G band of the unmodified MWCNT/CR nanocomposite. This was because of the disentanglement of the modified MWCNT and subsequent dispersion in the matrix due to the penetration of the CR into the bundles of MWCNTs during the mixing process [42]. Therefore, it can be deduced that the modification of MWCNTs with [Bmim][NTf_2] ionic liquid provides CR nanocomposites with moderate thermal properties and high tensile modulus, as well as good interaction.

Meanwhile, MWCNTs modified with [Bmim][NTf_2] ionic liquid have been employed for the fabrication of [Bmim][NTf_2]-modified MWCNT/NRL nanocomposites [51]. The thermal, mechanical, and chemical properties of the nanocomposites and their components were characterized by means of differential scanning calorimeter, thermogravimetric analyzer, universal tensile testing machine, and FTIR spectrometer. The thermal properties, such as the glass transition temperature, of the modified MWCNT/NRL nanocomposites increased by up to 7.9% in comparison to the unmodified MWCNT/NRL nanocomposite. This was caused by the good interaction between the modified MWCNT and NRL, which promoted the formation of three-dimensional MWCNTs networks and restricted the movement or flexibility of the molecular chains of NRL. This also indicated the partial miscibility of the nanocomposite components. However, the decomposition temperature of the modified MWCNT/NRL nanocomposites decreased owing to the high thermal conductivity of MWCNTs and the low molecular weight of [Bmim][NTf_2], which favored thermal decomposition of the nanocomposites. Furthermore, the mechanical properties, such as the tensile strength and elongation at break, of the modified MWCNT/NRL nanocomposites slightly decreased because of the small molecules of ionic liquid freely dispersed in the NRL matrix, which acted as defects that initiated failure. Nevertheless, the tensile modulus of the modified MWCNT/NRL nanocomposites increased by up to 73% compared to the unmodified MWCNT/NRL nanocomposite. This was attributed to the good dispersion of modified MWCNT, along with the formed MWCNTs networks in the matrix, assisted by interactions among ionic liquid, MWCNTs, and NRL. On the other hand, the chemical properties, such as the absorption band of the C−O stretching vibration, of the modified MWCNT/NRL nanocomposite shifted to a lower wavenumber region in comparison to the C−O stretching vibration of the unmodified MWCNT/NRL nanocomposite. This was due to the modified MWCNT bridging NRL and [Bmim][NTf_2] through π-π interactions in the nanocomposites [51]. Thus, it can be inferred that the modification of MWCNTs with [Bmim][NTf_2] ionic liquid gives NRL nanocomposites high glass transition temperature and tensile modulus, as well as good interaction.

MWCNTs modified with [Emim][NTf_2] ionic liquid have been employed for the fabrication of [Emim][NTf_2]-modified MWCNT/PVDF nanocomposites [54]. The thermal, mechanical, and chemical properties of the nanocomposites and their components were characterized using differential scanning calorimeter, micro universal testing machine, and FTIR spectrometer. The thermal properties, such as the glass transition temperature and melting temperature, of the modified MWCNT/PVDF nanocomposites decreased, which was attributed to the miscibility between the ionic liquid and PVDF. Moreover, the plasticizing effect of [Emim][NTf_2] on the PVDF matrix also decreased the thermal properties of the nanocomposites. On top of that, the mechanical properties, such as the tensile strength and elongation at break, of the modified MWCNT/PVDF nanocomposites significantly increased by up to 85% and 133%, respectively, compared to the unmodified MWCNT/PVDF nanocomposite. This was caused by the presence of [Emim][NTf_2] provided the good dispersion of MWCNTs in the PVDF matrix, which further enhanced the structural properties of the nanocomposites. Additionally, the modified MWCNT/PVDF nanocomposites were less brittle, with improved strength and ductility properties compared to unmodified MWCNT/PVDF nanocomposite. Nonetheless, the tensile modulus of the modified MWCNT/PVDF nanocomposites slightly decreased owing to the simultaneous increase in strength and elongation properties that could marginally reduce the rigidity of the nanocomposites. On the other hand, the chemical properties, such as the absorption band of the CH−CF−CH stretching vibration, of the modified MWCNT/PVDF nanocomposites shifted to a higher wavenumber region in comparison to the CH−CF−CH stretching vibration of the unmodified MWCNT/PVDF nanocomposite. This was due to the formation of an interaction between the modified MWCNT and PVDF via electrostatic interaction [54]. Hence, it can be concluded that the modification of MWCNTs with [Emim][NTf_2] ionic liquid grants PVDF nanocomposites high tensile strength and elongation properties, as well as good interaction.

5. Conclusions

Types of imidazolium-based ionic liquids, modification processes of nanofillers, and fabrication processes of ionic liquid-modified nanofiller/polymer nanocomposites were succinctly reviewed in this paper. The important properties, for example, thermal, mechanical, and chemical, of the nanocomposites were also described in this succinct review. Ionic liquids used for the modification of various types of nanofillers are mostly based on imidazolium cations combined with different counter anions. In addition, alkylimidazolium halide, alkylimidazolium hexafluorophosphate, alkylimidazolium tetrafluoroborate, and alkylimidazolium bistriflimide ionic liquids are the four most significant ionic liquids used for polymer nanocomposites. Alkylimidazolium halide ionic liquid is frequently employed in the modification of several types of nanofillers. Meanwhile, alkylimidazolium hexafluorophosphate, alkylimidazolium tetrafluoroborate, and alkylimidazolium bistriflimide ionic liquids are typically used in the modification of carbon-based nanofillers. The employment of ionic liquid-modified nanofillers can effectively improve the thermo-mechanico-chemical properties of polymer nanocomposites. Additionally, ionic liquid-modified nanofillers can form specific interactions with polymer matrices. Alkylimidazolium halide-modified nanofiller/polymer nanocomposites possess high mechanical properties and good interaction. On top of that, alkylimidazolium hexafluorophosphate-modified nanofiller/polymer nanocomposites exhibit high thermal and mechanical properties and good compatibility. Meanwhile, alkylimidazolium tetrafluoroborate-modified nanofiller/polymer nanocomposites demonstrate moderate thermal properties, high mechanical properties, and good interaction. Additionally, alkylimidazolium bistriflimide-modified nanofiller/polymer nanocomposites have moderate thermal properties, high mechanical properties, and good interaction. This succinct review may be valuable for the modification of nanofillers by using ionic liquids in the fabrication of polymer nanocomposites for a variety of usages.

Author Contributions: Conceptualization, A.A.S.; methodology, S.N.A.M.J.; validation, S.N.A.M.J. and K.A.; formal analysis, S.N.A.M.J.; investigation, A.A.S.; resources, K.A.; data curation, A.A.S.; writing—original draft preparation, A.A.S.; writing—review and editing, S.N.A.M.J.; project administration, K.A.; funding acquisition, K.A. All authors have read and agreed to the published version of the manuscript.

Funding: This succinct review was funded by the Universiti Putra Malaysia (vote number: 9001103).

Institutional Review Board Statement: Not applicable.

Informed Consent Statement: Not applicable.

Data Availability Statement: Not applicable.

Acknowledgments: The authors would like to thank M. Ali Aboudzadeh from the Centro de Física de Materiales, CSIC-UPV/EHU, and Shaghayegh Hamzehlou from the University of the Basque Country UPV-EHU for inviting the authors to write this succinct review.

Conflicts of Interest: The authors declare no conflict of interest. The funder had no role in the design of the review; in the collection, analyses, or interpretation of data; in the writing of the manuscript, or in the decision to publish the results.

References

1. Shamsuri, A.A.; Md Jamil, S.N.A. Compatibilization Effect of Ionic Liquid-Based Surfactants on Physicochemical Properties of PBS/Rice Starch Blends: An Initial Study. *Materials* **2020**, *13*, 1885. [CrossRef]
2. Shamsuri, A.A.; Md Jamil, S.N.A. Functional Properties of Biopolymer-Based Films Modified with Surfactants: A Brief Review. *Processes* **2020**, *8*, 1039. [CrossRef]
3. Shamsuri, A.A.; Daik, R. Mechanical and Thermal Properties of Nylon-6/LNR/MMT Nanocomposites Prepared Through Emulsion Dispersion Technique. *J. Adv. Res. Fluid Mech. Therm. Sci.* **2020**, *73*, 1–12. [CrossRef]
4. Prasad Sahoo, B.; Naskar, K.; Kumar Tripathy, D. Multiwalled carbon nanotube-filled ethylene acrylic elastomer nanocomposites: Influence of ionic liquids on the mechanical, dynamic mechanical, and dielectric properties. *Polym. Compos.* **2016**, *37*, 2568–2580. [CrossRef]
5. Zhang, X.; Xue, X.; Jia, H.; Wang, J.; Ji, Q.; Xu, Z. Influence of ionic liquid on the polymer–filler coupling and mechanical properties of nano-silica filled elastomer. *J. Appl. Polym. Sci.* **2017**, *134*, 44478. [CrossRef]
6. Yin, B.; Zhang, X.; Zhang, X.; Wang, J.; Wen, Y.; Jia, H.; Ji, Q.; Ding, L. Ionic liquid functionalized graphene oxide for enhancement of styrene-butadiene rubber nanocomposites. *Polym. Adv. Technol.* **2017**, *28*, 293–302. [CrossRef]
7. Sahu, G.; Tripathy, J.; Sahoo, B.P. Significant enhancement of dielectric properties of graphene oxide filled polyvinyl alcohol nanocomposites: Effect of ionic liquid and temperature. *Polym. Compos.* **2020**, *41*, 4411–4430. [CrossRef]
8. Qiu, M.; Zhang, B.; Wu, H.; Cao, L.; He, X.; Li, Y.; Li, J.; Xu, M.; Jiang, Z. Preparation of anion exchange membrane with enhanced conductivity and alkaline stability by incorporating ionic liquid modified carbon nanotubes. *J. Memb. Sci.* **2019**, *573*, 1–10. [CrossRef]
9. Huang, G.; Isfahani, A.P.; Muchtar, A.; Sakurai, K.; Shrestha, B.B.; Qin, D.; Yamaguchi, D.; Sivaniah, E.; Ghalei, B. Pebax/ionic liquid modified graphene oxide mixed matrix membranes for enhanced CO_2 capture. *J. Memb. Sci.* **2018**, *565*, 370–379. [CrossRef]

10. Thomas, E.; Parvathy, C.; Balachandran, N.; Bhuvaneswari, S.; Vijayalakshmi, K.P.; George, B.K. PVDF-ionic liquid modified clay nanocomposites: Phase changes and shish-kebab structure. *Polymer* **2017**, *115*, 70–76. [CrossRef]
11. Abraham, J.; Xavier, P.; Bose, S.; George, S.C.; Kalarikkal, N.; Thomas, S. Investigation into dielectric behaviour and electromagnetic interference shielding effectiveness of conducting styrene butadiene rubber composites containing ionic liquid modified MWCNT. *Polymer* **2017**, *112*, 102–115. [CrossRef]
12. Abraham, J.; Zachariah, A.K.; Wilson, R.; Ibarra-Gómez, R.; Muller, R.; George, S.C.; Kalarikkal, N.; Thomas, S. Effect of ionic liquid modified MWCNT on the rheological and microstructural developments in styrene butadiene rubber nanocomposites. *Rubber Chem. Technol.* **2019**, *92*, 531–545. [CrossRef]
13. Abraham, J.; Thomas, J.; Kalarikkal, N.; George, S.C.; Thomas, S. Static and Dynamic Mechanical Characteristics of Ionic Liquid Modified MWCNT-SBR Composites: Theoretical Perspectives for the Nanoscale Reinforcement Mechanism. *J. Phys. Chem. B* **2018**, *122*, 1525–1536. [CrossRef]
14. Abraham, J.; Jose, T.; Moni, G.; George, S.C.; Kalarikkal, N.; Thomas, S. Ionic liquid modified multiwalled carbon nanotube embedded styrene butadiene rubber membranes for the selective removal of toluene from toluene/methanol mixture via pervaporation. *J. Taiwan Inst. Chem. Eng.* **2019**, *95*, 594–601. [CrossRef]
15. George, S.; Abraham, J.; Jose, T.; Kalarikkal, N.; Thomas, S. Mechanics and Pervaporation Performance of Ionic Liquid Modified CNT Based SBR Membranes—A Case Study for the Separation of Toluene/Heptane Mixtures. *Int. J. Membr. Sci. Technol.* **2015**, *2*, 30–38. [CrossRef]
16. Abraham, J.; Sidhardhan Sisanth, K.; Zachariah, A.K.; Mariya, H.J.; George, S.C.; Kalarikkal, N.; Thomas, S. Transport and solvent sensing characteristics of styrene butadiene rubber nanocomposites containing imidazolium ionic liquid modified carbon nanotubes. *J. Appl. Polym. Sci.* **2020**, *137*, e49429. [CrossRef]
17. Sowińska, A.; Maciejewska, M.; Guo, L.; Delebecq, E. Effect of SILPs on the Vulcanization and Properties of Ethylene–Propylene–Diene Elastomer. *Polymers* **2020**, *12*, 1220. [CrossRef] [PubMed]
18. Le, H.H.; Wießner, S.; Das, A.; Fischer, D.; Auf Der Landwehr, M.; Do, Q.K.; Stöckelhuber, K.W.; Heinrich, G.; Radusch, H.J. Selective wetting of carbon nanotubes in rubber compounds—Effect of the ionic liquid as dispersing and coupling agent. *Eur. Polym. J.* **2016**, *75*, 13–24. [CrossRef]
19. Xu, P.; Cui, Z.P.; Ruan, G.; Ding, Y.S. Enhanced Crystallization Kinetics of PLLA by Ethoxycarbonyl Ionic Liquid Modified Graphene. *Chin. J. Polym. Sci.* **2019**, *37*, 243–252. [CrossRef]
20. Ge, Y.; Zhang, Q.; Zhang, Y.; Liu, F.; Han, J.; Wu, C. High-performance natural rubber latex composites developed by a green approach using ionic liquid-modified multiwalled carbon nanotubes. *J. Appl. Polym. Sci.* **2018**, *135*, 46588. [CrossRef]
21. Członka, S.; Strąkowska, A.; Strzelec, K.; Kairytė, A.; Kremensas, A. Melamine, silica, and ionic liquid as a novel flame retardant for rigid polyurethane foams with enhanced flame retardancy and mechanical properties. *Polym. Test.* **2020**, *87*, 106511. [CrossRef]
22. Bahader, A.; Haoguan, G.; HaoGoa, F.; Ping, W.; Shaojun, W.; Yunsheng, D. Preparation and characterization of poly(vinylidene fluoride) nanocomposites containing amphiphilic ionic liquid modified multiwalled carbon nanotubes. *J. Polym. Res.* **2016**, *23*, 184. [CrossRef]
23. Song, S.; Xia, S.; Liu, Y.; Lv, X.; Sun, S. Effect of Na+ MMT-ionic liquid synergy on electroactive, mechanical, dielectric and energy storage properties of transparent PVDF-based nanocomposites. *Chem. Eng. J.* **2020**, *384*, 123365. [CrossRef]
24. Shi, F.; Ma, Y.; Ma, J.; Wang, P.; Sun, W. Preparation and characterization of PVDF/TiO_2 hybrid membranes with ionic liquid modified nano-TiO2 particles. *J. Memb. Sci.* **2013**, *427*, 259–269. [CrossRef]
25. Pelit, F.O.; Pelit, L.; Dizdaş, T.N.; Aftafa, C.; Ertaş, H.; Yalçinkaya, E.E.; Türkmen, H.; Ertaş, F.N. A novel polythiophene—Ionic liquid modified clay composite solid phase microextraction fiber: Preparation, characterization and application to pesticide analysis. *Anal. Chim. Acta* **2015**, *859*, 37–45. [CrossRef]
26. Caldas, C.M.; Soares, B.G.; Indrusiak, T.; Barra, G.M.O. Ionic liquids as dispersing agents of graphene nanoplatelets in poly(methyl methacrylate) composites with microwave absorbing properties. *J. Appl. Polym. Sci.* **2021**, *138*, e49814. [CrossRef]
27. Mondal, T.; Basak, S.; Bhowmick, A.K. Ionic liquid modification of graphene oxide and its role towards controlling the porosity, and mechanical robustness of polyurethane foam. *Polymer* **2017**, *127*, 106–118. [CrossRef]
28. Chen, G.X.; Zhang, S.; Zhou, Z.; Li, Q. Dielectric properties of poly(vinylidene fluoride) composites based on bucky gels of carbon nanotubes with ionic liquids. *Polym. Compos.* **2015**, *36*, 94–101. [CrossRef]
29. Xiong, X.; Wang, J.; Jia, H.; Fang, E.; Ding, L. Structure, thermal conductivity, and thermal stability of bromobutyl rubber nanocomposites with ionic liquid modified graphene oxide. *Polym. Degrad. Stab.* **2013**, *98*, 2208–2214. [CrossRef]
30. Ke, K.; Pötschke, P.; Gao, S.; Voit, B. An Ionic Liquid as Interface Linker for Tuning Piezoresistive Sensitivity and Toughness in Poly(vinylidene fluoride)/Carbon Nanotube Composites. *ACS Appl. Mater. Interfaces* **2017**, *9*, 5437–5446. [CrossRef]
31. Członka, S.; Strąkowska, A.; Strzelec, K.; Kairytė, A.; Vaitkus, S. Composites of rigid polyurethane foams and silica powder filler enhanced with ionic liquid. *Polym. Test.* **2019**, *75*, 12–25. [CrossRef]
32. Chen, Y.; Tao, J.; Deng, L.; Li, L.; Li, J.; Yang, Y.; Khashab, N.M. Polyetherimide/bucky gels nanocomposites with superior conductivity and thermal stability. *ACS Appl. Mater. Interfaces* **2013**, *5*, 7478–7484. [CrossRef]
33. Zheng, X.; Li, D.; Feng, C.; Chen, X. Thermal properties and non-isothermal curing kinetics of carbon nanotubes/ionic liquid/epoxy resin systems. *Thermochim. Acta* **2015**, *618*, 18–25. [CrossRef]
34. Fang, D.; Zhou, C.; Liu, G.; Luo, G.; Gong, P.; Yang, Q.; Niu, Y.; Li, G. Effects of ionic liquids and thermal annealing on the rheological behavior and electrical properties of poly(methyl methacrylate)/carbon nanotubes composites. *Polymer* **2018**, *148*, 68–78. [CrossRef]

35. Shi, K.; Luo, J.; Huan, X.; Lin, S.; Liu, X.; Jia, X.; Zu, L.; Cai, Q.; Yang, X. Ionic Liquid-Graphene Oxide for Strengthening Microwave Curing Epoxy Composites. *ACS Appl. Nano Mater.* **2020**, *3*, 11955–11969. [CrossRef]
36. Damlin, P.; Suominen, M.; Heinonen, M.; Kvarnström, C. Non-covalent modification of graphene sheets in PEDOT composite materials by ionic liquids. *Carbon N. Y.* **2015**, *93*, 533–543. [CrossRef]
37. Lopes Pereira, E.C.; Soares, B.G. Conducting epoxy networks modified with non-covalently functionalized multi-walled carbon nanotube with imidazolium-based ionic liquid. *J. Appl. Polym. Sci.* **2016**, *133*, 43976. [CrossRef]
38. Rhyu, S.Y.; Cho, Y.; Kang, S.W. Nanocomposite membranes consisting of poly(ethylene oxide)/ionic liquid/ZnO for CO_2 separation. *J. Ind. Eng. Chem.* **2020**, *85*, 75–80. [CrossRef]
39. Ruan, H.; Zhang, Q.; Liao, W.; Li, Y.; Huang, X.; Xu, X.; Lu, S. Enhancing tribological, mechanical, and thermal properties of polyimide composites by the synergistic effect between graphene and ionic liquid. *Mater. Des.* **2020**, *189*, 108527. [CrossRef]
40. Cao, X.; Jin, M.; Liang, Y.; Li, Y. Synergistic effects of two types of ionic liquids on the dispersion of multiwalled carbon nanotubes in ethylene–vinyl acetate elastomer: Preparation and characterization of flexible conductive composites. *Polym. Int.* **2017**, *66*, 1708–1715. [CrossRef]
41. Xing, C.; Wang, Y.; Huang, X.; Li, Y.; Li, J. Poly(vinylidene fluoride) Nanocomposites with Simultaneous Organic Nanodomains and Inorganic Nanoparticles. *Macromolecules* **2016**, *49*, 1026–1035. [CrossRef]
42. Subramaniam, K.; Das, A.; Steinhauser, D.; Klüppel, M.; Heinrich, G. Effect of ionic liquid on dielectric, mechanical and dynamic mechanical properties of multi-walled carbon nanotubes/polychloroprene rubber composites. *Eur. Polym. J.* **2011**, *47*, 2234–2243. [CrossRef]
43. Subramaniam, K.; Das, A.; Heinrich, G. Development of conducting polychloroprene rubber using imidazolium based ionic liquid modified multi-walled carbon nanotubes. *Compos. Sci. Technol.* **2011**, *71*, 1441–1449. [CrossRef]
44. Subramaniam, K.; Das, A.; Häußler, L.; Harnisch, C.; Stöckelhuber, K.W.; Heinrich, G. Enhanced thermal stability of polychloroprene rubber composites with ionic liquid modified MWCNTs. *Polym. Degrad. Stab.* **2012**, *97*, 776–785. [CrossRef]
45. Subramaniam, K.; Das, A.; Simon, F.; Heinrich, G. Networking of ionic liquid modified CNTs in SSBR. *Eur. Polym. J.* **2013**, *49*, 345–352. [CrossRef]
46. Subramaniam, K.; Das, A.; Heinrich, G. Highly conducting polychloroprene composites based on multi-walled carbon nanotubes and 1-butyl 3-methyl imidazolium bis(trifluoromethylsulphonyl)imide. *KGK Kautsch. Gummi Kunstst.* **2012**, *65*, 44–46.
47. Steinhauser, D.; Subramaniam, K.; Das, A.; Heinrich, G.; Klüppel, M. Influence of ionic liquids on the dielectric relaxation behavior of CNT based elastomer nanocomposites. *Express Polym. Lett.* **2012**, *6*, 927–936. [CrossRef]
48. Le, H.H.; Hoang, X.T.; Das, A.; Gohs, U.; Stoeckelhuber, K.W.; Boldt, R.; Heinrich, G.; Adhikari, R.; Radusch, H.J. Kinetics of filler wetting and dispersion in carbon nanotube/rubber composites. *Carbon N. Y.* **2012**, *50*, 4543–4556. [CrossRef]
49. Subramaniam, K.; Das, A.; Heinrich, G. Improved oxidation resistance of conducting polychloroprene composites. *Compos. Sci. Technol.* **2013**, *74*, 14–19. [CrossRef]
50. Semeriyanov, F.F.; Chervanyov, A.I.; Jurk, R.; Subramaniam, K.; König, S.; Roscher, M.; Das, A.; Stöckelhuber, K.W.; Heinrich, G. Non-monotonic dependence of the conductivity of carbon nanotube-filled elastomers subjected to uniaxial compression/decompression. *J. Appl. Phys.* **2013**, *113*, 103706. [CrossRef]
51. Krainoi, A.; Kummerlöwe, C.; Nakaramontri, Y.; Wisunthorn, S.; Vennemann, N.; Pichaiyut, S.; Kiatkamjornwong, S.; Nakason, C. Influence of carbon nanotube and ionic liquid on properties of natural rubber nanocomposites. *Express Polym. Lett.* **2019**, *13*, 327–348. [CrossRef]
52. Jiang, G.; Song, S.; Zhai, Y.; Feng, C.; Zhang, Y. Improving the filler dispersion of polychloroprene/carboxylated multi-walled carbon nanotubes composites by non-covalent functionalization of carboxylated ionic liquid. *Compos. Sci. Technol.* **2016**, *123*, 171–178. [CrossRef]
53. Abraham, J.; Mohammed Arif, P.; Kailas, L.; Kalarikkal, N.; George, S.C.; Thomas, S. Developing highly conducting and mechanically durable styrene butadiene rubber composites with tailored microstructural properties by a green approach using ionic liquid modified MWCNTs. *RSC Adv.* **2016**, *6*, 32493. [CrossRef]
54. Sharma, M.; Sharma, S.; Abraham, J.; Thomas, S.; Madras, G.; Bose, S. Flexible EMI shielding materials derived by melt blending PVDF and ionic liquid modified MWNTs. *Mater. Res. Express* **2014**, *1*, 035003. [CrossRef]
55. Hassouneh, S.S.; Yu, L.; Skov, A.L.; Daugaard, A.E. Soft and flexible conductive PDMS/MWCNT composites. *J. Appl. Polym. Sci.* **2017**, *134*, 44767. [CrossRef]
56. Shamsuri, A.A.; Jamil, S.N.A.M. A Short Review on the Effect of Surfactants on the Mechanico-Thermal Properties of Polymer Nanocomposites. *Appl. Sci.* **2020**, *10*, 4867. [CrossRef]
57. Shamsuri, A.A.; Abdullah, D.K. Protonation and Complexation Approaches for Production of Protic Eutectic Ionic Liquids. *J. Phys. Sci.* **2010**, *21*, 15–28.
58. Shamsuri, A.A.; Abdan, K.; Kaneko, T. A Concise Review on the Physicochemical Properties of Biopolymer Blends Prepared in Ionic Liquids. *Molecules* **2021**, *26*, 216. [CrossRef]
59. Shamsuri, A.A.; Abdullah, D.K. Synthesizing of ionic liquids from different chemical pathways. *Int. J. Appl. Chem.* **2011**, *7*, 15–24.
60. Hassanshahi, N.; Hu, G.; Li, J. Application of Ionic Liquids for Chemical Demulsification: A Review. *Molecules* **2020**, *25*, 4915. [CrossRef]

processes

MDPI

Article

Mesoscale Morphologies of Nafion-Based Blend Membranes by Dissipative Particle Dynamics

Unal Sen [1,2,*], Mehmet Ozdemir [3], Mustafa Erkartal [4], Alaattin Metin Kaya [5], Abdullah A. Manda [3], Ali Reza Oveisi [6], M. Ali Aboudzadeh [7,8,*] and Takashi Tokumasu [2,*]

1 Department of Materials Science and Engineering, Faculty of Engineering, Eskisehir Technical University, Eskisehir 26555, Turkey
2 Institute of Fluid Science, Tohoku University, 2-1-1 Aoba-ku, Sendai, Miyagi 980-8577, Japan
3 Basic Engineering Sciences, College of Engineering, Imam Abdulrahman Bin Faisal University, Dammam 31451, Saudi Arabia; miozdemir@iau.edu.sa (M.O.); aamanda@iau.edu.sa (A.A.M.)
4 Department of Materials Science and Nanotechnology Engineering, Abdullah Gul University, Kayseri 38080, Turkey; merkartal@mail.com
5 Department of Mechanical Engineering, Faculty of Engineering, Bursa Uludag University, Bursa 16059, Turkey; alaattinkaya@uludag.edu.tr
6 Department of Chemistry, Faculty of Science, University of Zabol, Zabol P.O. Box 98615-538, Iran; aroveisi@uoz.ac.ir
7 Centro de Física de Materiales, CSIC-UPV/EHU, Paseo Manuel Lardizábal 5, 20018 Donostia-San Sebastián, Spain
8 Donostia International Physics Center (DIPC), Paseo Manuel Lardizábal 4, 20018 Donostia-San Sebastián, Spain
* Correspondence: unalsen@eskisehir.edu.tr (U.S.); mohammadali.aboudzadeh@ehu.eus (M.A.A.); tokumasu@ifs.tohoku.ac.jp (T.T.)

Citation: Sen, U.; Ozdemir, M.; Erkartal, M.; Kaya, A.M.; Manda, A.A.; Oveisi, A.R.; Aboudzadeh, M.A.; Tokumasu, T. Mesoscale Morphologies of Nafion-Based Blend Membranes by Dissipative Particle Dynamics. *Processes* **2021**, *9*, 984. https://doi.org/10.3390/pr9060984

Academic Editor: Domenico Frattini

Received: 1 May 2021
Accepted: 28 May 2021
Published: 2 June 2021

Abstract: Polymer electrolyte membrane (PEM) composed of polymer or polymer blend is a vital element in PEM fuel cell that allows proton transport and serves as a barrier between fuel and oxygen. Understanding the microscopic phase behavior in polymer blends is very crucial to design alternative cost-effective proton-conducting materials. In this study, the mesoscale morphologies of Nafion/poly(1-vinyl-1,2,4-triazole) (Nafion-PVTri) and Nafion/poly(vinyl phosphonic acid) (Nafion-PVPA) blend membranes were studied by dissipative particle dynamics (DPD) simulation technique. Simulation results indicate that both blend membranes can form a phase-separated microstructure due to the different hydrophobic and hydrophilic character of different polymer chains and different segments in the same polymer chain. There is a strong, attractive interaction between the phosphonic acid and sulfonic acid groups and a very strong repulsive interaction between the fluorinated and phosphonic acid groups in the Nafion-PVPA blend membrane. By increasing the PVPA content in the blend membrane, the PVPA clusters' size gradually increases and forms a continuous phase. On the other hand, repulsive interaction between fluorinated and triazole units in the Nafion-PVTri blend is not very strong compared to the Nafion-PVPA blend, which results in different phase behavior in Nafion-PVTri blend membrane. This relatively lower repulsive interaction causes Nafion-PVTri blend membrane to have non-continuous phases regardless of the composition.

Keywords: dissipative particle dynamics; Nafion; mesoscale morphology; poly(1-vinyl-1,2,4-triazole); poly(vinylphosphonic acid)

1. Introduction

Polymer electrolyte membrane (PEM) fuel cell, being more efficient, environmentally friendly, and modular, is one of the most outstanding candidates to replace the internal combustion engine [1–4]. Fluorinated or hydrocarbon-based proton exchange polymeric ionomers serve as the central component in PEM fuel cells owing to their crucial roles. In order to use them in viable technological applications, these materials need to have superior properties like high proton conductivity, thermal and chemical stability, impermeability to gases, and easier compatibility with electrodes [5–7]. Morphological peculiarities of these materials mostly stem from the micro-phase separation of hydrophobic and hydrophilic parts of the macromolecules [8]. Many studies have displayed that the conduction of protons takes place via hydrophilic domains that are resulted from micro-phase separation [9]. Morphological attributes of these membrane materials are very much affected by water and the backbone's chemistry.

High-temperature (100–200 °C) PEM fuel cells have been of great interest to researchers because of their outstanding benefits in terms of higher catalytic activity, CO poisoning elimination, low cost, device cooling, and water management. Several authors have studied novel polyelectrolyte systems with high conductivity at temperatures from 100 °C to 200 °C as an alternative solution to the commercial perfluorinated polyelectrolytes (e.g., Nafion), which show high conductivities only through water-assisted proton conduction [10–12] and operate at temperatures below 100 °C [13]. To date, building strong acid-base complexes among functionalities connected to the polymeric backbones has been one of the most well-known methods that enable high proton conductivity under low humidity or even anhydrous conditions [14–20]. In these materials, a homogeneous polymeric blend is formed by strong multiple acid-base interactions and hydrogen bonding networks, allowing proton conductivity by Bronsted acid-base pairs.

Understanding the process of micro-phase separation in various ionomers is important and leads researchers to employ different modeling methods. It is surmised that molecular features and processing conditions result in mesoscale alterations at different length scales in morphology. Atomic simulations by molecular dynamics (MD) and Monte Carlo (MC) simulation not only help to understand the molecular structures of small systems, but they are also helpful in analyzing the aggression of the sulfonic acid groups and local transport of protons in perfluorinated sulfonic acid (PFSA) ionomers. On the other hand, these simulations are not useful to analyze the nanoscale morphology in complicated inhomogeneous phases because they require more atoms and excessive time for equilibration. For this reason, many modeling methods are used on a comparable scale with experimental SAXS (small-angle X-ray scattering) data in these systems [21–23]. The coarse-grained and continuum are mainly two methods for convenient modeling of intermolecular interactions. Individual conceptual sites with appropriate non-bonded interaction parameters can be put in place of chemical groups of a complex polymer in particle-based models. However, local concentration fields with collective variables are used to figure out self-organizing structures for field-based methods. The length is scaled in 10–103 nm, and time is scaled up to milliseconds (even seconds), enabling the simulation of larger size systems and getting results comparable to experimental data. The morphology evolution and eventual phase separation of several ionomer systems have been investigated via mesoscale modeling using dissipative particle dynamics (DPD) simulations [24–27]. Moreover, self-consistent mean-field (SCMF) simulations are also employed in order to study how temperature and water content affect phase separation and morphology in PFSA membranes [28].

There are various studies on PFSA-based proton conducting blend membranes as an alternative to costly fluorinated membranes. Although mesoscale morphologies of hydrated PFSA-based membranes were investigated both experimentally and computationally, very few studies have investigated the mesoscale phase behavior of blend membranes for polymer electrolyte fuel cells [26]. The present study is complementary to our previous experimental studies on proton-conducting PFSA-based polymer blends. In these studies, Nafion/poly(vinyl phosphonic acid) (Nafion-PVPA) [16] and Nafion/poly(1-vinyl-1,2,4-triazole) (Nafion-PVTri) [29] blend membranes with significant anhydrous proton conductivity were fabricated. However, it is not possible to experimentally analyze the mesoscale morphologies of these blends. In this work, mesoscale phase morphologies of Nafion-PVTri and Nafion-PVPA blend membranes with different compositions were investigated using the DPD simulation method. Simulation results show that both blend membranes can form a highly separated microstructure due to the hydrophobic and hydrophilic character of different polymer chains forming the blend of different segments in the same polymer chain. This study uses dissipative particle dynamics to construct the mesoscale phase morphologies of Nafion-PVPA and Nafion-PVTri blend membranes. Herein, we establish a connection between the simulated morphologies and experimental properties.

2. Theoretical Background

In order to understand complex fluid dynamics, mesoscopic simulation methods are significant and have been widely used. Hoogerbrugge and Koelman were the first to create the DPD method, which is later improved by Groot and Warren [30–32]. In DPD, Fluid material's mesoscopic regions with similar chemical features are acted as fundamental particles called "beads. It is unlike MD that these particles are assumed to have no atomic properties. This is because all degrees of freedom that are greater than a bead radius in the DPD unit are neglected. Thus, coarse-grained interactions among beads are computed, and all atomic details are gone.

Simulations are done on a combination of particles that interacts by Newton's equations of motion.

$$\frac{d\mathbf{r}_i}{dt} = \mathbf{v}_i, \ \frac{d\mathbf{v}_i}{dt} = \mathbf{f}_i \tag{1}$$

The masses are adjusted to 1; thus, the acceleration equals the force acting on a particle. The sum of three pairwise additive components gives the force $\mathbf{f}_i$:

$$\mathbf{f}_i = \sum_{j \neq i} (\mathbf{F}_{ij}^{\mathrm{C}} + \mathbf{F}_{ij}^{\mathrm{D}} + \mathbf{F}_{ij}^{\mathrm{R}}) \tag{2}$$

The summation runs over all other particles j in a given cutoff radius $\mathbf{r}_c$ for particle i. It is this short-range cutoff that makes the interactions local. The cutoff radius is assumed to be unity for simplicity as this is the only length–scale in the system. So, forces beyond the bead radius are not taken into account for all beads. Soft repulsion interactions are taken into consideration, whereas the excluded-volume effect is not functional.

$\mathbf{F}_{ij}^{\mathrm{C}}$ is the conservative force, a soft repulsive interaction between the particles' centers i and j. $\mathbf{F}_{ij}^{\mathrm{D}}$ is the drag force (or dissipative force), and it depends on the relative velocities of the beads i and j. It normalizes velocities to decrease the relative momentum of the particles. $\mathbf{F}_{ij}^{R}$ denotes a random force stabilizing the temperature.

The repulsive conservative force $\mathbf{F}_{ij}^{\mathrm{C}}$ exists between the centers of the ith and jth particles, and it is given as:

$$\mathbf{F}_{ij}^{\mathrm{C}} = \begin{cases} a_{ij}(1 - r_{ij})\hat{\mathbf{r}}_{ij} & (r_{ij} \leq 1) \\ 0 & (r_{ij} \geq 1) \end{cases}, \tag{3}$$

a_{ij} stands for the maximum repulsion force between the ith and jth. particles; and $\boldsymbol{r_{ij}}$ is given as $\mathbf{r}_{ij} = \mathbf{r}_i - \mathbf{r}_j$, $r_{ij} = |\mathbf{r}_{ij}|$, $\hat{\mathbf{r}}_{ij} = \mathbf{r}_{ij}/|\mathbf{r}_{ij}|$.

$\mathbf{F}_{ij}^{\mathrm{D}}$, the dissipative force acting to reduce the relative momentum of two beads depends on the relative velocity of these beads, and it is given as:

$$\mathbf{F}_{ij}^{\mathrm{D}} = -\gamma\varpi^{\mathrm{D}}(r_{ij})(\hat{\mathbf{r}}_{ij}\mathbf{v}_{ij})\hat{\mathbf{r}}_{ij}. \tag{4}$$

γ is a friction coefficient, $\varpi^{D}(r_{ij})$ is a short-range weight function, and $\mathbf{v}_{ij} = \mathbf{v}_i - \mathbf{v}_j$. Due to the dissipative force structure chosen, the total momenta for any pair of particles and hence the system overall is conserved.

For the random force given below, a different distance-dependent $\varpi^{\mathrm{R}}(r_{ij})$ is employed. This force is between all pairs of beads with a similar short-range cutoff:

$$\mathbf{F}_{ij}^{\mathrm{R}} = \sigma\varpi^{\mathrm{R}}(r_{ij})\theta_{ij}\hat{\mathbf{r}}_{ij} \tag{5}$$

$\theta_{ij}(t)$ is a random variable with Gaussian statistics satisfying $\langle \theta_{ij}(t) \rangle = 0$, and $\langle \theta_{ij}(t)\theta_{kl}(t') \rangle = \left(\delta_{ik}\delta_{jl} + \delta_{il}\delta_{jk}\right)\delta_{ij}(t - t')$.

It has been proved that one can define only one of the two weight functions $\mathbf{F}_{ij}^{\mathrm{R}}$ or $\mathbf{F}_{ij}^{\mathrm{D}}$, and the other one is defined accordingly [33].

In DPD simulations, beads move freely. Some molecules like polymers and surfactants are represented by two or more beads. Therefore, to assure the connectivity of the chain, an additional force between successive beads must be used:

$$\mathbf{F}_{i,i+1}^{\mathrm{spring}} = K\mathbf{r}_{i,i+1} \tag{6}$$

In the DPD method, the particles refer to beads rather than molecules or atoms, and beads contain coarse-grained groups of atoms. Bead by bead interactions take the places of atomic-level interactions. The engagement of springs between two successive beads along the chain further enables the use of this system for polymers. The random and dissipative forces perform in unison. Their total effect is a thermostat. This fine conservative repulsive force reveals the system's leading chemistry. The parameters represented by a_{ij} are named bead–bead repulsion parameters. It is because these parameters are dependent on underlying atomistic interactions that they are also referred to as DPD interaction parameters.

The simulations of liquid-liquid and liquid-solid interfaces can be done by employing the DPD method. Therefore, it is very similar to the mean-field theory of Flory–Huggins for polymer chains and can be considered a continuous version of the lattice model established therein [34–36]. The mean-field theory is used to describe polymer's miscibility in a given solvent, and this is done via comparing the free energy of the system before and after mixing. Description of thermodynamics of polymer blends is done with the help of a similar theory, and block copolymers and their blends with homopolymers can also be described with some modifications. The mixing energy is directly related to the dimensionless Flory–Huggins interaction parameter χ, which represents the change in energy in units of kT when a bead of **A** is drawn from an environment of bulk **A** and surrounded with a bead **B** in an environment of bulk **B**. **A** and **B** would be polymer or molecules.

The only computational parameter that the beads' chemistry is involved in the conservative force as noise couples with dissipation. The relationship between the repulsion parameter a_{ij} acting between neighboring beads and the Flory–Huggins interaction parameter χ is described by Groot and Warren [31]. The repulsion parameter and the energy of mixing are proportional to each other.

The Flory–Huggins model in binary systems is the most well-known theory employed for the thermodynamics of phase separation and mixing. The general equation for the system's free energy of mixing (ΔG_{mix}) with bead types A and B can be expressed as:

$$\frac{\Delta G_{\mathrm{mix}}}{RT} = \left(\frac{\varphi_A}{N_A}\right) \ln \varphi_A + \left(\frac{\varphi_B}{N_B}\right) \ln \varphi_B + \chi_{AB}\varphi_A\varphi_B \tag{7}$$

φ_A and φ_B stand for the volume fraction for type A and B beads, respectively. In Equation (7), the first two quantities refer to the entropy of mixing, and these quantities vary with respect to the number of distinct ways in which the chains can come together on a lattice, and the last term is the free energy of interaction. The positive interaction parameter χ leads to various phase diagrams.

The Flory–Huggins interaction parameters can be defined as:

$$\chi = \frac{E_{\mathrm{mix}}}{RT} \tag{8}$$

E_{mix} is the energy of mixing describing the change in free energy because of the interaction between the pure and mixed state. It can be calculated using the traditional Flory–Huggins model:

$$E_{mix} = \frac{1}{2}Z(E_{AB} + E_{AB} - E_{AA} - E_{BB}) \quad (9)$$

Z denotes coordination number and E_{ij} is the binding energy for the chosen ith and jth components.

χ, the Flory–Huggins parameter can also be calculated from Hildebrand solubility parameters, which is calculated from cohesive energy density values:

$$\chi_{AB} = \frac{V_{bead}}{kT}(\delta_A - \delta_B)^2 \quad (10)$$

$$\delta = \left(\frac{\Delta E_v}{V_m}\right)^{1/2} = (CED)^{1/2} \quad (11)$$

where V_{bead} is the beads' average molar volume, δ_A and δ_B denote the parameters for solubility for beads A and B, respectively, and ΔE_v is the molar enthalpy for vaporization. Parameters for solubility vary depending on the species' chemical nature in question.

3. Computational Methodology

Chemical structures for Nafion, PVPA, and PVTri are shown in Figure 1. To construct systems for the DPD, copolymer Nafion is divided into three beads (A, B, and S), and homopolymers of PVPA and PVTri are represented by a single bead (P for PVPA and T for PVTri). Since the fluorine-containing backbone of Nafion is chemically different from the ether and the sulfonic acid-containing side chain, the backbone is constructed as a single bead (A) while the side chain is built as two different beads (B and S). By attaching extra fluorine and hydrogen atoms to the connection points of Nafion and homopolymer beads, respectively, consistency is obtained.

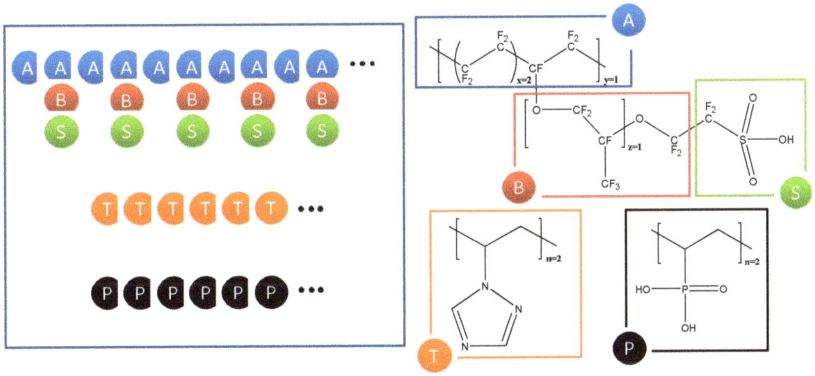

Figure 1. Chemical structures of Nafion, poly(1-vinyl-1,2,4-triazole) (PVTri), and poly(vinyl phosphonic acid) (PVPA) and structure of beads of polymers used in dissipative particle dynamics (DPD) simulations.

The DPD modules of the Material Studio suite of programs were used in all DPD simulations carried out. The optimization of the particles' structures was done via molecular mechanics along with COMPASS force field parameters by employing the Forcite module in Materials Studio. The number of particles per unit, namely, the simulated density value, was set at $\rho = 3$. Calculation of interaction radius ($r_c = 4.5$ A) was done using the beads' average size. The Blends and Synthia modules in Materials Studio are used in order to calculate the Flory–Huggins parameters corresponding to bead–bead interactions. Interaction parameters of DPD are computed by using the parameter χ. Groot established a relationship between the parameter χ_{ij} and the repulsion parameter a_{ij} between the beads i and j [31]. The density affects this relationship, and the density is set at 3 in our computations:

$$a_{ij} = \frac{\chi_{ij}}{0.306} + 25 \tag{12}$$

In a simulation box whose edges are 36 nm, about 192,000 DPD beads were positioned in arbitrary locations at the beginning. It is equilibrated for a period of 6 ns. The examination of the blend morphologies is done via homopolymer density contour plots, which are formed from the equilibrated mesostructures.

4. Results and Discussion

4.1. Flory–Huggins Parameters and Interaction Energies

Conventional PFSA-based membranes' proton conductivity highly depends on the water content in their protonated form. This dependency limits the fuel cells' working temperature to below 100 °C due to the membranes' dehydration from hydrophilic cavities at elevated temperatures. These membranes typically show phase separation where the network of hydrophilic nanopores and nanochannels are embedded in a hydrophobic phase domain. The hydrophilic nanoporous phase contains water and acidic sites, facilitating proton conductivity by transporting free protons. The hydrophobic phase serves as a mechanical force to stabilize the morphology of the membrane. Due to the drawbacks of low-temperature PEM fuel cells, developing high-temperature proton conducting membranes is needed. Recently, having thermal and chemical stability, acid-base polymer blends have been examined as alternative membranes for PEM fuel cells operating above 100 °C. Among them, Nafion-PVPA and Nafion-PVTri show promising properties and considerable proton conductivities.

Flory–Huggins interaction parameters χ for chosen components i and j are calculated through Equation (8), using the binding energies and coordination number. The χ parameters (Table 1) give pairwise information about how the beads will interact with each other. The backbone segment of Nafion is immiscible with all other beads except for the bead B. The molar energy of vaporization for the nonpolar materials should be less because of their weak intermolecular energies. The polar parts result in cohesive energy density, which is what is calculated in this study for the fluorine-containing groups. A highly positive χ indicates the contacts of the bead A with the beads S, P, and T. The contact of the bead B with the beads S and P are less favorable when compared with the contact of bead A with bead B and bead B with bead T. A negative value of χ indicates the interactions of the bead S with P and T is highly preferred due to complexation tendency between corresponding segments.

The various trends investigating the solubility parameters of different beads can be elucidated by partitioning short-range interaction energy into hydrogen-bonding, London dispersion (or van der Waals type), and dipole-dipole types in the Hansen solubility parameters. There is only the former one for the case of nonpolar molecules. Therefore, It is due to due to weaker intermolecular interactions that cohesive energy density or molar energy of vaporization for the nonpolar beads is less than those for polar beads. Hence, parameters of the solubility for the nonpolar fluorinated segments and polar triazole and oxygen moieties are placed on distinct solubility spectrum sites. For the molecules' consistency, attaching hydrogen atoms to the head and tail points might cause little variation in the computation of solubility parameters, and this might be discarded in this study.

Table 1. Calculated Flory–Huggins (χ) parameters and corresponding repulsion parameters (a_{ij}) of DPD simulation for each pair in (**a**) Nafion–PVPA and (**b**) Nafion–PVTri blends.

	(a)	
Pair	χ	a_{ij} $(k_B T)$
A–B	0.75	27.61
A–S	4.53	40.87
A–P	10.05	60.19
B–S	2.86	35.02
B–P	4.35	40.24
S–P	−1.49	20.12
	(b)	
Pair	χ	a_{ij} $(k_B T)$
A–B	0.75	27.61
A–S	4.53	40.87
A–T	1.89	31.60
B–S	2.86	35.02
B–T	0.14	25.49
S–T	−0.55	23.21

In theory, liquids having similar cohesive energy densities are expected to carry similar characteristics for solubility. The parameter χ indicates that segments containing fluorine are immiscible with other segments. The smaller solubility parameter of fluorine-containing bead has a smaller solubility than other beads, which shows that by increasing the structure's fluorine content, the immiscibility of fluorine-containing segments can be increased. The other beads, on the other hand, have larger solubility parameters due to their chemical structure.

4.2. Morphology

There have been many computational studies to understand the phase morphologies of hydrated polymer electrolytes. The control of phase behavior is significant to improve membranes' proton conductivity. In this work, the effects of the molar composition of Nafion/PVPA and Nafion/PVTri on the mesoscopic morphologies of blend membranes were investigated. As the concentration of PVPA increases, PVPA clusters become continuous after a 1:2 molar ratio (Figure 2). Proton conductivity increases with PVPA content. On the other hand, Nafion-PVPA blend membranes are not stable with high PVPA content. In our experimental work, maximum proton conductivity was obtained with 1:3 composition [16]. In that composition, it was clear that the PVPA phase was continuous enough to promote proton conductivity through interaction with sulfonic acid groups of Nafion.

Herein, Nafion, having an amphiphilic character because of incompatible polar and nonpolar segments, was considered. These segments tend to reside away from each other due to having distinct interaction parameters, which results in phase separation of chains. However, they are joined together due to a covalent bond that increases their involuntary compatibility, enabling only microscopic phase separation. While this transition from disorder to order causes high order continuous phases in the case of Nafion-PVPA blends, there is a comparatively low-order phase of cylinders in Nafion-PVTri blends (Figure 3), which is, in fact, consistent with the previous observations done for Nafion-containing blends.

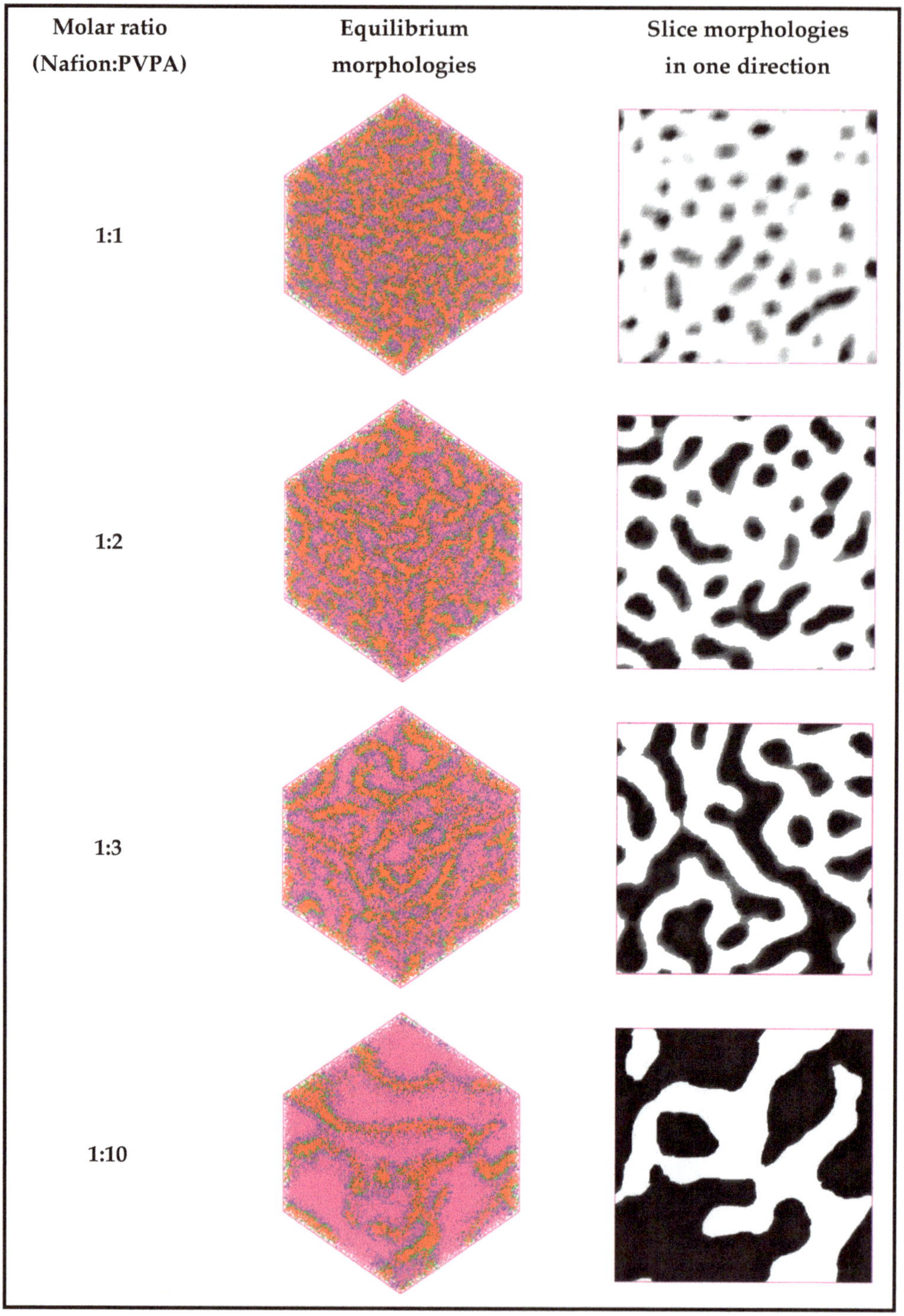

Figure 2. Comparison of equilibrium morphologies and slice morphologies of Nafion/PVPA blend membranes with various molar compositions.

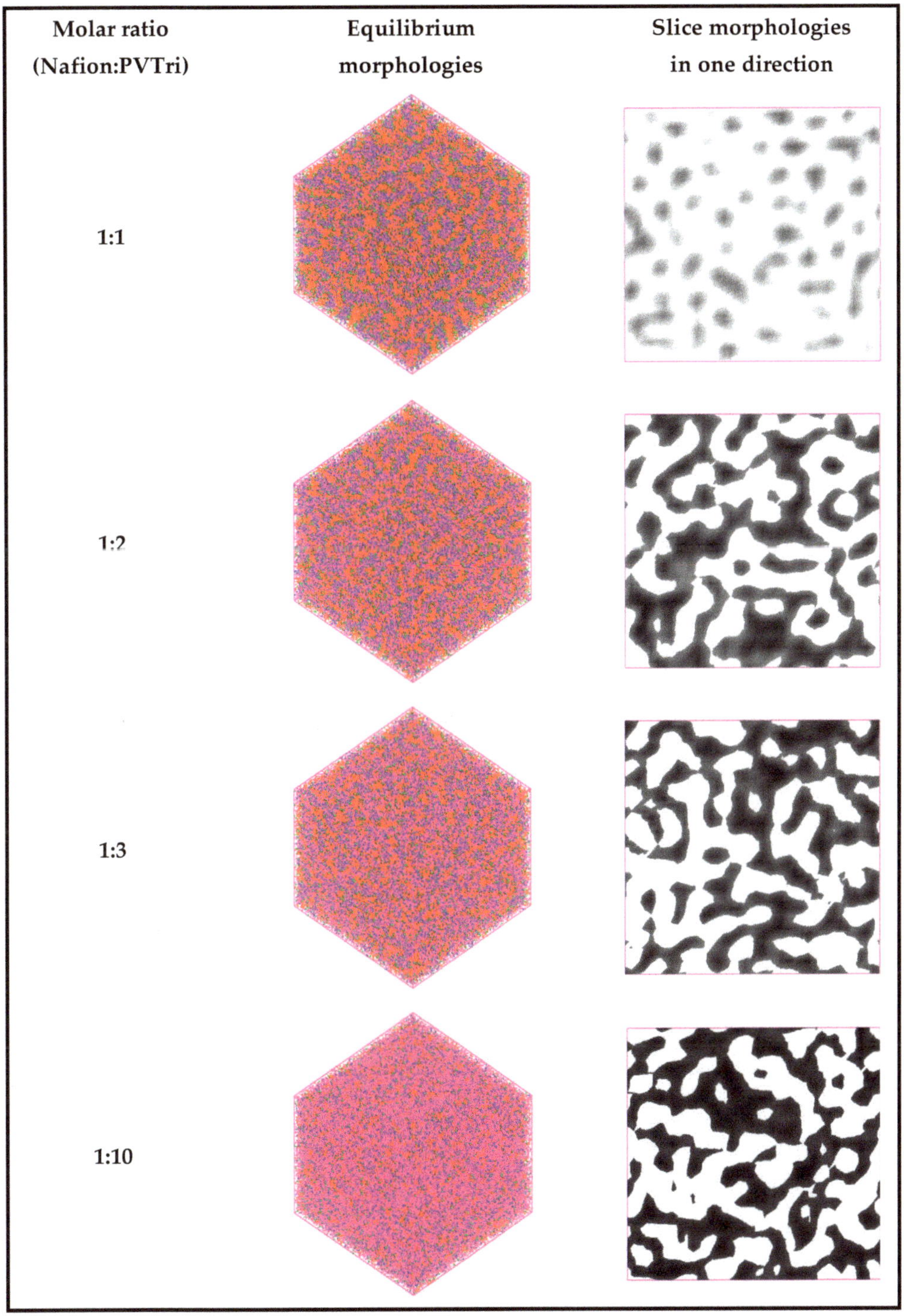

Figure 3. Comparison of equilibrium morphologies and slice morphologies of Nafion/PVTri blend membranes with various molar compositions.

4.3. End-to-End Distance of Blend Membranes

A single polymer chain can appear in many possible conformations ranging from a tight coil to a straight chain. The probability of a single polymer chain having a specific end-to-end distance rises as the number of possible conformations that can have this end-to-end distance rises. A straight chain can be produced by only one conformation; however, more possible conformations exist as the molecule coils up more. Therefore, a polymer chain is virtually inclined to coil up. A model that assumes that molecules are composed of many segments can predict the end-to-end distance of a polymer chain. All the segments in this model are rigid, but they are freely jointed at both ends and hence capable of making any angle with the next segment. Therefore, a molecule model can be constructed by connecting each consecutive segment at an arbitrary angle. The name of this process is a random walk. In the present study, the end-to-end distance for Nafion chains, an important structural property of polymeric materials, is computed according to PVPA and PVTri by employing DPD simulations. Since there is a high interaction parameter between sulfonic and phosphonic acid groups, root means square (RMS) of end-to-end distances of polymer chains in Nafion-PVPA are smaller than those in Nafion-PVTri (Figure 4).

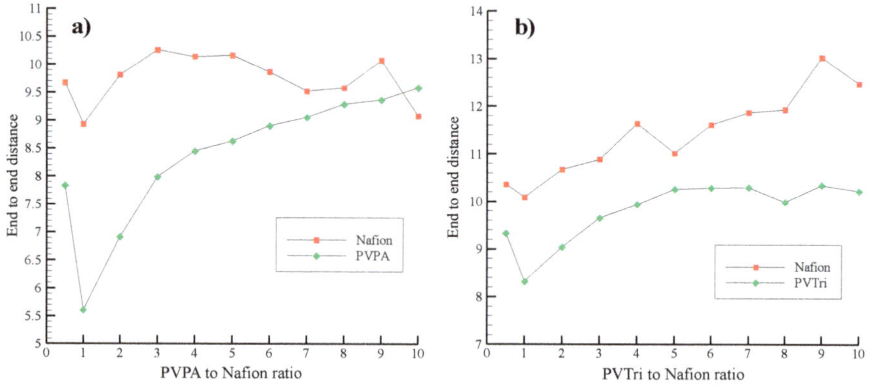

Figure 4. Average RMS end-to-end distances of polymer chains in (**a**) Nafion–PVPA and (**b**) Nafion–PVTri blends at different compositions.

RMS end-to-end distance is highly dependent on the type and composition of the blends. The smallest RMS end-to-end distance is obtained in 1:1 composition in both systems, which may be attributed to the fact that polymer chains in the exact one-to-one ratio prefer tight coil conformation. The length of individual chains in the system does not change in any environment, while their relative positioning in the bulk of the system changes to lead to diverse morphologies. As PVPA content increases RMS end-to-end distance increases, this is because the continuous PVPA phase starts. However, RMS end-to-end distance does not vary significantly through changing the composition of the Nafion-PVTri blend system. It is because DPD is a coarse-grained method, and the only parameter that carries the atomistic information is the soft repulsion parameter that these results should be confirmed using other conformational search methods (e.g., high-temperature Molecular Dynamics).

4.4. Diffusion Rates of Blend Membranes

Diffusivity rates of beads can be seen in Figure 5. Diffusivity of the bead C is the highest among other bead diffusivities. Since C is located on the polymer's side chains, it is relatively more mobile than the other beads like A on the chain, making its diffusivity the highest.

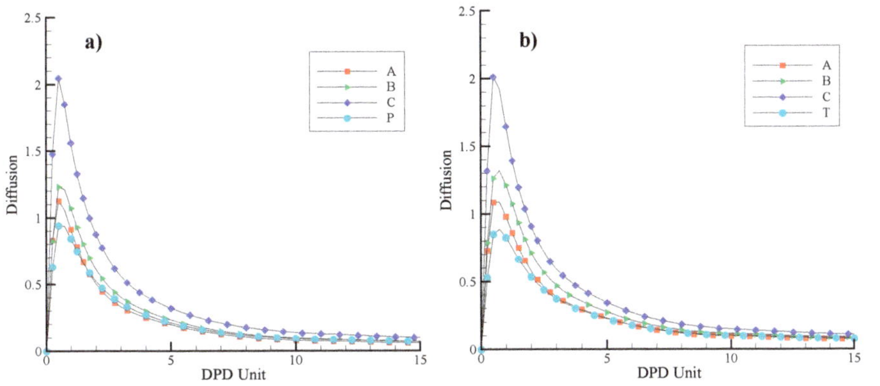

Figure 5. Diffusion rates of the beads for (**a**) Nafion/PVPA blend membranes, (**b**) Nafion/PVTri blend membranes.

4.5. Density Profiles of Blend Membranes

In the x-direction, the density profiles for the beads A and P in the Nafion-PVPA system and the density profiles for the beads A and T in the Nafion-PVTri system are presented in Figure 6a,b, respectively. The results are in line with the previously conducted studies demonstrating morphology. Both P and T are highly immiscible with A that densities of them will decrease as the density of A increases and vice versa. However, the degree of repulsive interaction parameter directly affects the density profiles of the blend membranes. The density variation in Nafion-PVPA is higher than in Nafion-PVTri due to higher repulsion between the beads A and P.

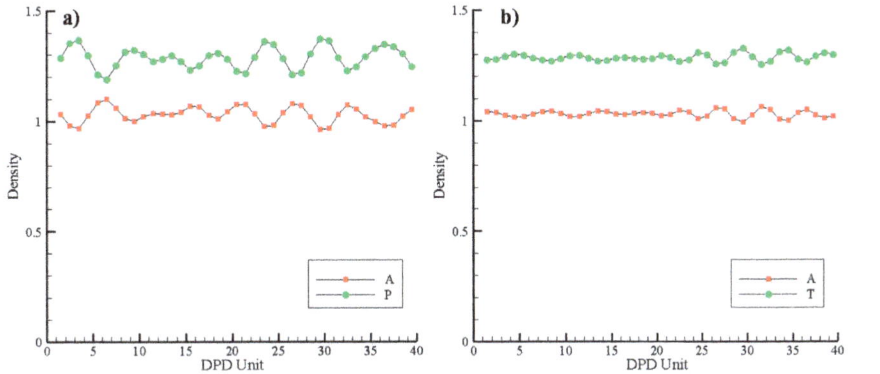

Figure 6. Comparison of densities of beads for (**a**) Nafion/PVPA blend membranes, (**b**) Nafion/PVTri blend membranes.

5. Conclusions

Understanding the process of micro-phase separation in various ionomers is of enormous importance, which leads researchers to employ different modeling methods. It is surmised that molecular features and processing conditions result in mesoscale alterations at different length scales in morphology. In this work, mesoscale phase morphologies of Nafion-PVTri and Nafion-PVPA blend membranes with different compositions were investigated using the DPD simulation method. The present study is complementary to our previous experimental study on proton-conducting PFSA-based polymer blends. In our earlier study, Nafion-PVPA and Nafion-PVTri blend membranes with significant anhydrous proton conductivity were fabricated. Simulation results show that both blend membranes can form a highly separated microstructure due to the hydrophobic and hydrophilic character of different polymer chains forming the blend and different segments in the same polymer chain. There is a strong, attractive interaction between the sulfonic and phosphonic acid groups and a very strong repulsive interaction between the fluorinated segment and phosphonic acid groups in the Nafion-PVPA blend membrane. With the increase of the PVPA content in the blend membrane, the PVPA clusters' size gradually increases and forms a continuous phase in the membrane. On the other hand, the repulsive interaction between fluorinated segment and triazole unit in Nafion-PVTri blend is not very strong compared to that in the Nafion-PVPA blend, which results in different phase behavior in Nafion-PVTri blend membrane. This relatively lower repulsive interaction leads to the Nafion-PVTri blend membrane having non-continuous phases regardless of the composition.

Author Contributions: Conceptualization, U.S. and T.T.; methodology, U.S. and M.E.; software, M.O. and M.E.; validation, A.M.K.; writing—original draft preparation, U.S. and M.O.; writing—review and editing, M.A.A., A.M.K., A.A.M. and A.R.O.; supervision, M.A.A. and T.T.; funding acquisition, U.S. and T.T. All authors have read and agreed to the published version of the manuscript.

Funding: This research received no external funding.

Institutional Review Board Statement: Not applicable.

Informed Consent Statement: Not applicable.

Data Availability Statement: Not applicable.

Acknowledgments: U. Sen was supported by JSPS RONPAKU (Dissertation PhD) program with ID No. TUR 11008.

Conflicts of Interest: The authors declare no conflict of interest.

References

1. Chandan, A.; Hattenberger, M.; El-kharouf, A.; Du, S.; Dhir, A.; Self, V.; Pollet, B.G.; Ingram, A.; Bujalski, W. High temperature (HT) polymer electrolyte membrane fuel cells (PEMFC)—A review. *J. Power Sources* **2013**, *231*, 264–278. [CrossRef]
2. Shin, D.W.; Guiver, M.D.; Lee, Y.M. Hydrocarbon-Based Polymer Electrolyte Membranes: Importance of Morphology on Ion Transport and Membrane Stability. *Chem. Rev.* **2017**, *117*, 4759–4805. [CrossRef]
3. Wang, Y.; Chen, K.S.; Mishler, J.; Cho, S.C.; Adroher, X.C. A review of polymer electrolyte membrane fuel cells: Technology, applications, and needs on fundamental research. *Appl. Energy* **2011**, *88*, 981–1007. [CrossRef]
4. Zhang, H.; Shen, P.K. Recent development of polymer electrolyte membranes for fuel cells. *Chem. Rev.* **2012**, *112*, 2780–2832. [CrossRef] [PubMed]
5. Chen, J.; Asano, M.; Maekawa, Y.; Yoshida, M. Suitability of some fluoropolymers used as base films for preparation of polymer electrolyte fuel cell membranes. *J. Membr. Sci.* **2006**, *277*, 249–257. [CrossRef]
6. Iulianelli, A.; Basile, A. Sulfonated PEEK-based polymers in PEMFC and DMFC applications: A review. *Int. J. Hydrogen Energy* **2012**, *37*, 15241–15255. [CrossRef]
7. Savadogo, O. Emerging membranes for electrochemical systems: Part II. High temperature composite membranes for polymer electrolyte fuel cell (PEFC) applications. *J. Power Sources* **2004**, *127*, 135–161. [CrossRef]
8. Yang, Y.; Siu, A.; Peckham, T.J.; Holdcroft, S. Structural and morphological features of acid-bearing polymers for PEM fuel cells. In *Advances in Polymer Science*; Springer: Berlin/Heidelberg, Germany, 2008; pp. 55–126.
9. Paddison, S.J. Proton Conduction Mechanisms at Low Degrees of Hydration in Sulfonic Acid-Based Polymer Electrolyte Membranes. *Annu. Rev. Mater. Res.* **2003**, *33*, 289–319. [CrossRef]
10. Araya, S.S.; Zhou, F.; Liso, V.; Sahlin, S.L.; Vang, J.R.; Thomas, S.; Thomas, S.; Thomas, S.; Kær, S.K. A comprehensive review of PBI-based high temperature PEM fuel cells. *Int. J. Hydrogen Energy* **2016**, *41*, 21310–21344. [CrossRef]
11. Li, Q.; Aili, D.; Jensen, J.O. High temperature polymer electrolyte membrane fuel cells: Approaches, status, and perspectives. In *High Temperature Polymer Electrolyte Membrane Fuel Cells: Approaches, Status, and Perspectives*; Springer: Berlin/Heidelberg, Germany, 2016; pp. 1–545.
12. Rosli, R.E.; Sulong, A.B.; Daud, W.R.W.; Zulkifley, M.A.; Husaini, T.; Rosli, M.I.; Majlan, E.H.; Haque, M.A. A review of high-temperature proton exchange membrane fuel cell (HT-PEMFC) system. *Int. J. Hydrogen Energy* **2017**, *42*, 9293–9314. [CrossRef]
13. Mauritz, K.A.; Moore, R.B. State of understanding of Nafion. *Chem. Rev.* **2004**, *104*, 4535–4585. [CrossRef] [PubMed]
14. Erkartal, M.; Aslan, A.; Dadi, S.; Erkilic, U.; Yazaydin, O.; Usta, H.; Sen, U. Anhydrous proton conducting poly(vinyl alcohol) (PVA)/ poly(2-acrylamido-2-methylpropane sulfonic acid) (PAMPS)/1,2,4-triazole composite membrane. *Int. J. Hydrogen Energy* **2016**, *41*, 11321–11330. [CrossRef]

15. Sen, U.; Acar, O.; Bozkurt, A.; Ata, A. Proton conducting polymer blends from poly(2,5-benzimidazole) and poly(2-acrylamido-2-methyl-1-propanesulfonic acid). *J. Appl. Polym. Sci.* **2011**, *120*, 1193–1198. [CrossRef]
16. Sen, U.; Acar, O.; Celik, S.U.; Bozkurt, A.; Ata, A.; Tokumasu, T.; Miyamoto, A. Proton-conducting blend membranes of Nafion/poly(vinylphosphonic acid) for proton exchange membrane fuel cells. *J. Polym. Res.* **2013**, *20*, 217. [CrossRef]
17. Sen, U.; Usta, H.; Acar, O.; Citir, M.; Canlier, A.; Bozkurt, A.; Ata, A. Enhancement of anhydrous Proton conductivity of poly(vinylphosphonic acid)-poly(2,5-benzimidazole) membranes via in situ polymerization. *Macromol. Chem. Phys.* **2015**, *216*, 106–112. [CrossRef]
18. Ahn, M.K.; Lee, B.; Jang, J.; Min, C.-M.; Lee, S.-B.; Pak, C.; Lee, J.-S. Facile preparation of blend proton exchange membranes with highly sulfonated poly(arylene ether) and poly(arylene ether sulfone) bearing dense triazoles. *J. Membr. Sci.* **2018**, *560*, 58–66. [CrossRef]
19. Giffin, G.A.; Galbiati, S.; Walter, M.; Aniol, K.; Aniol, K.; Kerres, J.; Zeis, R. Interplay between structure and properties in acid-base blend PBI-based membranes for HT-PEM fuel cells. *J. Membr. Sci.* **2017**, *535*, 122–131. [CrossRef]
20. Sui, Y.; Du, Y.; Hu, H.; Qian, J.; Zhang, X. Do acid-base interactions really improve the ion conduction in a proton exchange membrane?-a study on the effect of basic groups. *J. Mater. Chem. A* **2019**, *7*, 19820–19830. [CrossRef]
21. Komarov, P.V.; Veselov, L.N.; Chu, P.P.; Khalatur, P.G. Mesoscale simulation of polymer electrolyte membranes based on sulfonated poly(ether ether ketone) and Nafion. *Soft Matter* **2010**, *6*, 3939–3956. [CrossRef]
22. Ryan, E.M.; Mukherjee, P.P. Mesoscale modeling in electrochemical devices—A critical perspective. *Prog. Energy Combust. Sci.* **2019**, *71*, 118–142. [CrossRef]
23. Wescott, J.T. Mesoscale simulation of morphology in hydrated perfluorosulfonic acid membranes. *J. Chem. Phys.* **2006**, *124*, 134702. [CrossRef]
24. Morohoshi, K.; Hayashi, T. Modeling and simulation for fuel cell polymer electrolyte membrane. *Polymers* **2013**, *5*, 56–76. [CrossRef]
25. Roy, S.; Markova, D.; Kumar, A.; Klapper, M.; Müller-Plathe, F. Morphology of phosphonic acid-functionalized block copolymers studied by dissipative particle dynamics. *Macromolecules* **2009**, *42*, 841–848. [CrossRef]
26. Sun, D.L.; Zhou, J. Dissipative particle dynamics simulations on mesoscopic structures of Nafion and PVA/Nafion blend membranes. *Wuli Huaxue Xuebao/Acta Phys. Chim. Sin.* **2012**, *28*, 909–916.
27. Wu, D.; Paddison, S.J.; Elliott, J.A. A comparative study of the hydrated morphologies of perfluorosulfonic acid fuel cell membranes with mesoscopic simulations. *Energy Environ. Sci.* **2008**, *1*, 284–293. [CrossRef]
28. Krueger, J.J.; Simon, P.P.; Ploehn, H.J. Phase behavior and microdomain structure in perfluorosulfonated ionomers via self-consistent mean field theory. *Macromolecules* **2002**, *35*, 5630–5639. [CrossRef]
29. Sen, U.; Bozkurt, A.; Ata, A. Nafion/poly(1-vinyl-1,2,4-triazole) blends as proton conducting membranes for polymer electrolyte membrane fuel cells. *J. Power Sources* **2010**, *195*, 7720–7726. [CrossRef]
30. Hoogerbrugge, P.J.; Koelman, J.M.V.A. Simulating microscopic hydrodynamic phenomena with dissipative particle dynamics. *Europhys. Lett.* **1992**, *19*, 155–160. [CrossRef]
31. Groot, R.D.; Warren, P.B. Dissipative particle dynamics: Bridging the gap between atomistic and mesoscopic simulation. *J. Chem. Phys.* **1997**, *107*, 4423–4435. [CrossRef]
32. Groot, R.D.; Madden, T.J. Dynamic simulation of diblock copolymer microphase separation. *J. Chem. Phys.* **1998**, *108*, 8713–8724. [CrossRef]
33. Español, P.; Warren, P. Statistical Mechanics of Dissipative Particle Dynamics. *Europhys. Lett.* **1995**, *30*, 191. [CrossRef]
34. Hu, Y.; Ying, X.; Wu, D.T.; Prausnitz, J.M. Molecular thermodynamics of polymer solutions. *Fluid Phase Equilibria* **1993**, *83*, 289–300. [CrossRef]
35. Klenin, V.J. Chapter 3-Polymer+flow-molecular-liquid system. Mean field approaches. Liquid-liquid phase separation. In *Thermodynamics of Systems Containing Flexible-Chain Polymers*; Klenin, V.J., Ed.; Elsevier Science B.V.: Amsterdam, The Netherlands, 1999; pp. 253–508.
36. Olemskoi, A.; Savelyev, A. Theory of microphase separation in homopolymer–oligomer mixtures. *Phys. Rep.* **2005**, *419*, 145–205. [CrossRef]

processes MDPI

Article

Experimental Study on the Effect of Basalt Fiber and Sodium Alginate in Polymer Concrete Exposed to Elevated Temperature

Seyed Esmaeil Mohammadyan-Yasouj [1,*], Hossein Abbastabar Ahangar [2], Narges Ahevani Oskoei [1], Hoofar Shokravi [3], Seyed Saeid Rahimian Koloor [4] and Michal Petrů [5]

1 Department of Civil Engineering, Najafabad Branch, Islamic Azad University, Najafabad 8514143131, Iran; narges.oskoei@yahoo.com
2 Department of Chemistry, Najafabad Branch, Islamic Azad University, Najafabad 8514143131, Iran; abbastabar@pmt.iaun.ac.ir
3 Department of Structures and Materials, School of Civil Engineering, Faculty of Engineering, Universiti Teknologi Malaysia, Skudai 81310, Johor, Malaysia; shokravihoofar@utm.my
4 Institute for Nanomaterials, Advanced Technologies and Innovation (CXI), Technical University of Liberec (TUL), Studentska 2, 461 17 Liberec, Czech Republic; s.s.r.koloor@gmail.com
5 Technical University of Liberec (TUL), Studentska 2, 461 17 Liberec, Czech Republic; michal.petru@tul.cz
* Correspondence: semy2016@pci.iaun.ac.ir

Abstract: Polymer concrete contains aggregates and a polymeric binder such as epoxy, polyester, vinyl ester, or normal epoxy mixture. Since polymer binders in polymer concrete are made of organic materials, they have a very low heat and fire resistance compared to minerals. This paper investigates the effect of basalt fibers (BF) and alginate on the compressive strength of polymer concrete. An extensive literature review was completed, then two experimental phases including the preliminary phase to set the appropriate mix design, and the main phase to investigate the compressive strength of samples after exposure to elevated temperatures of 100 °C, 150 °C, and 180 °C were conducted. The addition of BF and/or alginate decreases concrete compressive strength under room temperature, but the addition of BF and alginate each alone leads to compressive strength increase during exposure to heat and increase in the temperature to 180 °C showed almost positive on the compressive strength. The addition of BF and alginate both together increases the rate of strength growth of polymer concrete under heat from 100 °C to 180 °C. In conclusion, BF and alginate decrease the compressive strength of polymer concretes under room temperature, but they improve the resistance against raised temperatures.

Keywords: concrete; basalt fiber; epoxy resin; alginate; raised temperature; compressive strength

Citation: Mohammadyan-Yasouj, S.E.; Ahangar, H.A.; Oskoei, N.A.; Shokravi, H.; Koloor, S.S.R.; Petrů, M. Experimental Study on the Effect of Basalt Fiber and Sodium Alginate in Polymer Concrete Exposed to Elevated Temperature. *Processes* **2021**, *9*, 510. https://doi.org/10.3390/pr9030510

Academic Editor: Shaghayegh Hamzehlou

Received: 11 February 2021
Accepted: 8 March 2021
Published: 11 March 2021

1. Introduction

Concrete is an old building material, in use for centuries. Concrete is widely used in the construction of civil engineering structures such as buildings [1,2], bridges [3,4], and other civil structures [5,6]. Reinforced concrete structures are vulnerable to high-temperature conditions, which greatly shortens the service life and hinders its applications [7]. The properties of the mix design parameters have a significant effect on the performance of concrete exposed to elevated temperature [8–11]. Several research studies have been carried out to improve the thermal stability of concrete exposed to elevated temperatures using various materials, such as fillers, nanoparticles, and fibers [12–15].

Sodium alginate (SA) is a well-known natural polymer that is extracted from cell walls of brown algae [16,17]. Recently, SA has received much attention in concrete applications due to self-healing behavior [18,19]. The influence of the SA on mechanical and flexural properties is still unclear and different studies achieved different results. Abbas and Mohsen [20] indicated that the addition of SA increased the fresh and mechanical properties of concrete. However, Heidari et al. [21] reported a decrease in mechanical properties by using SA in the mix design. Mignon et al. [22] indicated that the addition of 1% calcium alginate can result in a 15% reduction in compression strength of concrete while it was 28% for SA. SA also improves the temperature resistance of concrete and enhances flame-, fire- and heat-resistance of cementitious composites [23]. Moreover, it was stated that the properties of the produced alginate gel are greatly affected by the polymer concentration and type [16,19]. Mignon et al. [22] showed that alginate can improve the internal curing of concrete effectively. Ouwerx et al. [19] showed that the concentration of SA in concrete can limit the elasticity of beads. Pathak et al. [16] prepared and compared different metal alginates and alginic acid exploited from fresh algae using the extraction method. It was reported that decomposition of cobalt alginate takes place at a higher temperature compared to sodium and calcium alginate. The distribution and pore size were significantly influenced by the presence of metal ions. The type of metal alginates may lead to variation in thermal behavior of the compounds. Table 1 shows a summary of the studies on the effect of alginates on mechanical parameters of concrete.

Table 1. A summary of the studies on the effect of alginates on mechanical parameters of concrete.

Reference.	Alginate Type	Weight Fraction (% of Cement)	Remarks
Mohammadyan-Yasouj et al. [24]	NaAlg	0.1	• Alginate decreases the flexural, tensile, and compressive strengths of concrete.
Ouwerx et al. [19]	Alginate gel bead	-	• Properties of alginate significantly depend on polymer type and concentration.
Pathak et al. [16]	Alginic acid and metal alginates	-	• Presence of metal ions impacted on the pore size distribution of alginates. • Different thermal behavior was observed in different metal alginates due to structural differences.
Heidari et al. [21]	Alginic acid	0.5 and 1	• Alginate decreases the flexural, tensile, and compressive strengths of concrete. • Alginate diminished the fresh properties of SCC.
Mignon et al. [22]	NaAlg, CaAlg	0.5 and 1	• Alginate improves internal curing • CaAlg beads of 1% reduced 15% of the compressive strength. • NaAlg of 1% resulted in a 28% decrease in compressive strength due to increasing in the water uptake.

Polymeric concrete is a kind of concrete in which natural aggregates such as silica sand and gravel are bound together in a matrix with a polymer binder as a supplement or replacement of cement in concrete [25]. Polymeric concrete has higher mechanical strengths, chemical resistance, and ductility, compared with ordinary concrete (OC) [26,27]. Currently, three distinct types of polymeric concrete are studied that include polymer-Portland cement concrete (PPCC), polymer impregnated concrete (PIC), and polymer concrete (PC) [27].

According to American Concrete Institute (ACI 548.3R), in PC, the polymer is served as the sole binding material present in concrete [28]. However, filler materials such as silica fume and fly ash can be utilized along with aggregates to enhance mechanical properties and minimize building cost. PC can achieve about 80% of its 28 day compressive strength in one day, and compressive strength of 100 MPa can be attained [26,29]. Higher ductility after 60 °C [30] and no visible failure under compression for mixes containing epoxy or epoxy and filler of 30% of the total volume is also reported for PC [31]. An investigation by Golestaneh et al. [32] indicated an optimal amount of epoxy to reach the maximum compressive and flexural strength was 15% of the total volume, while Elalaoui et al. [33] suggested 13% to reach the highest physical and mechanical properties at the lowest cost. However, the higher value for the suggested optimal amount by Golestaneh et al. [32], even with 20% epoxy, may be due to the utilization of silica fume as filler to get the maximum tensile strength.

The results indicate that, compared to OC, polymeric concrete exhibited fewer cracks that are probably due to the bound effect between fiber and matrix [34]. PC has higher strengths than OC [34–36] and, similar to the OC, with the addition of fiber, reduction in its workability is unavoidable [34]. Acceptable interface adhesion between fiber and polymeric matrix for fiber loading of 15% of the total volume resulted in high tensile and flexural strengths [37]. Understanding the behavior of concrete exposed to various temperatures is considered an important factor in meeting the safety and service life objectives in which structures are intended and designed. Three different temperature ranges are generally used in concrete studies that include low temperature (<0 °C), medium temperature (0–50 °C), and high or elevated temperature (<50 °C). The thermal response of concrete to elevated temperature mainly depends on constituents' characteristics and the mix composition [38]. Based on the applications and intended exposure condition, previous researches have considered various temperature ranges to evaluate the thermal response of concrete which can be classified as temperatures below 200 °C [39–41], below 600 °C [42–44], and below 1000 °C [45–47]. The value of 600 °C is generally selected because the interior of the concrete members will not exceed 600 °C in a short time. Higher temperatures (e.g., 1000 °C) are considered for the peak gas temperature of a fire that may reach a higher value [48], while the temperature lower than 200 °C is generally considered for members exposed to indirect fire heat.

The duration of exposure and heating rate are two other important factors that have been investigated in many research studies. Different heating durations are generally based on the recommendation by different specifications for maximum fire-resistance rating. For example, the fire resistance for column members in China fire design specifications is defined as 180 min (3 h) [48]. Alharbi et al. [49] indicated that duration of exposure has an effect on degradation of the stiffness while prolonged exposures can significantly reduce the strength of concrete. The heating speed caused by exposure to real fire is generally specified by fire standards (e.g., ISO 834 standard fire).

Investigation on the behavior of PC with basalt fiber has revealed that BF >1% reduces concrete strength [50]. On the other hand, the studies on epoxy/basalt polymer concrete showed enhancement of mechanical properties and more ductility under increasing temperature up to 100 °C [36]. Reis [30] observed that the epoxy mortars had higher sensitivity to temperature variations due to the heat distortion temperature of the resins [19]. Niaki et al. [51] demonstrated that basalt fiber improved the mechanical properties and increased the thermal stability of the PC subjected to different temperatures (up to 250 °C). Several research studies have been conducted to improve the durability of polymer concrete exposed to elevated temperatures [9]. Elalaoui et al. [52] proposed ammonium polyphosphate to be used as a constituent of the PC to improve the temperature resistance of the concrete. Niaki et al. [36] reported that adding crushed basalt aggregates into PC led to decreases in the highest maximum stress rapidly, but increases the yield displacement significantly due to temperature increase. Gorninski et al. [53] produced PC composites using waste alumina and showed that the addition of alumina flame retardant waste can efficiently improve the temperature resistance of the PC.

The effect of using BF on concrete has been carried out by many researchers, and several experiments were conducted to study the influence of adding different proportions, diameter, and length of fibers in concrete. Jiang et al. [54] demonstrated that BF addition can considerably improve the tensile and flexural strength and toughness index of fiber reinforce concrete. It was indicated that adding BF had no obvious increase in compressive strength. Compared to other types of fibers such as polypropylene and glass fiber (GF) higher performance for splitting tensile, flexural strengths, crack resistance, ductility post cracking flexural response have been reported for BF [54–57]. Almost all previous researchers are agreed upon the fact that an increase in the fiber content and/or length decreases the workability of concrete and increases porosity [58]. Kabay [59] showed that increasing the length and/or volume of fibers reduces concrete workability. It was observed that void content has a higher impact on abrasion compared to compressive and flexural strength. The inclusion of fibers reduced compression strength and abrasive wear of concrete. The highest flexure strength was achieved in mixtures with 60% w/c. Dias and Thaumaturgo [34] studied the effect of BF in geopolymer concrete. It was indicated that BF in geopolymer concretes improved the flexural and splitting tensile strength compared to geopolymer concretes; however, the addition of 1.0% BF resulted in a 26.4% reduction in compressive strengths of cement concretes.

The length and weight fraction of BF are two interactive factors that influence on mechanical and fresh properties of cementitious composites. Khan and Cao [60] investigated the mechanical properties of basalt-fiber-reinforced cementitious composites with four different fiber lengths (3, 6, 12, and 20 mm) and also with a combination of various lengths and contents. The results indicated that the compressive strength of basalt-fiber-reinforced cementitious composites containing fiber with the length of 6 and 12 mm is higher than that with 3 and 20 mm length. It was observed that short fibers in the matrix play a bridging role to limit the expansion of microcracks while longer fibers provided additional capacity to restrict the development of macrocracks. Amuthakkannan et al. [61] studied the effect of basalt fiber length and content on the mechanical properties of basalt-fiber-reinforced polymer matrix composites of the fabricated composites. It was shown that basalt fiber with the length of 10 mm and 50 mm provided better tensile strength than 4 mm and 21 mm in polymer matrix composites. It was indicated that shorter fiber lengths will create more fiber ends that act as stress concentration points, resulting in a faster failure at these points.

Almost all studies conducted on properties of fiber-reinforced cementitious materials indicated that increasing the fiber contents and/or length harms the workability and porosity of the fresh mix [58]. Reduction in the workability due to an increase in the fiber content requires higher water-to-cement ratio, which leads to higher porosity in the hardened concrete. Accordingly, fiber content is also a main parameter considered by previous researchers to control the fiber effect on the concrete properties. However, selecting an appropriate amount of fiber content to reach proper workability with expected mechanical properties for concrete is necessary. As shown in Table 2, the length of BF is varied from 3 to 30 mm while the fiber content is ranged from 0.05% to 5% of the volume fraction in the mixes. As it can be seen in the table, 2% is the most frequently used weight fraction in studies related to fresh and hardened concrete properties. Chopped fibers with shorter length are characterized with a more uniform distribution while longer lengths can better contribute with crack control.

Table 2. A summary of the studies on the effect of basalt fiber (BF) on mechanical parameters of concrete.

Reference.	Fiber Type	Length (mm)	Weight Fraction (%)	Remarks
Jiang et al. [54]	Basalt Polypropylene	12 and 22 4–19	0.05, 0.1, 0.3, & 0.5	- Compressive strength ↑ - Toughness & splitting tensile strength ↑ - Flexural strength (0.3% BF) and (0.5% BF) - Tensile and flexure strength (22 mm length of BF)
Kabay [59]	Basalt	12 and 24	0.07 and 0.14	- Length and/or volume of BF ↑ ⟹ workability ↓ - Compression strength ↓ - Abrasive wear (in the range of 2–18%) ↓ - Flexure strength ↑
Ayoub et al. [62]	Basalt	25	1, 2, and 3	- BF contents ↑ ⟹ compressive strength ↓ - Tension strength ↑ - No correlation between BF and E-value
Kizilkanat et al. [57]	Basalt Glass	12 12	0.25, 0.5, 0.75, and 1	- BF contents ↑ ⟹ fracture energy ↑ - Compressive strength ↑ (>0.25% fiber content) - Crack avoidance, tensile strength, and ductility ↑ - GF ↑ ⟹ flexure strength
Fenu et al. [56]	Basalt Glass	12 12	3 and 5	No significant impact on dynamic in-crease factor - Energy absorption ↑ - Static flexural strength and post-peak behavior ↑
Shafiqh et al. [55]	Basalt Polyvinyl Alcohol	25 30	1, 2, and 3	- Strain attaining capacity ↑ - PVA fibers ↑ ⟹ ductility, post-cracking flexural response, and toughness ↑ - BF does not correlate with post-peak flex-ural behavior
Girgin [63]	Basalt Glass	24 24	2	- Fibers ↑ ⟹ fracture energy ↑; workability ↓ wand E-value ↓ - Compressive strength ↑ (0.5% and 0.75% fiber content) - Crack avoidance ↑, tensile strength ↑, and ductility ↑
Afroz et al. [9]	Basalt (Chinese) Basalt (Russian)	3 25	0.5	- Stability in alkaline medium ↑ - Protection against alkaline ion - Mechanical properties ↓ - Long-term flexural strengths and splitting tensile ↑
Zhao et al. [64]	Basalt	18	1, 1.5, 2 & 2.5	- Anti-impact deformation characteristics ↑
Katkhuda and Shatarat [65]	Basalt	18	0.1, 0.3, 0.5, 1, and 1.5	- Flexural and splitting tensile compressive strength ↑
Sun et al. [66]	Basalt	6 and 12	1, 2, 3, 4, & 5	- Compressive and splitting tensile strengths ↑; bending strength increase ↓

Ayoub et al. [62] stated that an increase in BF contents of the concrete mix leads to reduced compressive strength. However, 0.5% and 0.75% of BF content can slightly improve the compressive strength. The addition of 10% silica and BF can result in a higher tensile strength in high-performance concrete. BF content does not correlate with E-value. Fenu et al. [56] reported that the addition of GF and BF increased dynamic energy absorption at a high strain rate. Furthermore, it enhanced the post-peak behavior and static flexural strength of concrete. Shafiqh et al. [55] indicated that the addition of GF and BF enhanced strain capacity attaining of the concrete. Significant improvement was achieved in flexural response, ductility, and toughness by the inclusion of PVA fibers. However, no relation was found between BF addition and post-peak flexural behavior of concrete. Girgin 2016 [63] stated that the addition of GF resulted in a 35% reduction in the average strain capacity of concrete compared to the control specimens. The flexural strain capacity of BF was influenced by cement hydration in the matrix–fiber interface. Afroz et al. [67] studied the influence of BF fibers for sulfate and chloride resistance in concrete for long-term durability applications. It was shown that modified BF had better stability in the alkaline medium compared to non-modified ones. However, a slight reduction in the mechanical properties was reported for concretes reinforced with treated fibers.

Since polymer binders in polymer concrete are made of organic materials, they have a very low heat and fire resistance compared to minerals. Hence improving the thermal performance of the polymer concretes is of utmost importance for the construction industry. The limited information that exists in the literature indicates that application of SA in the concrete mixture could associate with self-healing of cracks and improving the fire, flame, and heat resistance of concrete while the inclusion of BF could enhance the energy absorption and ductile behavior of concrete that can prevent brittle fracture of structures. Several pieces of research have been conducted to study the effect of using SA and BF in cement concrete; however, such studies are limited when it comes to their applications in PC. To the author's knowledge, this study is the first that investigated the strength characteristic of SA-based BF-reinforced PC subjected to elevated temperature. After an extensive literature review, two experimental phases, including the preliminary phase to set the appropriate mix design, and the main phase to investigate the compressive strength of samples after exposure to elevated temperatures of 100 °C, 150 °C, and 180 °C, were conducted. A summary of the research on PC/PPCC with and without fiber reinforcement is presented in Tables 3 and 4, respecrively.

Table 3. Polymeric concrete.

Remarks	Filler and Polymer			Codes & Sample Size (mm)	Test Age (Day)	Conducted Tests and Sample Monitoring	Ref.
	Amount (% of the Total Volume (Vol%) or the Total Weight (wt %))	Details	Name				
• 15% epoxy & 200% filler (15% fine silica, 25% medium-size silica, & 60% coarse silica powder) had maximum compressive & flexural strengths • Tensile strength was maximized with 20% resin & 200% filler • Mechanical strength of polymer concrete was 4–5-folds higher than cementitious concrete	100, 150, & 200 Vol%	50–60, 600, & 1100 µm Size	Silica Fume	➢ ASTM (C33-03, C579-01, C293-02, C496/C496/M, & C496-04) • 50 × 50 × 50 cube • 76.2 × 152.4 cylinder		• Uniaxial compressive test • Uniaxial tensile test • Flexural compressive test	Golestaneh et al. 2010 [32]
	10, 15, & 20 wt %	Epoxy, Bisphenol A	Epoxy Resin				
• Investigate the effect of temperature on the performance of epoxy and unsaturated polyester polymer mortars • Higher ductility after 60 °C • Epoxy more sensitive than polyester • From room temperature to 60 °C, flexural and compressive strength decreases drastically as temperature increases; a loss of more than 50% is reported • The epoxy mortars are more sensitive to temperature changes than the unsaturated polyester ones	88 Vol%	245 µm	Only Coarse Aggregate	➢ RILEM TC113/PC-2, RILEM TC113/PCM-8, ASTM (C39-05, C348-02) • 40 × 40 × 160 prism • 50 × 100 cylinder	7	• Uniaxial compressive test • Flexural compressive test	Reis 2012 [34]
	12 wt %	Diglycidyl ether bisphenol A & an aliphatic amineHardener, RR515	Epoxy Resin				
• Mechanical and physical properties of epoxy polymer concrete after exposure to temperatures up to 250 °C • Optimum polymer amount is 13% • 50% reduction in compressive strength after exposure to 250 °C • No change in flexure strength • The optimal polymer content is 13% which leads to obtain the highest physical and mechanical properties at the lowest cost	-	$(0–4) \times 10^3$ µm	Sand	➢ RILEM CPT (PCM 2, PCM 8), NF EN (1097-6, 197-1), EN (12390-3, 206-1, 934-2) • 50 × 50 × 305 prism • 70 × 70 × 280 prism • 50 × 100 cylinder • 150 × 300 cylinder • 50 × 150 cylinder	7 & 28	• Uniaxial compressive test • Flexural compressive test • Heating • Porosity • Scanning electron microscopy • E-value calculation	Elalaoui et al. 2012 [33]
	-	$(4–10) \times 10^3$ µm	Gravel				
	6, 9, 13, & 16 wt %	Bisphenol A diglycidyl ether resin, Eponal 371	Epoxy resin				
• Effect of fly ash on the behavior of polymer concrete with different types of resin was investigated • Improve in ductility & increase in E-value by decreasing in fly ash content for all the mixes of PC concrete • Compressive strength as high as 100 MPa can be achieved using resins in polymer concrete • About 80% of the 28-day compressive strength can be achieved in 7 days for polymer concrete • Split tensile and flexural strengths decreased with the increasing fly ash content for all the mixes	0, 3, 10, 13, 20, 21, & 23 Vol%	15 µm (Type F)	Fly Ash	➢ ASTM (C128, D695, D3967), ISO 178 • 9 × 16 × 160 prism • 50 × 100 cylinder	3, 7, 14, 21, & 28	• Uniaxial compressive test • Uniaxial tensile test • Flexural compressive test • E-value calculation	Lokuge & Aravinthan 2013 [26]
	20, 22, 30, 40, & 43 Vol%	Orthophthalic	Polyester Resin				
		Bisphenol	Vinylester Resin				
		Thixotropic	Epoxy Resin				
• Increase in density of polymer matrix by an increase in filler content which resulted in more & bigger voids • An increase in filler amount to 60% reduced the flexural strength by 70%. • No visible failure under compression was observed for the mixes with up to 30% filler. • An increase in filler amount leads to increased compressive modulus of elasticity. • Adding filler could help preserve structural performance through absorbing or blocking UV radiation before reaching chromophores on which the color of a polymer matrix is dependent • The mix with 30% filler has better mechanical properties while the mix with 50% filler would be the most economical choice. Both 30% and 50% filler mix have high durability for UV radiation	0, 10, 20, 30, 40, 50, & 60 Vol%	Fire Retardant, Hollow Microsphere, Fly Ash	light-weight Filler	➢ ASTM (D7028, D4065, C905, C580, & C579) • 60 × 14 × 4 prism • 80 × 10 × 10 prism • 25 × 25 cylinder		• Uniaxial compressive test • Uniaxial tensile test • Flexural compressive test • Porosity • Ultraviolet • Density • Heat Generation • Glass transition temperature	Ferdous et al. 2016 [31]

Table 4. Polymer concrete (PC)/polymer-Portland cement concrete (PPCC) with fiber.

Remarks	Basalt and Polymer: Amount (% of the Total Volume (Vol%) or the Total Weight (wt %))	Basalt and Polymer: Etra details	Basalt and Polymer: Diameter (μm)	Basalt and Polymer: Length (mm)	Basalt and Polymer: Name	Codes & Sample Size (mm)	Test Age (Day)	Conducted Tests and Sample Monitoring	Ref.
• Geopolymer concrete showed higher strengths than Portland cement concretes • BF of 1.0% resulted in 26.4% & 12% reductions in compressive and splitting tensile strengths for Portland cement concretes • BF in geopolymer concretes resulted in higher flexural & splitting tensile strength than geopolymer concretes without fibers • Addition of fibers to all concretes tested caused increases in the VeBe time • Geopolymer concretes exhibited less cracks that is probably due to the bond between fiber and matrix	(0, 0.5, 1) Vol%	-	9	45	Basalt Fibre (BF)	➢ NBR (5738/93, 5739/94, 7222/94, 7222/94), ASTM (1993), ESIS (1992), Rilem Draft Recommendations (1989) • 150 × 300 cylinder • 150 × 150 × 500 prism (with 30 mm depth and 3 mm thickness notch)	28	• Uniaxial compressive test • Uniaxial tensile test • Flexural compressive test • VeBe consistometer test	Dias & Thaumaturgo 2005 [34]
	A/B = 1 (13.85 wt %) & 47.7 wt %	A: SiO_2/Al_2O_3 ratio = 5.35, $(Na_2O + K_2O)/SiO_2$ = 0.209, B:SiO_2/Na_2O = 2.24, KOH = 14M	-	-	Geopolymer (Poly(Siloxo-Sialate))				
• Comparison between epoxy polymer concrete plain, reinforced with CF and GF • Increase in compressive strength of epoxy polymer (27.5–45.4% by GF & 36.1–55.1% by CF was observed • Slightly ductile failure of reinforced polymer, while unreinforced polymer showed a brittle failure • Comparing plain polymer concrete to ordinary concrete, compressive strength is 85% higher • When polymer concrete is reinforced, more than 100% increase in compression strength was observed • Epoxy polymer concrete (with and without fiber) proved to be an excellent alternative to concretes	1 & 2 Vol%	-	Chopped	6	Carbon Fiber (CF)	➢ RILEM, ASTM (C469-94, C617-98) • 50 × 100 cylinder	7	• Uniaxial compressive test	Reis 2005 [35]
				6	Glass Fiber (GF)				
	20% wt %	Eposil 551 (Silicem®), Based on a diglycidyl-ether of bisphenol A, Aliphatic amine hardener	-	-	Epoxy Polymer				
• Maximum tensile and flexural strengths of BF/polyester poly butylene succinate (BF/PBS) were achieved at a fiber loading of 15 vol% and a good interface adhesion between the fiber and matrix observed in the composites • Impact strength of BF/PBS decreased with fiber addition primarily and then increased by increasing fiber • Thermal deflection temperature (HDT) of BF/PBS composite was significantly higher than the HDT of PBS resin	(0.0–15) Vol%	-	10	continuous twistless roving	BF (coated by silane coupling agent)	➢ GB/T(1040-1992, 9341, 1843, 1633, 1634) • Dumbbell shape • 4 × 10 × 80 prism (with 2 mm depth notch)	-	• Uniaxial tensile test • Flexural compressive test • Impact test • Thermal stability • Scanning electron microscopy	Zhang et al. 2012 [37]
	20–30 wt %	Diglycidyl ether of bisphenol A resin (DGEBA), Low viscosity (CY 184), Polyaminehardener (Aradur®2965)	-	-	Resin				
• BF was exposed to water, styrene butadiene rubber and superplasticizer that prevails in cementitious materials • BF >1% reduced concrete strength & basalt was degraded and degenerated by biochemical changes • In increased alkaline conditions, spalling of fibers was observed when exposed to the pH of 12 & further disintegration of fibers were noticed when exposed to polycarboxylate based superplasticizer and polymer matrix of styrene butadyene rubber at pH 12 • Further studies are required to enhance the structural stability of BF in the concrete matrix	(0.5, 1, 1.5, & 2) Vol%	-	13	24	BF	➢ - • 100 × 100 × 100 cube • 150 × 300 cylinder	14 & 28	• Compression • Uniaxial tensile test • X-ray diffraction test • Fourier transform infrared • Scanning electron microscopy • Energy dispersive X-ray	Sarayu et al. 2016 [30]
	97–85 vol%	Mp 114 °C, Vicat softening temperature 96 °C	-	-	Polymer/Styrene Butadiene Rubber (SBR)				
• Mechanical properties of an epoxy/basalt polymer concrete at 24, 50, 75, & 100 °C were investigated • Increase in the amount of epoxy resin to 25 wt % enhanced mechanical properties of this concrete • Increase in temperature resulted in more ductility, but mechanical properties deterioration and stiffness loss • Decrease in aggregate size resulted in higher flexural & splitting tensile & lower compressive strengths	(0, 70, 72.5, 75, 77.5, & 80) Vol%	-	1–5	-	Crushed Basalt (aggregate)	• ASTM (C39/C39M, C580, C496/C496M) • 50 × 100 cylinder • 25 × 25 × 54 prism	7 & 28	• Uniaxial compressive test • Uniaxial tensile test • Flexural compressive test Scanning electron microscopy	Niaki et al. 2017 [36]

2. Materials and Methods

2.1. Materials

Materials including aggregates, BF, SA, and two-part epoxy resin for this experiment were provided by a local supplier. Basalt fiber, SA, and epoxy resin were productions originally from the U.S., Iran, and Germany, respectively. Triplicate cubic concrete samples of 50 × 50 × 50 mm were tested for the compressive strength of each mix composition. According to ASTM C117 and E11, the aggregates for the mortar were sieved through sieve number 4 to provide relatively fine aggregates of size less than 4.75 mm. The fineness modulus of the used aggregates was 2.72. Size distribution curve for aggregate used in the present research is shown in Figure 1. The review of the literature presented in Table 2 showed that the length of the BF considered in the table for production of polymer and cementitious composites is ranged from 6 to 45 mm. Shorter chopped lengths fibers are more effective for appropriate fiber distribution while longer chopped lengths are more effective in crack bridging action [68].

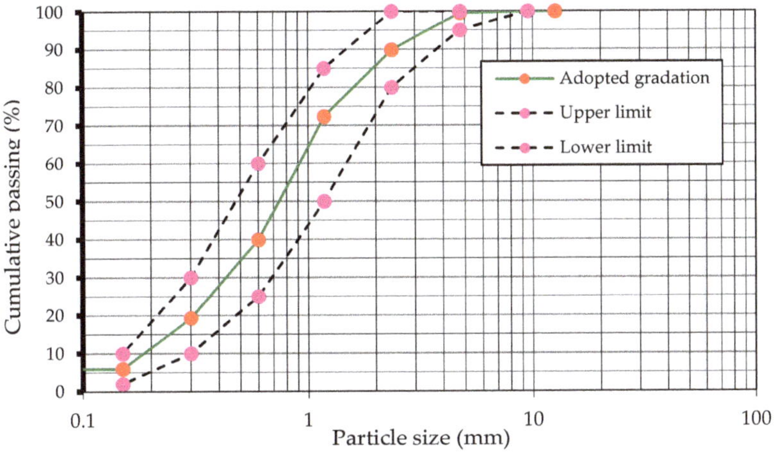

Figure 1. Size distribution curve for aggregate used in the present research.

In order to achieve the objectives of this study and in light of the findings reported in previous studies, together with the literature review discussed in the introduction section, the average length of the chopped BF in the mortar mixtures is considered 10 mm for this study. Moreover, the amount of investigated fiber content reported in previous studies and literature review is ranged between 0.15% and 2% of the total volume of the PC. In this study, the content of the chopped BF in the mortar mixtures was considered 2% of the total volume, which is consistent with the literature. BF is another admixture in the composite to improve the thermal resistance of the composite. Nonetheless, the addition of BF into the mixtures may improve the fresh properties of the composite, it reduces the compressive strength of the hardened mortar specimens. In order to restrict the adverse effect and take advantage of BF in improving thermal resistance and rhetorical properties, a lower proportion of SA compared to that used in literature was incorporated. BF amount used in this study occupied only 0.1% of the total mortar volume. The geometrical and mechanical properties of the BF used in the present study are shown in Table 5. The fine fraction of aggregates 0–150 µm were excluded.

Table 5. Properties of BF used in the experiment.

Cutting Length (mm)	10
Diameter (µm)	17
Density (gr/cm3)	2.65
Elastic modulus (GPa)	93–110
Tensile strength (MPa)	4100–4800
Elongation (%)	1.3–3.2
Softening point (°C)	1050
Water absorption (%)	<0.5

Two-part epoxy resin with a specific weight of 1.1 gr/cm^3 was used in the experiment. In this process, a low viscous epoxy, known as saturant resin (Part A) and hardener (Part B), was mixed to obtain a homogeneous mixture. Epoxy resins are the most widely used PC compounds [69]. Epoxy concrete is composed of two parts of the epoxy and aggregate blend. The curing of the epoxy resin systems is an exothermic reaction. The aggregates were checked to be dry and free of dirt, debris, and organic materials using sieving and oven drying. The mix proportion of the saturant resin and hardener with the trading name of Araks FK20 was as per the guidelines of the manufacturer. Generally, epoxy formulations are skin sensitizers, and handling precautions should be taken carefully [69]. In order to determine the general proportions of selected materials in the experimental study, two amounts of 13% and 15% of the total mortar volume were used for epoxy content in the mix design stage. SA occupied 0.1% of the total mortar volume.

The curing time and temperature for PC are affected by parameters such as the epoxy formulation, mix proportions of PC composition, and mass volume. Generally, PC mortar cures in hours after casting while complete curing of the cement-based materials takes days or weeks.

2.2. *Mix Designs and Preparation of Specimens*

The experimental study was conducted in two phases, one aimed to obtain appropriate proportions and the other is to investigate the main objective of this research. In the first phase of the experimental program, two epoxy concrete mixes made of aggregate and epoxy resin without the inclusion of BF and SA were tested. The results of the compressive strength for the epoxy resin contents of 13% and 15% were 41.9 MPa and 39.7 MPa. For the second phase, different mix designs of PC were determined to have aggregate, epoxy resin, FB, and SA in the composition of the mix. The resin amount, in the PC mix designs, was constant and equal to 13% of the total concrete volume. The mix design for SA-based BF-reinforced mortar subjected to 0 °C and elevated temperatures of 100 °C, 150 °C, and 180 °C is presented in Table 6. Zero temperature in these tables that is room-curing temperature without heating was a reference before heating of samples. The hardened mortar specimens were exposed to three elevated temperatures of 100 °C, 150 °C and 180 °C for a duration of three hours. Then, the compressive strength of samples was tested and compared for each mix design under raised temperature conditions. To test compressive strength, PC samples were placed in the oven for three hours under the intended temperature after 7 days of curing. In order to prevent cooling shock, samples were left at room temperature to cool down gradually before applying compressive load in the testing machine.

Table 6. Details of PC in the main study.

Name **	BF (%) *	SA (%) *	Evaluation Temperature (°C)
MPT_0	-	-	0
MPT_{100}	-	-	100
MPT_{150}	-	-	150
MPT_{180}	-	-	180
$MPBT_0$	0.2	-	0
$MPBT_{100}$	0.2	-	100
$MPBT_{150}$	0.2	-	150
$MPBT_{180}$	0.2	-	180
$MPAT_0$	-	0.1	0
$MPAT_{100}$	-	0.1	100
$MPAT_{150}$	-	0.1	150
$MPAT_{180}$	-	0.1	180
$MPBAT_0$	0.2	0.1	0
$MPBAT_{100}$	0.2	0.1	100
$MPBAT_{150}$	0.2	0.1	150
$MPBAT_{180}$	0.2	0.1	180

* of the total weight; ** MPT, MPBT, MPAT, MPBAT are PC with no SA and BF, with BF only, with SA only, and with SA and BF, respectively.

2.3. *Compressive Strength Testing*

The compressive strength tests of the specimens were carried out according to ASTM C109-08. For each mixture, three 50 × 50 × 50 mm cubic specimens of each mix were tested after 3 days. In order to determine the ultimate load-carrying capacity under pure compression, the axial compressive load was applied continuously with a rate of 2.4 kN/sec according to ASTM C31 by a compression machine with 2000 kN capacity. The load was applied until the rupture of the specimen which means that the failure of the specimen at the point, the reading of the load in kN, and stress in MPa were recorded from the dial gauge.

3. Result and Discussions

Compressive tests were performed to investigate the temperature resistance of PC mortar and the effect of the inclusion of chopped BF and SA on the mechanical strength of PC. The results of the ultimate load-carrying capacity of the mortar specimens for different mix designs are shown in Table 7. The result shows that incorporation of BF and/or SA into PC reduced the compressive strength to 33.8, 34.3, and 35.3, which is consistent with the observation published by Dias and Thaumaturgo [34] and Niaki et al. [36], that BF reduces the compressive strength of PC.

Table 7. Compressive strength for the PC with/without SA and BF.

Name	f_c(MPa) 7-Day	Variation to MPT_0 (%)	Variation after Heating (%)	
MPT_0	43.7	-	-	Compared to MPT_0
MPT_{100}	47.33	+8.31	+8.31	
MPT_{150}	46.3	+5.95	+5.95	
MPT_{180}	39.85	−8.81	−8.81	
$MPBT_0$	33.8	−21.10	-	Compared to $MPBT_0$
$MPBT_{100}$	39.33	−10.00	+16.36	
$MPBT_{150}$	41.5	−5.03	+22.78	
$MPBT_{180}$	41.15	−5.83	+21.75	
$MPAT_0$	34.3	−21.51	-	Compared to $MPAT_0$
$MPAT_{100}$	41.53	−4.97	+21.08	
$MPAT_{150}$	44.4	+1.60	+29.45	
$MPAT_{180}$	44.8	+2.52	+30.61	
$MPBAT_0$	35.3	−19.22	-	Compared to $MPBAT_0$
$MPBAT_{100}$	38.3	−12.36	+8.50	
$MPBAT_{150}$	42.9	−1.83	+21.53	
$MPBAT_{180}$	44.5	+1.83	+26.06	

Figure 2 shows the variation in the compressive strength of different mix designs compared to the reference state. The analysis of the results obtained by the compressive strength test of the samples maintained at room temperature revealed that the addition of BF and SA to the reference mix design decreased compressive strength. A lower decrease in compressive strength was observed for the mortar mix containing sodium alginate (SA) compared to those with only alginate. The samples with sodium alginate pose lower compressive strength due to the amount and size of voids in the structure of the hardened composite compared to the control mortar [24]. This weakening of compressive strength may ascribe to forming voids by the addition of the fibers into the matrix, which is in agreement with test results by Reis [30], Kabay [59], and Sun et al. [66]. However, the observation during uniaxial compressive loading indicated that the inclusion of FB leads to the absorption of a large amount of plastic energy before failure that is indicated also in other research studies. He and Lu [70] and Donker and Obonyo [71]. The analysis of the results obtained by the compressive strength test for polymer mortar containing BF and SA revealed that the addition of BF and SA to the reference mix design enhanced compressive strength growth in elevated temperatures. The cure of an epoxy resin is an exothermic process; however, the observation of an experimental investigation on curing temperature of epoxy resin conducted by Lahouar et al. [72] reported absorption of energy between 20 °C and 80 °C representing an endothermic peak due to possible phase change in curing state of the resin. Variations in the compressive strength after heating of the PC containing either BF or SA alone or both together indicate improving the effectiveness of these additives. The obtained results are not in agreement with those of either two works of Reis et al. [35] and [30], indicating a sharp increase (up to 50%) or drastic drop of compressive strength due to temperature elevation.

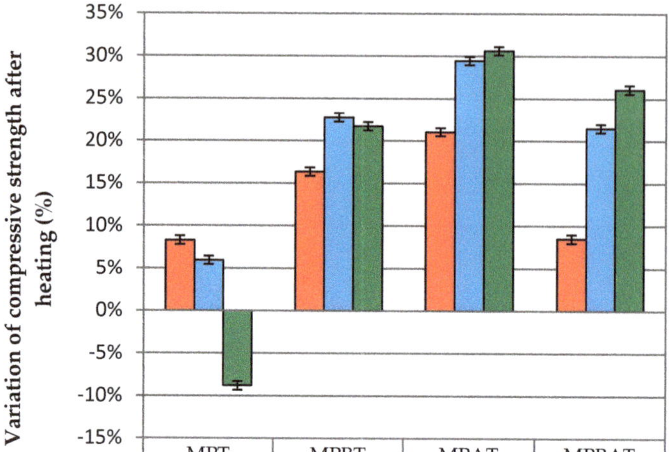

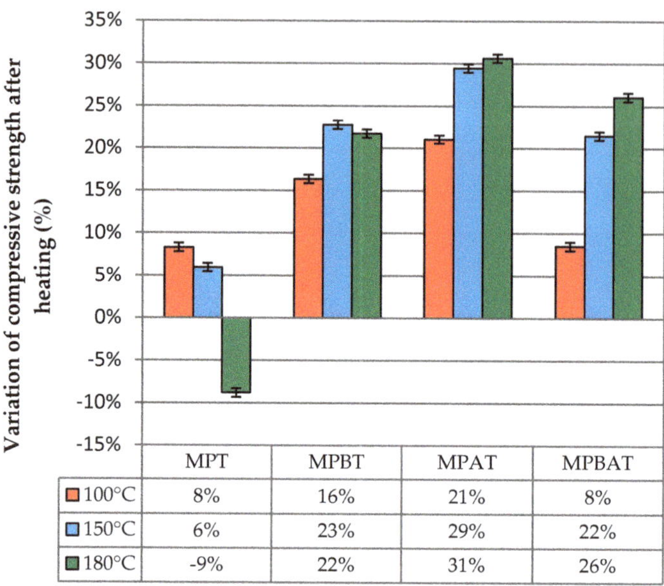

	MPT	MPBT	MPAT	MPBAT
100°C	8%	16%	21%	8%
150°C	6%	23%	29%	22%
180°C	-9%	22%	31%	26%

Figure 2. Failure of samples under compression test (all tests were conducted in triplicate).

The negative percentage for variation of compressive strength of MPT specimens is in agreement with previous studies that indicated the adverse effect of elevated temperature on the hardened specimens. The increase in the variation of compressive strength in MPBT, MPAT, and MPBAT supports the finding of this research that the addition of BF and SA enhances the compressive strength of PC. The highest positive influence by raised temperature is corresponding to the mixture including SA alone that with an increase in the temperature from room temperature to 100 °C then 150 °C and finally to 180 °C, growth in the compressive strength became 21.08%, 29.45%, and 30.61%, respectively. A lower increase of the compressive strength of the BF-reinforced PC is due to the entrapped voids because of fiber addition. The obtained results about enhanced temperature resistance of SA-based mortar specimens also conform to the finding of DeBrouse [23].

Utilization of both BF and SA together in the PC has been with a lower increase from room temperature to 100 °C until 180 °C, compared to the individual application of these materials. However, it can be seen that heating of PC for 3 h under 180 °C resulted in strength reduction, while together or individual effectiveness of BF and SA in the PC is remarkable on the improvement in the compressive strength after heating for all the three temperatures. From Table 6, the highest compressive strength for samples of PC after 3 days of air curing was higher than the cement concrete samples. A comparison between the obtained compressive strength of polymer and cement concretes with BF and SA indicated that the elevated temperature had a greater influence on polymer concrete than that of cement concrete. The rate of increase was around 20% for polymer concrete, while it was around 5% for cement concrete for temperature rise from room temperature to 180 °C.

Polymer concrete samples after failure under compression test are shown in Figure 3. It should be demonstrated that PC samples exhibited no obvious crack on the surface when the applied load was released after failure under compression. The produced polymer concrete can be considered as a composition of two phases of polymer mortars and aggregates. Through applying the load displacement into the specimens, it was observed that the failure was initiated by the creation of interphase microcracks between polymer mortars and aggregates. By increasing the compressive load, longitudinal cracks were propagated along the jaws' direction, which was similar to those that happen in normal concrete. It was observed that the time between crack initiation and total failure of the specimens was lower than that of cementitious concrete specimens [24,73].

(**a**) PC (**b**) BF-reinforced PC (**c**) SA-based PC (**d**) SA-based BF-reforced PC

Figure 3. Failure of samples under compression test.

The compressive load was applied until the mortar specimens of each mix reached their ultimate load-bearing capacity and failure occurred. High plastic deformation of the polymer concrete was observed corresponding to the applied load.

4. Conclusions

In order to investigate the effectiveness of the addition of BF with the natural source and/or SA as a waste material to reduce the strength degradation of polymer concretes after exposure to raised temperatures of 100 °C, 150 °C, and 180 °C, an experimental program was conducted in two phases including primary study and main study. The addition of BF and/or SA decreases concrete compressive strength under room temperature, but the addition of BF and SA each alone leads to compressive strength increase during exposure to heat. An increase in the temperature to 180 °C showed an almost positive effect on the compressive strength. The highest positive influence by raised temperature is corresponding to the mixture including SA alone that, with an increase in the temperature from room temperature to 100 °C then 150 °C and finally to 180 °C, growth in the compressive strength became 21.08%, 29.45%, and 30.61%, respectively. The addition of BF and SA both together increases the rate of strength growth of polymer concrete under heat from 100 °C to 180 °C. In view of these conclusions, the addition of BF and SA results in a decrease of compressive strength of polymer concretes under room temperature, but it is obvious that these materials are positive to improve the resistance of PC. However, the most important outcome of this research may be a natural source of BF and waste material SA to prevent PC compressive strength degradation under elevated temperature. Increase of compressive strength up to 30%, that when noticed to the cement amount with the energy-wasting process to be produced, is a very valuable finding.

Author Contributions: Conceptualization, N.A.O., S.E.M.-Y. and H.A.A.; methodology, N.A.O., S.E.M.-Y. and H.A.A.; software, N.A.O.; validation, N.A.O., S.E.M.-Y. and H.A.A.; formal analysis, N.A.O., S.E.M.-Y. and H.A.A.; investigation, N.A.O., S.E.M.-Y. and H.A.A.; resources, N.A.O., S.E.M.-Y., H.A.A., H.S., S.S.R.K., and M.P.; data curation, N.A.O.; writing—original draft preparation, N.A.O., S.E.M.-Y., H.A.A. and H.S.; writing—review and editing, N.A.O., S.E.M.-Y., H.A.A., H.S., M.P. and S.S.R.K.; visualization, N.A.O., S.E.M.-Y., H.A.A., H.S., S.S.R.K., and M.P.; supervision, S.E.M.-Y., H.A.A.; project administration, N.A.O., S.E.M.-Y., H.A.A., H.S., S.S.R.K., and M.P. funding acquisition, N.A.O., S.E.M.-Y., H.A.A., H.S., S.S.R.K., and M.P. All authors have read and agreed to the published version of the manuscript.

Funding: The research was supported by the Islamic Azad University, Najafabad Branch, and the Ministry of Education, Youth, and Sports of the Czech Republic and the European Union (European Structural and Investment Funds Operational Program Research, Development, and Education) in the framework of the project "Modular platform for autonomous chassis of specialized electric vehicles for freight and equipment transportation", Reg. No. CZ.02.1.01/0.0/0.0/16_025/0007293.

Conflicts of Interest: The authors declare no conflict of interest.

References

1. Mohammadyan-Yasouj, S.E.; Marsono, A.K.; Abdullah, R.; Moghadasi, M. Wide beam shear behavior with diverse types of reinforcement. *ACI Struct. J.* **2015**, *112*, 199–208. [CrossRef]
2. Moghadasi, M.; Marsono, A.K.; Mohammadyan-Yasouj, S.E. A study on rotational behaviour of a new industrialised building system connection. *Steel Compos. Struct.* **2017**, *25*, 245–255.
3. Shokravi, H.; Shokravi, H.; Bakhary, N.; Koloor, S.S.R.; Petrů, M. A Comparative Study of the Data-driven Stochastic Subspace Methods for Health Monitoring of Structures: A Bridge Case Study. *Appl. Sci.* **2020**, *10*, 3132. [CrossRef]
4. Shokravi, H.; Shokravi, H.; Bakhary, N.; Heidarrezaei, M.; Koloor, S.S.R.; Petru, M. Vehicle-assisted techniques for health monitoring of bridges. *Sensors* **2020**, *20*, 3460. [CrossRef] [PubMed]
5. Shokravi, H.; Shokravi, H.; Bakhary, N.; Heidarrezaei, M.; Rahimian Koloor, S.S.; Petrů, M. Application of the Subspace-Based Methods in Health Monitoring of Civil Structures: A Systematic Review and Meta-Analysis. *Appl. Sci.* **2020**, *10*, 3607. [CrossRef]
6. Shokravi, H.; Shokravi, H.; Bakhary, N.; Koloor, S.S.R.; Petrů, M. Health Monitoring of Civil Infrastructures by Subspace System Identification Method: An Overview. *Appl. Sci.* **2020**, *10*, 2786. [CrossRef]
7. Aslani, F.; Samali, B. Predicting the bond between concrete and reinforcing steel at elevated temperatures. *Struct. Eng. Mech.* **2013**, *48*, 643–660. [CrossRef]
8. Shokravi, H.; Mohammadyan-Yasouj, S.E.; Rahimian Koloor, S.S.; Petrů, M.; Heidarrezaei, M. Effect of alumina additives on mechanical and fresh properties of self-compacting concrete: A review. *Processes* **2021**. (under review).
9. Mohammadyan-Yasouj, S.E.; Heidari, N.; Shokravi, H. Influence of Waste Alumina Powder on Self-compacting Concrete Resistance under Elevated Temperature. *J. Build. Eng.* **2021**, *14*. (under review).
10. Awal, A.S.M.A.; Shehu, I.A. Performance evaluation of concrete containing high volume palm oil fuel ash exposed to elevated temperature. *Constr. Build. Mater.* **2015**, *76*, 214–220. [CrossRef]
11. Ahmad, S.; Sallam, Y.S.; Al-Hawas, M.A. Effects of key factors on compressive and tensile strengths of concrete exposed to elevated temperatures. *Arab. J. Sci. Eng.* **2014**, *39*, 4507–4513. [CrossRef]
12. Thirumurugan, S.; Anandan, S. Residual strength characteristics of polymer fibre concrete exposed to elevated temperature. *Eng. J.* **2015**, *19*, 117–131. [CrossRef]
13. Fraternali, F.; Ciancia, V.; Chechile, R.; Rizzano, G.; Feo, L.; Incarnato, L. Experimental study of the thermo-mechanical properties of recycled PET fiber-reinforced concrete. *Compos. Struct.* **2011**, *93*, 2368–2374. [CrossRef]
14. Rabehi, B.; Ghernouti, Y.; Boumchedda, K.; Li, A.; Drir, A. Durability and thermal stability of ultra high-performance fibre-reinforced concrete (UHPFRC) incorporating calcined clay. *Eur. J. Environ. Civ. Eng.* **2017**, *21*, 594–611. [CrossRef]
15. Lee, S.-J.; Kim, S.-H.; Won, J.-P. Strength and fire resistance of a high-strength nano-polymer modified cementitious composite. *Compos. Struct.* **2017**, *173*, 96–105. [CrossRef]
16. Pathak, T.S.; San Kim, J.; Lee, S.-J.; Baek, D.-J.; Paeng, K.-J. Preparation of alginic acid and metal alginate from algae and their comparative study. *J. Polym. Environ.* **2008**, *16*, 198–204. [CrossRef]
17. Engbert, A.; Gruber, S.; Plank, J. The effect of alginates on the hydration of calcium aluminate cement. *Carbohydr. Polym.* **2020**, *236*, 116038. [CrossRef]
18. Wang, J.; Mignon, A.; Snoeck, D.; Wiktor, V.; Van Vliergerghe, S.; Boon, N.; De Belie, N. Application of modified-alginate encapsulated carbonate producing bacteria in concrete: A promising strategy for crack self-healing. *Front. Microbiol.* **2015**, *6*, 1088. [CrossRef]
19. Ouwerx, C.; Velings, N.; Mestdagh, M.M.; Axelos, M.A.V. Physico-chemical properties and rheology of alginate gel beads formed with various divalent cations. *Polym. Gels Netw.* **1998**, *6*, 393–408. [CrossRef]
20. Abbas, W.A.; Mohsen, H.M. Effect of Biopolymer Alginate on some properties of concrete. *J. Eng.* **2020**, *26*, 121–131.
21. Heidari, A.; Ghaffari, F.; Ahmadvand, H. Properties of Self Compacting Concrete Incorporating Alginate and Nano Silica. *Asian J. Civil Eng. Build. Hous.* **2015**, *16*, 1–16.
22. Mignon, A.; Snoeck, D.; D'Halluin, K.; Balcaen, L.; Vanhaecke, F.; Dubruel, P.; Van Vlierberghe, S.; De Belie, N. Alginate biopolymers: Counteracting the impact of superabsorbent polymers on mortar strength. *Constr. Build. Mater.* **2016**, *110*, 169–174. [CrossRef]
23. DeBrouse, D.R. *Alginate-Based Building Materials*; World Intellectual Property Organization: Geneva, Switzerland, 2012.
24. Mohammadyan-Yasouj, S.E.; Ahangar, H.A.; Oskoei, N.A.; Shokravi, H.; Koloor, S.S.R.; Petrů, M. Thermal Performance of Alginate Concrete Reinforced with Basalt Fiber. *Crystals* **2020**, *10*, 779. [CrossRef]
25. Barbuta, M.; Rujanu, M.; Nicuta, A. Characterization of polymer concrete with different wastes additions. *Procedia Technol.* **2016**, *22*, 407–412. [CrossRef]

26. Lokuge, W.; Aravinthan, T. Effect of fly ash on the behaviour of polymer concrete with different types of resin. *Mater. Des.* **2013**, *51*, 175–181. [CrossRef]
27. Venkatesh, B.; Student, U.G. Review on performance of polymer concrete with resins and its applications. *Int. J. Pure Appl. Math.* **2018**, *119*, 175–184.
28. Abd_Elmoaty, A.M. Self-healing of polymer modified concrete. *Alexandria Eng. J.* **2011**, *50*, 171–178. [CrossRef]
29. Rebeiz, K.S. Precast use of polymer concrete using unsaturated polyester resin based on recycled PET waste. *Constr. Build. Mater.* **1996**, *10*, 215–220. [CrossRef]
30. Reis, J.M.L. Effect of temperature on the mechanical properties of polymer mortars. *Mater. Res.* **2012**, *15*, 645–649. [CrossRef]
31. Ferdous, W.; Manalo, A.; Aravinthan, T.; Van Erp, G. Properties of epoxy polymer concrete matrix: Effect of resin-to-filler ratio and determination of optimal mix for composite railway sleepers. *Constr. Build. Mater.* **2016**, *124*, 287–300. [CrossRef]
32. Golestaneh, M.; Amini, G.; Najafpour, G.D.; Beygi, M.A. Evaluation of mechanical strength of epoxy polymer concrete with silica powder as filler. *World Appl. Sci. J.* **2010**, *9*, 216–220.
33. Elalaoui, O.; Ghorbel, E.; Mignot, V.; Ouezdou, M. Ben Mechanical and physical properties of epoxy polymer concrete after exposure to temperatures up to 250 C. *Constr. Build. Mater.* **2012**, *27*, 415–424. [CrossRef]
34. Dias, D.P.; Thaumaturgo, C. Fracture toughness of geopolymeric concretes reinforced with basalt fibers. *Cem. Concr. Compos.* **2005**, *27*, 49–54. [CrossRef]
35. Reis, J.M.L. Mechanical characterization of fiber reinforced polymer concrete. *Mater. Res.* **2005**, *8*, 357–360. [CrossRef]
36. Hassani Niaki, M.; Fereidoon, A.; Ghorbanzadeh Ahangari, M. Mechanical properties of epoxy/basalt polymer concrete: Experimental and analytical study. *Struct. Concr.* **2018**, *19*, 366–373. [CrossRef]
37. Zhang, Y.; Yu, C.; Chu, P.K.; Lv, F.; Zhang, C.; Ji, J.; Zhang, R.; Wang, H. Mechanical and thermal properties of basalt fiber reinforced poly (butylene succinate) composites. *Mater. Chem. Phys.* **2012**, *133*, 845–849. [CrossRef]
38. Mohamad, S.A.; Al-Hamd, R.K.S.; Khaled, T.T. Investigating the effect of elevated temperatures on the properties of mortar produced with volcanic ash. *Innov. Infrastruct. Solut.* **2020**, *5*, 1–11. [CrossRef]
39. Abdel-Fattah, H.; El Hawary, M.M.; Falah, A. Effect of elevated temperatures on the residual fracture toughness of epoxy modified concrete. *Kuwait J. Sci. Eng.* **2000**, *27*, 27–39.
40. Al-Salloum, Y.A.; Elsanadedy, H.M.; Abadel, A.A. Behavior of FRP-confined concrete after high temperature exposure. *Constr. Build. Mater.* **2011**, *25*, 838–850. [CrossRef]
41. Behnood, A.; Ziari, H. Effects of silica fume addition and water to cement ratio on the properties of high-strength concrete after exposure to high temperatures. *Cem. Concr. Compos.* **2008**, *30*, 106–112. [CrossRef]
42. Amin, M.; Tayeh, B.A. Investigating the mechanical and microstructure properties of fibre-reinforced lightweight concrete under elevated temperatures. *Case Stud. Constr. Mater.* **2020**, *13*, e00459. [CrossRef]
43. Tang, Y.; Feng, W.; Feng, W.; Chen, J.; Bao, D.; Li, L. Compressive properties of rubber-modified recycled aggregate concrete subjected to elevated temperatures. *Constr. Build. Mater.* **2021**, *268*, 121181. [CrossRef]
44. De Larissa, C.A.; dos Anjos, M.A.S.; de Sa, M.V.V.A.; de Souza, N.S.L.; de Farias, E.C. Effect of high temperatures on self-compacting concrete with high levels of sugarcane bagasse ash and metakaolin. *Constr. Build. Mater.* **2020**, *248*, 118715.
45. Wang, W.-C.; Wang, H.-Y.; Chang, K.-H.; Wang, S.-Y. Effect of high temperature on the strength and thermal conductivity of glass fiber concrete. *Constr. Build. Mater.* **2020**, *245*, 118387. [CrossRef]
46. Zhou, J.; Lu, D.; Yang, Y.; Gong, Y.; Ma, X.; Yu, B.; Yan, B. Physical and Mechanical Properties of HighStrength Concrete Modified with Supplementary Cementitious Materials after Exposure to Elevated Temperature up to 1000 °C. *Materials* **2020**, *13*, 532. [CrossRef] [PubMed]
47. Zhang, H.; Li, L.; Yuan, C.; Wang, Q.; Sarker, P.K.; Shi, X. Deterioration of ambient-cured and heat-cured fly ash geopolymer concrete by high temperature exposure and prediction of its residual compressive strength. *Constr. Build. Mater.* **2020**, *262*, 120924. [CrossRef]
48. Wang, Y.; Xu, T.; Liu, Z.; Li, G.; Jiang, J. Seismic behavior of steel reinforced concrete cross-shaped columns after exposure to high temperatures. *Eng. Struct.* **2021**, *230*, 111723. [CrossRef]
49. Alharbi, Y.R.; Abadel, A.A.; Elsayed, N.; Mayhoub, O.; Kohail, M. Mechanical properties of EAFS concrete after subjected to elevated temperature. *Ain Shams Eng. J.* **2020**. [CrossRef]
50. Sarayu, K.; Gopinath, S.; Ramachandra, M.A.; Iyer, N.R. Structural stability of basalt fibers with varying biochemical conditions-A invitro and invivo study. *J. Build. Eng.* **2016**, *7*, 38–45. [CrossRef]
51. Niaki, M.H.; Fereidoon, A.; Ahangari, M.G. Experimental study on the mechanical and thermal properties of basalt fiber and nanoclay reinforced polymer concrete. *Compos. Struct.* **2018**, *191*, 231–238. [CrossRef]
52. Elalaoui, O.; Ghorbel, E.; Ouezdou, M. Ben Influence of flame retardant addition on the durability of epoxy based polymer concrete after exposition to elevated temperature. *Constr. Build. Mater.* **2018**, *192*, 233–239. [CrossRef]
53. Gorninski, J.P.; Tonet, K.G.; Sokołowska, J.J. Use of polishing alumina as flame retardant in orthophthalic polyester resin matrix composites. In *Proceedings of the Advanced Materials Research*; Trans Tech Publications: Zürich, Switzerland, 2015; Volume 1129, pp. 209–216.
54. Jiang, C.; Fan, K.; Wu, F.; Chen, D. Experimental study on the mechanical properties and microstructure of chopped basalt fibre reinforced concrete. *Mater. Des.* **2014**, *58*, 187–193. [CrossRef]

55. Shafiq, N.; Ayub, T.; Khan, S.U. Investigating the performance of PVA and basalt fibre reinforced beams subjected to flexural action. *Compos. Struct.* **2016**, *153*, 30–41. [CrossRef]
56. Fenu, L.; Forni, D.; Cadoni, E. Dynamic behaviour of cement mortars reinforced with glass and basalt fibres. *Compos. Part B Eng.* **2016**, *92*, 142–150. [CrossRef]
57. Kizilkanat, A.B.; Kabay, N.; Akyüncü, V.; Chowdhury, S.; Akça, A.H. Mechanical properties and fracture behavior of basalt and glass fiber reinforced concrete: An experimental study. *Constr. Build. Mater.* **2015**, *100*, 218–224. [CrossRef]
58. Sarkar, A.; Hajihosseini, M. Feasibility of Improving the Mechanical Properties of Concrete Pavement Using Basalt Fibers. *J. Test. Eval.* **2020**, *48*, 2908–2917. [CrossRef]
59. Kabay, N. Abrasion resistance and fracture energy of concretes with basalt fiber. *Constr. Build. Mater.* **2014**, *50*, 95–101. [CrossRef]
60. Khan, M.; Cao, M. Effect of hybrid basalt fibre length and content on properties of cementitious composites. *Mag. Concr. Res.* **2020**, 1–12. [CrossRef]
61. Amuthakkannan, P.; Manikandan, V.; Jappes, J.T.W.; Uthayakumar, M. Effect of fibre length and fibre content on mechanical properties of short basalt fibre reinforced polymer matrix composites. *Mater. Phys. Mech.* **2013**, *16*, 107–117.
62. Ayub, T.; Shafiq, N.; Nuruddin, M.F. Mechanical properties of high-performance concrete reinforced with basalt fibers. *Procedia Eng.* **2014**, *77*, 131–139. [CrossRef]
63. Girgin, Z.C.; Yıldırım, M.T. Usability of basalt fibres in fibre reinforced cement composites. *Mater. Struct.* **2016**, *49*, 3309–3319. [CrossRef]
64. Zhao, Y.-R.; Wang, L.; Lei, Z.-K.; Han, X.-F.; Xing, Y.-M. Experimental study on dynamic mechanical properties of the basalt fiber reinforced concrete after the freeze-thaw based on the digital image correlation method. *Constr. Build. Mater.* **2017**, *147*, 194–202. [CrossRef]
65. Katkhuda, H.; Shatarat, N. Improving the mechanical properties of recycled concrete aggregate using chopped basalt fibers and acid treatment. *Constr. Build. Mater.* **2017**, *140*, 328–335. [CrossRef]
66. Sun, X.; Gao, Z.; Cao, P.; Zhou, C. Mechanical properties tests and multiscale numerical simulations for basalt fiber reinforced concrete. *Constr. Build. Mater.* **2019**, *202*, 58–72. [CrossRef]
67. Afroz, M.; Patnaikuni, I.; Venkatesan, S. Chemical durability and performance of modified basalt fiber in concrete medium. *Constr. Build. Mater.* **2017**, *154*, 191–203. [CrossRef]
68. Bi, J.; Huo, L.; Zhao, Y.; Qiao, H. Modified the smeared crack constitutive model of fiber reinforced concrete under uniaxial loading. *Constr. Build. Mater.* **2020**, *250*, 118916. [CrossRef]
69. Committee, A.C.I. Guide for the use of polymers in concrete. *ACI Comm. Rep.* **2009**, *548*, 1–100.
70. He, D.; Lu, Z. Experimental Study on Mechanical Properties of Chopped Basalt Fiber Reinforced Concrete. *J. Henan Univ. Nat. Sci.* **2009**, *3*, 320–322.
71. Donkor, P.; Obonyo, E. Earthen construction materials: Assessing the feasibility of improving strength and deformability of compressed earth blocks using polypropylene fibers. *Mater. Des.* **2015**, *83*, 813–819. [CrossRef]
72. Lahouar, M.A.; Caron, J.-F.; Pinoteau, N.; Forêt, G.; Benzarti, K. Mechanical behavior of adhesive anchors under high temperature exposure: Experimental investigation. *Int. J. Adhes. Adhes.* **2017**, *78*, 200–211. [CrossRef]
73. Joshani, M.; Koloor, S.S.R.; Abdullah, R. Damage Mechanics Model for Fracture Process of Steel-concrete Composite Slabs. In *Applied Mechanics and Materials*; Trans Tech Publications Ltd.: Stafa-Zurich, Switzerland, 2012; pp. 339–345.

processes

MDPI

Article

Development of Poly(L-Lactic Acid)/Chitosan/Basil Oil Active Packaging Films via a Melt-Extrusion Process Using Novel Chitosan/Basil Oil Blends

Constantinos E. Salmas [1,*], Aris E. Giannakas [2,*], Maria Baikousi [1], Areti Leontiou [3], Zoe Siasou [2] and Michael A. Karakassides [1,4]

1 Department of Materials Science and Engineering, University of Ioannina, 45110 Ioannina, Greece; mariabaikousi@gmail.com (M.B.); mkarakas@uoi.gr (M.A.K.)
2 Department of Food Science and Technology, University of Patras, 30100 Agrinio, Greece; zoesiasou@gmail.com
3 Department of Business Administration of Food and Agricultural Enterprises, University of Patras, 30100 Agrinio, Greece; aleontiu@upatras.gr
4 Institute of Materials Science and Computing, University Research Center of Ioannina (URCI), 45110 Ioannina, Greece
* Correspondence: ksalmas@uoi.gr (C.E.S.); agiannakas@upatras.gr (A.E.G.)

Abstract: Following the global trend toward a cyclic economy, the development of a fully biodegradable active packaging film is the target of this work. An innovative process to improve the mechanical, antioxidant, and barrier properties of Poly(L-Lactic Acid)/Chitosan films is presented using essential basil oil extract. A Chitosan/Basil oil blend was prepared via a green evaporation/adsorption method as a precursor for the development of the Poly(L-Lactic Acid)/Chitosan/Basil Oil active packaging film. This Chitosan/Basil Oil blend was incorporated directly in the Poly(L-Lactic Acid) matrix with various concentrations. Modification of the chitosan with the Basil Oil improves the blending with the Poly(L-Lactic Acid) matrix via a melt-extrusion process. The obtained Poly(L-Lactic Acid)/Chitosan/Basil Oil composite films exhibited advanced food packaging properties compared to those of the Poly(L-Lactic Acid)/Chitosan films without Basil Oil addition. The films with 5%wt and 10%wt Chitosan/Basil Oil loadings exhibited better thermal, mechanical, and barrier behavior and significant antioxidant activity. Thus, PLLA/CS/BO5 and PLLA/CS/BO10 are the most promising films to potentially be used for active packaging applications.

Keywords: PLLA; chitosan; basil oil; active packaging; films; barrier properties; antioxidant properties

Citation: Salmas, C.E.; Giannakas, A.E.; Baikousi, M.; Leontiou, A.; Siasou, Z.; Karakassides, M.A. Development of Poly(L-Lactic Acid)/Chitosan/Basil Oil Active Packaging Films via a Melt-Extrusion Process Using Novel Chitosan/Basil Oil Blends. *Processes* **2021**, *9*, 88. https://doi.org/10.3390/pr9010088

Received: 18 December 2020
Accepted: 29 December 2020
Published: 3 January 2021

1. Introduction

Nowadays, the global trend towards a cyclic economy, sustainability, green economy, and nanotechnology suggests the use of by-products, biomass, and/or bio-wastes which have zero environmental fingerprints as raw materials for the development of novel biodegradable active packaging materials. One of the most promising and widely used bio-based polymers which has already been commercialized, is Poly(L-Lactic Acid) (PLLA). PLLA is produced by the polymerization of the L-Lactic Acid monomer, which is produced through fermentation of sugars. Such sugars are extracted from biomass sources (e.g., corn starch, tapioca, sugar cane, or sugar beet) [1,2]. Compared with petroleum-based polymer production, the PLLA production process requires about 50% less energy consumption [3]. Moreover, PLLA is not only bio-based but also compostable and biodegradable through hydrolysis by microorganisms [4]. Thus, PLLA could be a green alternative to some common thermoplastic polymers such as polyethylene (PE), polypropylene (PP), polystyrene (PS), and polyethylene terephthalate (PET) [5,6]. Its mechanical properties make this polymer suitable for use in various production processes. However, its high elastic modulus makes its plastic deformation limited, and thus, it becomes brittle.

Chitosan (CS) is another biopolymer promising for food packaging applications. CS is a linear polysaccharide that is produced by treating a food by-product, i.e., the chitin shells of crustaceans [7]. CS exhibits high barrier values, significant antioxidant properties, and antimicrobial activity. Such properties give this material great potential for use as packaging material [8]. The main disadvantages of CS are the weak tensile properties and the disability to blend in extruders with other materials that are commonly used in the industry for packaging film production.

Researchers developed composite membranes to overcome the disadvantages of various biopolymers. Blending Poly(L-Lactic Acid) (PLLA) with CS to produce a composite membrane is an attractive procedure for the production of new polymeric materials with controlled properties [9–15]. One more advantage of the polymer blending procedure is the production of materials exhibiting versatility, simplicity, and cost-effectiveness. The physicochemical and mechanical properties of the material which is produced after blending are dependent on the state of the mixture (i.e., solid or liquid) and the miscibility of the components.

On the other hand, the recent trend in food production processes is the development of active/bioactive packaging films using the advantages of nanotechnology [16–19]. In this way, researchers are using various methods to intrude natural antioxidants, such as essential oils [16] or natural extracts [19], in packaging films. The replacement of synthetic antioxidants (e.g., butyl-hydroxytoluene (BHT), butyl-hydroxyanisole (BHA), and tert-butyl hydroxyquinone (TBHQ) which exhibit bad effects on consumers health) with natural antioxidants in active/bioactive packaging films enhances the shelf-life of food and increases the positive environmental fingerprint.

The novel biodegradable composite active packaging film developed during this work was made with (1) PLLA, (2) CS, and (3) Basil essential Oil (BO). Modification of CS with BO via an easy green method before mixing with PLLA was the main innovation of the proposed process. This modification led to a CS/BO blend. Sequentially, this novel CS/BO blend was mixed with the PLLA polymeric matrix in four different %*w*/*w* concentrations (i.e., 5, 10, 20, and 30%*w*/*w*) and melt-extruded. The CS/BO blend and all the PLLA/CS/BO active packaging films were characterized via various techniques. More specifically, parameters such as tensile properties, water and oxygen barrier properties, water sorption, and antioxidant activity were measured for the obtained packaging films.

2. Materials and Methods

2.1. Materials

PLLA was supplied by NatureWorks, Minnetonka, MN, US, with the trade name IngeoTM Biopolymer 3052D. CS with medium molecular weight and deacetylation degree at 90% was supplied from Flurochem, Hadfield, Derbyshire, United Kingdom (cat. no. FCB051814). Basil essential oil (BO) was purchased from Esperis spa., Binda, Milano, Italy and according to the attached safety data sheets, the % mass composition was 70–80% estragole, 7.5–10% linalool, 1–3% eucalyptol, 0.5–1.0% eugenol, and 0.5–1.0% D-limonene.

2.2. Preparation Methods

2.2.1. Preparation of CS/BO Hybrid Blend

The CS/BO blend was prepared according to the adsorption/evaporation process which is described in our previous publication [20]. Briefly, 5 g of CS was spread in an aluminum beaker. A small quartz beaker was placed in the middle of the aluminum beaker and was filled with 5 g of BO. The whole "apparatus" was sealed and put in an oven at 130 °C for 24 h. Under these conditions, the volatile components of the BO were evaporated and adsorbed into the CS. Following this innovative preparation method, we successfully encapsulated the most volatile, the most active, and the less toxic fraction of extraction oils. The obtained CS/BO hybrid blend could be easily mixed as a component for active packaging film production.

2.2.2. Preparation of PLLA/CS/BO Films

PLLA/CS/BO films of 300–350 μm average thickness were developed by the melt-mixing method (see Figure 1) using a minilab twin co-rotating extruder (D516 mm, L/D524). PLLA/CS films were also prepared for comparison reasons. PLLA pellets were dried in an oven under vacuum at 98 °C for 2 h before their use. For 5 min of total melt process time, the temperature was kept stable at 170 °C, and the screw rotation speed was 100 rpm. Samples were prepared by the addition of the CS/BO hybrid blend into PLLA. The CS/BO blend composition was fixed at 5, 10, 20, and 30% *w*/*w* Different samples were also prepared by the addition of pure CS into PLLA with the same previous compositions. The melted strands were exported from the extruder machine and cut into small granules with a granulated machine. Films were produced using a hydraulic press with heated platens. Hot-pressing of approximately 1 g of the obtained granules at 110 °C under 2.0 MPa constant pressure for 3 min provided the final product. These products were kept in a desiccator under 25 °C temperature and relative humidity of 50% RH. Finally, for comparison reasons, a "blank" film of PLLA without the addition of pure CS or CS/BO was also prepared following the same procedure. The code names of all samples are listed in Table 1. The amount of PLLA and CS/BO used in the preparation stage of the PLLA/CS/BO active films as well as the process conditions at the melting extrusion stage are listed in Table 1.

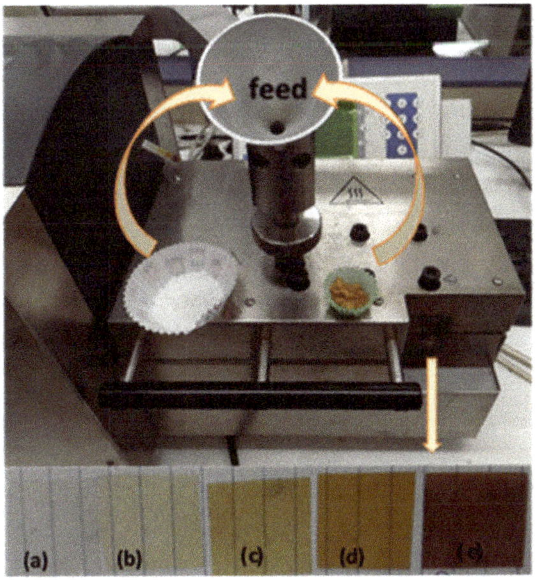

Figure 1. Scheme with the extrusion process used for the preparation of Poly(L-Lactic Acid) (PLLA)/chitosan (CS) and PLLA/CS/BO composites films: (**a**) "blank" PLLA, (**b**) PLLA/CS/BO5, (**c**) PLLA/CS/BO10, (**d**) PLLA/CS/BO20, and (**e**) PLLA/CS/BO30.

2.3. XRD Analysis

The morphological evaluation of the PLLA/CS and PLLA/CS/BO films was investigated via XRD patterns obtained using a Brüker D8 Advance X-ray diffractometer (Bruker, Analytical Instruments, S.A., Athens, Greece) equipped with a LINXEYE XE High-Resolution Energy-Dispersive detector. Typical scanning parameters were set as follows: two theta range 2–30°, increment 0.03°, and PSD 0.764.

Table 1. Code names, amounts of PLLA and CS/BO used, and extrusion processing conditions for all prepared active films.

Code Name	PLLA(g)	CS (g)	CS/BO (g)	Extruder Temperature (°C)	Extruder Rotation Speed (rpm)	Extruder Total Processing Time (min)
PLLA	5.00		-	170	150	5
PLLA/CS5	4.75	0.25	-	170	150	5
PLLA/CS10	4.50	0.50	-	170	150	5
PLLA/CS20	3.00	1.00	-	170	150	5
PLLA/CS30	3.50	1.50	-	170	150	5
PLLA/CS/BO5	4.75	-	0.25	170	150	5
PLLA/CS/BO10	4.50	-	0.50	170	150	5
PLLA/CS/BO20	3.00	-	1.00	170	150	5
PLLA/CS/BO30	3.50	-	1.50	170	150	5

2.4. FT-IR Spectrometry

The chemical structure of the PLLA/CS and PLLA/CS/BO films were investigated via IR spectra measurements. Infrared (FT-IR) spectra, which were the average of 32 scans at 2 cm^{-1} resolution, were measured using an FT/IR-6000 JASCO Fourier transform spectrometer (JASCO, Interlab, S.A., Athens, Greece). Scans were carried out in the frequency range 4000–400 cm^{-1}.

2.5. Thermogravimetric/Differential Thermal Analysis (TG-DTA)

Thermogravimetric (TGA) and differential thermal analysis (DTA) measurements were performed on PLLA/CS/BO samples. The results were compared with the same type of measurements on PLLA/CS samples. A Perkin-Elmer Pyris Diamond TGA/DTA instrument (Interlab, S.A., Athens, Greece) was used for such measurements. Samples of approximately 5 mg were heated under nitrogen atmosphere from 25 to 700 °C and with an increasing temperature rate of 5 °C/min.

2.6. Tensile Properties

Tensile measurements were carried out according to the ASTM D638 method on PLLA/CS and PLLA/CS/BO films as well as on a "blank" PLLA film (used for comparison reasons). Measurements were performed using a Simantzü AX-G 5kNt instrument (Simantzu. Asteriadis, S.A., Athens, Greece). Three to five samples of each film were tensioned at an across head speed of 2 mm/min. The samples were cut in dumb-bell shape with gauge dimensions of 10 mm × 3 mm × 0.22 mm. Force (N) and deformation (mm) were recorded during the test. Stress, stain, and modulus of elasticity were calculated based on these measurements and the gauge dimensions.

2.7. Water Sorption

Films were cut into small pieces (12 mm × 12 mm), desiccated overnight under vacuum, and weighed to determine their dry mass. The weighed films were placed in closed beakers containing 30 mL of water (pH = 7) and stored at T = 25 °C. The sorption versus time plots were developed by periodical weighting of the samples until equilibrium was reached and according to the following equation:

$$W.S. (\%) = (m_{Wet} \times m_{Dry})/m_{Dry} \times 100 \quad (1)$$

where m_{Wet} and m_{Dry} are the weights of the wet and dry films, respectively, and W.S. is the water sorption.

2.8. Water Vapor Permeability (WVTR)

Water vapor permeability of the PLLA/CS/BO films as well as of the "blank" PLLA film were determined at 38 °C and 50% RH according to the ASTM E96/E 96M-05 method. Measurements were carried out using a handmade apparatus and following the methodology described extensively in our previous publications [21–23].

2.9. Oxygen Permeability (OP)

The Oxygen Transition Rates (OTRs) of the PLLA/CS/BO films as well as of the "blank" PLLA film were measured using an oxygen permeation analyzer (8001, Systech Illinois Instruments Co., Johnsburg, IL, USA). Tests were carried out at 23 °C and 0% RH according to the ASTM D 3985 method. OTR values were measured in cc $O_2/m^2/day$. The OP values of the tested samples were calculated by multiplying the OTR values with the average film thickness, which was approximately 300–350 μm. The mean OTR value for each kind of film resulted from the measurements of three samples.

2.10. Antioxidant Activity

The antioxidant activity of films was evaluated using 500 mg of small pieces (approximately 3 mm × 3 mm) of each film. The sample was placed in a dark-colored glass bottle with a plastic screw cap and filled with 10 mL of DPPH ethanolic solution at 50 ppm concentration. After incubation for 24 h at 25 °C in darkness, the % antioxidant activity of the films was calculated according to Equation (2):

$$\%\ \text{Antioxidant activity} = (\text{Abs}_{\text{control}} \times \text{Abs}_{\text{sample}})/\text{Abs}_{\text{control}}) \times 100 \quad (2)$$

2.11. Statistical Analysis

All measurements were carried out measuring three samples at least for each kind of film. The statistical analysis was performed using the Statistical Software SPSS 20 for windows (SPSS Inc., Chicago, IL, USA). The results for mean values and standard deviation are presented below in Tables 2 and 3. A detailed discussion for the statistical analysis is presented in the Results section.

Table 2. Modulus of elasticity (E), tensile strength (σ_{uts}), and % elongation at break (ε_b) of all the tested PLLA/CS and PLLA/CS/Basil essential Oil (BO) films as well as the "blank" PLLA film.

Code Name	Tensile E (St. Dev.) (MPa)	σ_{uts} (MPa) (St. Dev.)	%ε (St. Dev.)
PLLA	2891.3(61.9)	33.9(4.9)	1.4(0.5)
PLLA/CS5	2877.3(125.3)	32.2(5.4)	1.3(0.3)
PLLA/CS10	2557.5(137.8)	28.2(6.0)	1.1(0.3)
PLLA/CS20	2006.3(225.4)	16.1(6.3)	0.8(0.2)
PLLA/CS30	1446.7(298.5)	6.1(2.4)	0.5(0.1)
PLLA/CS/BO5	2690.7(285.3)	40.3(7.9)	2.0(0.4)
PLLA/CS/BO10	2913.4(121.7)	36.3(3.7)	1.6(0.3)
PLLA/CS/BO20	3060.2(184.5)	21.5(4.0)	0.8(0.1)
PLLA/CS/BO30	3226.5(188.6)	20.9(4.4)	0.7(0.3)

Table 3. Water vapor transmission rate (WVTR), Oxygen Permeability (OP), water sorption, and antioxidant activity values of all tested PLLA/CS and PLLA/CS/BO composite films.

Code Name	WVTR (St. Dev.) ($g/m^2 \cdot day$)	OP (St. Dev.) $cm^3 \cdot mm/m^2 \cdot day$	% Water Sorption (St. Dev.)	Antioxidant Activity after 24 h (St. Dev.)
PLLA	13.80(3.5)	541.6(2.4)	1.85(0.22)	n.d.
PLLA/CS5	14.2(2.8)	538.7(2.8)	2.92(0.25)	2.6(0.9)
PLLA/CS10	16.3(3.1)	548.3(3.5)	5.04(1.65)	5.8(1.2)
PLLA/CS20	24.4(3.4)	569.5(4.3)	9.23(1.55)	11.3(1.4)
PLLA/CS30	35.1(3.8)	586.4(5.2)	26.57(2.96)	16.4(1.5)
PLLA/CS/BO5	12.74(2.5)	468.2(1.5)	1.81(0.35)	7.8(0.9)
PLLA/CS/BO10	13.54(3.2)	518.5(2.1)	1.94(1.75)	12.8(1.2)
PLLA/CS/BO20	20.22(2.8)	552.7(2.9)	6.23(1.85)	24.4(1.4)
PLLA/CS/BO30	30.41(4.2)	562.4(3.5)	20.57(2.86)	34.6(1.5)

3. Results

3.1. XRD Analysis

Figure 2 presents the XRD plots of the PLLA/CS and the PLLA/CS/BO films as well as of the "blank" PLLA film. The most intense diffraction peaks of the XRD measurement of the PLLA film is at 2 θ; values of 16.6° and 18.96°. This result is in agreement with previous reports [24,25]. By increasing the CS and CS/BO contents in the PLLA/CS and PLLA/CS/BO mixtures, the corresponding PLLA characteristic peaks of the FT-IR plots are decreased and shifted in smaller angles. Moreover, at a more careful glance, it is observed that the characteristic broad peak of CS at around 20° is increased by increasing the CS and CS/BO contents. These observations indicate effective blending of the PLLA chains with both the CS and CS/BO blends.

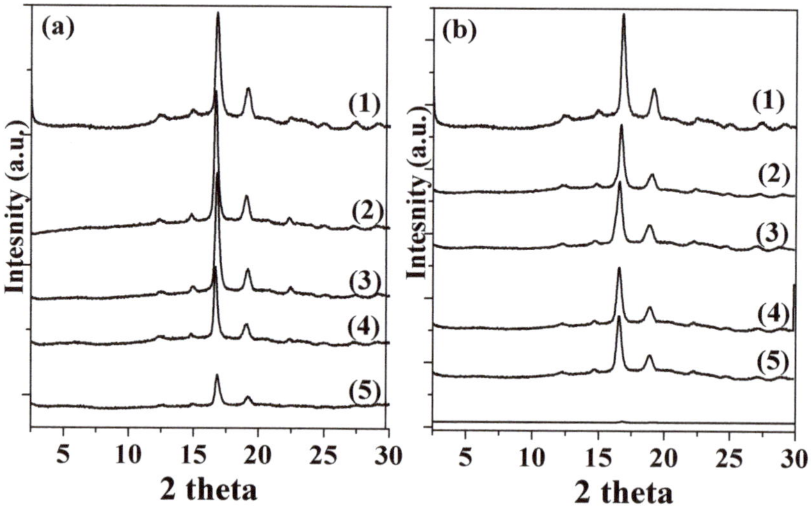

Figure 2. XRD plots of (**a**) (2) PLLA/CS5, (3) PLLA/CS10, (4) PLLA/CS20, and (5) PLLA/CS30 as well as (**b**) (2) PLLA/CS/BO5, (3) PLLA/CS/BO10, (4) PLLA/CS/BO20, and (5) PLLA/CSB/O30: in the upper part of both graphs (line (1)) is the corresponding "blank" PLLA XRD plot for comparison.

3.2. FT-IR

Figure 3 presents the FT-IR spectra of the PLLA/CS and the PLLA/CS/BO films as well as the FT-IR spectra of the "blank" PLLA film. Figure 3c presents the magnified spectra of the PLLA, PLLA/CS30, and PLLA/CS/BO30 films. The main characteristic peaks of both the PLLA and the CS material are observed in all spectra. For the PLLA/CS/BO30 film (see Figure 3c, spectrum (3)), the band at approximately 1511 cm^{-1} can be attributed to the BO vibrations [20] and indicates effective mixing of the CS/BO blend with the PLLA chains. As it is denoted with the dot lines (see Figure 3a–c), the OH stretch band of the PLLA is presented at the wavenumber value 3510 cm^{-1}. At wavenumber values of 2941 and 2993 cm^{-1}, the CH and CH_3 stretching bands are visible, whereas the characteristic band of the CO ester group is observed at 1750 cm^{-1} [26].

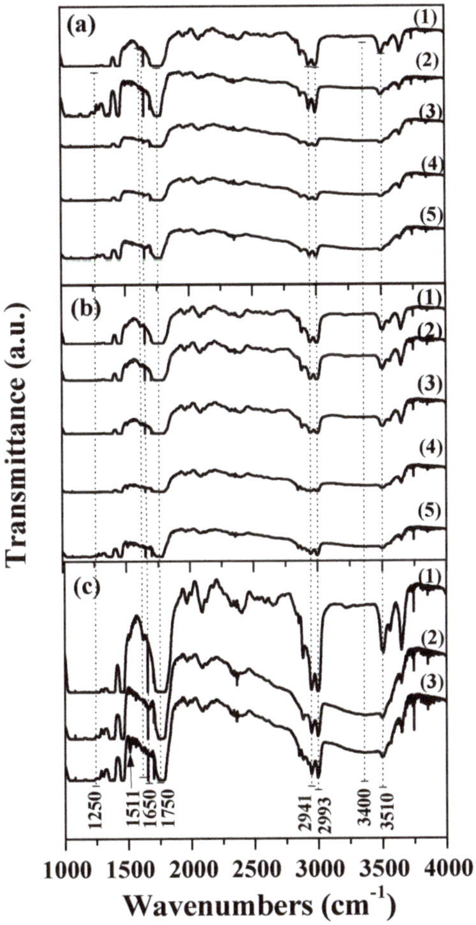

Figure 3. FT-IR spectra plots of (**a**) (1) PLLA (2) PLLA/CS5, (3) PLLA/CS10, (4) PLLA/CS20, and (5) PLLA/CS30; (**b**) (1) PLLA (2) PLLA/CS/BO5, (3) PLLA/CS/BO10, (4) PLLA/CS/BO20, and (5) PLLA/CS/BO30; and (**c**) (1) PLLA, (2) PLLA/CS30, and (3) PLLA/CS/BO30.

Furthermore, the broadband at the wavenumber value around 3400 cm^{-1} corresponds to the OH and NH stretch of CS. For the CS material, the band at the wavenumber value 1650 cm^{-1} corresponds to the amide I band. On the other hand, the amide III band [26,27] was not detected. This happens probably because of the low transmittance of the films in the wavenumber range 1050–1300 cm^{-1}. By increasing the CS and the CS/BO contents in the PLLA/CS and PLLA/CS/BO films, the characteristic peaks of the FT-IR plots, which correspond to the PLLA material, decreased. This fact indicated effective blending of the PLLA chains with the CS chains and the CS/BO blend. Furthermore, no PLLA peak shift is observed in PLLA/CS or in PLLA/CS/BO films. This fact indicates that no interaction exists between PLLA and CS chemical groups [28].

3.3. TG-DTA

Figure 4a shows the TG plots for pure PLLA and CS and for all the PLLA/CS blends. Figure 4b shows the TG plots for pure PLLA and CS/BO and for all the PLLA/CS/BO blends. Pure CS (see Figure 4a) and the CS/BO blend (see Figure 4b) exhibit two weight-loss steps. The first weight-loss step starts at around 100 °C and ends at around 200 °C and occurs due to elimination of the adsorbed moisture. The second weight-loss step starts at approximately 230 °C, ends at approximately 550 °C, and occurs because of decomposition of the CS chains [29]. Thus, the modification of CS with BO does not cause remarkable changes to the thermal behavior of the obtained CS/BO blend.

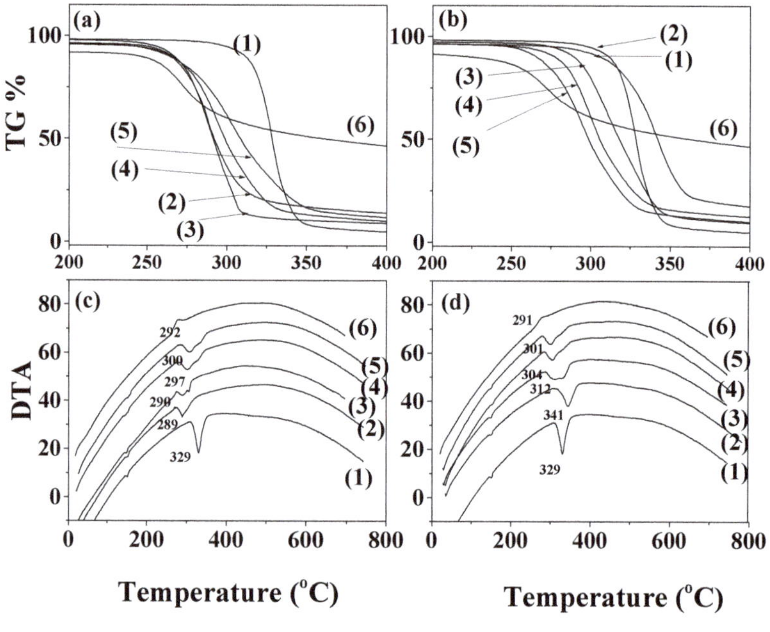

Figure 4. Thermogravimetric (TG) (**a**,**b**) and Differential Thermal Analysis (DTA) (**c**,**d**) plots of all PLLA/CS (**a**,**c**) and PLLA/CS/BO composite films as well as blank PLLA film, and pure CS and modified CS/BO powders: in Figure 4a,c (1) PLLA, (2) PLLA/CS5, (3) PLLA/CS10, (4) PLLA/CS20, (5) PLLA/CS30, and (6) CS are shown, and in Figure 4b,d, (1) PLLA, (2) PLLA/CS/BO5, (3) PLLA/CS/BO10, (4) PLLA/CS/BO20, (5) PLLA/CS/BO30, and (6) CS/BO are shown.

The TG plot of the pure PLLA (see Figure 4a,b line 1) shows one weight-loss step, which starts at around 310 °C [28,30]. Incorporation of pure CS in the PLLA chains causes a decrease in the decomposition temperature compared to that of the pure PLLA. After such mixing, the decomposition starting step occurs at lower temperature values. The same observation occurs in cases of 10%wt, 20%wt, and 30%wt CS/BO loading in the PLLA matrix. On the contrary, the incorporation of 5%wt of the CS/BO blend in the PLLA matrix shifts the decomposition temperature to higher values. These decomposition temperatures are better depicted in the DTA plots of the PLLA/CS samples (see Figure 4c) and the PLLA/CS/BO samples (see Figure 4d). According to these values, the PLLA/CS/BO films exhibit higher thermal stability than the corresponding PLLA/CS films. The PLLA/CS/BO5 sample exhibits the highest decomposition temperature value, which is also higher that of the "blank" PLLA film. Concluding, we could say that the thermogravimetric analysis experiments showed an enhancement of the thermal stability of the PLLA/CS/BO composite films compared to the thermal stability of the PLLA/CS composite films. The PLLA/CS/BO films with a composition of 5%w/w CS/BO blend in the PLLA matrix exhibited the highest thermal stability.

3.4. Tensile Properties

Figure 5a,b shows the stress–strain curves of the PLLA/CS and the PLLA/CS/BO composite films correspondingly. The average value, the standard deviation of Young's Modulus (E) as well as of the tensile strength (σ_{uts}), and the elongation at break (εb) were calculated. The results are presented in Table 2. A decrease in Young's Modulus and in tensile strength, and the elongation at break values are observed for all the PLLA/CS films compared to the respective values of the "blank" PLLA films. More specifically, as the CS content increased in the PLLA/CS films, the obtained Young's Modulus, the tensile strength, and the elongation at break values decreased further. This result is consistent with previous reports where the CS was blended with PLLA [9,11]. This happens because mixing of the thermoplastic polymer PLLA and CS is thermodynamically prohibited. Moreover, the two polymers are inherently incompatible [11,27].

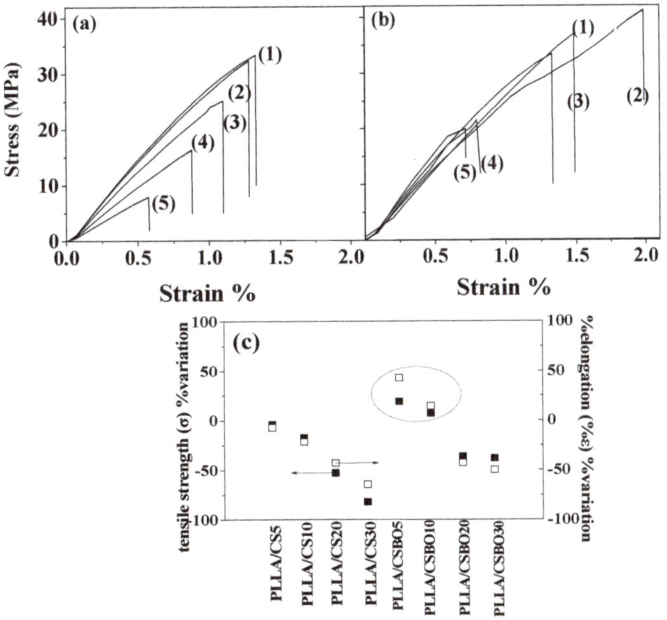

Figure 5. Strain–stress curves of (**a**) (1) PLLA, (2) PLLA/CS5, (3) PLLA/CS10, (4) PLLA/CS20, and (5) PLLA/CS30 and (**b**) (1) PLLA, (2) PLLA/CS/BO5, (3) PLLA/CS/BO10, (4) PLLA/CS/BO20, and (5) PLLA/CS/BO30, and (**c**) %variation of tensile strength (σ) values and %elongation at break—(%ε) values are for all tested PLLA/CS and PLLA/CS/BO active packaging films.

On the contrary, the modified CS/BO blend exhibited better compatibility with PLLA. For PLLA/CS/BO5 and PLLA/CS/BO10 films (see Table 2), higher values of Young's Modulus, tensile strength, and elongation at break were obtained compared to the respective values of the "blank" PLLA films. Comparing the mechanical properties of the PLLA/CS/BO20 and the PLLA/CS/BO30 films with the relevant properties of the "blank" PLLA film, Young's Modulus values are still higher but the tensile strength and the elongation at break values decreased. To the best of our knowledge, it is the first reference that PLLA/CS blends exhibit enhanced tensile strength and elongation at break values. This result is illustrated better in Figure 5c, where the %variation of tensile strength and elongation at break values of all PLLA/CS, PLLA/CS/BO, and "blank" PLLA films are plotted. From this figure, it is obvious that the CS/BO blends exhibited higher compatibility with the PLLA matrix in the cases of 5%*w*/*w* and 10%*w*/*w* nominal concentrations. This result indicates the use of the novel CS/BO blend as a compatibilizer agent.

3.5. Water Sorption

Calculated water sorption values are listed in Table 3. The addition of both CS and CS/BO in the PLLA matrixes increased the water sorption values because of the higher hydrophilicity of CS [9,11]. Hydrophobic modification of the CS with BO led to lower water sorption values for PLLA/CS/BO films as compared to the respective water sorption values of the PLLA/CS films. The lowest water sorption values were obtained for the PLLLA/CS/BO5 and PLLA/CS/BO10 samples.

3.6. Water Barrier Properties

Calculated WVTR values are shown in Table 3. The PLLA/CS films exhibited higher WVTR values than the "blank" PLLA film, which is consistent with previous studies, [9,11]. This phenomenon is caused by the greater hydrophilicity of CS, which allows the diffusion of water molecules through the film. As the CS content increased, the WVTR values of the obtained PLLA/CS films increased further. On the contrary, the PLLA/CS/BO films exhibited lower WVTR values than the PLLA/CS films. The hydrophobic behavior of the BO makes the PLLA/CS/BO5 and PLLA/CS/BO10 films show lower WVTR values compared to the "blank" PLLA film. This result indicates that the hydrophobicity of the CS/BO blend prevails over the hydrophilicity of the pure CS material. This is supported by the TG and FTIR experiments which were discussed above. Thus, the BO in the modified CS/BO blend acts as a compatibilizer between the PLLA and the CS chains. Moreover, it acts as a hydrophobic barrier agent.

3.7. Oxygen Permeability (OP)

The calculated OP values for all the tested PLLA/CS, PLLA/CS/BO, and "blank" PLLA samples are listed in Table 3. The OP values exhibited a similar trend to that of the WVTR values. The PLLA/CS/BO films exhibited lower OP values than the PLLA/CS films. This result indicates that CS modification with BO led to more effective blending of the CS/BO blend with the PLLA matrix. Furthermore, it enhances the tortuosity of paths which are followed by the oxygen molecules inside the PLLA/CS/BO composite matrix. The PLLA/CS/BO5 and PLLA/CS/BO10 samples exhibited lower OP values compared to the respective values of the PLLA film.

This result in combination with the WVTR values and the % water sorption values could support the conclusion that 5%wt and 10%*w*/*w* CS/BO loading is optimal for such PLLA/CS/BO composite films.

3.8. Antioxidant Activity

Antioxidant activity values for the tested composite films are listed in Table 3. Considering the PLLA/CS films, such values were found to increase when the CS content increased. The PLLA/CS5 material exhibited values around 7.8%, the PLLA/CS10 exhibited values around 12.8%, and the PLLA/CS20 as well as the PLLA/CS30 exhibited values around 16.4%. It is known [30] that CS inhibits reactive oxygen species and prevents lipid oxidation of food. Antioxidant activity values increased further when the CS/BO blend was added to the PLLA matrix. The antioxidant activity values in such cases were around 2.6% for PLLA/CS5, around 5.8% for PLLA/CS10, around 24.4% for the PLLA/CS20, and around 34.6% for the PLLA/CS30 sample. This result indicates successful use of the BO molecules as an antioxidant agent in the modified CS/BO blend.

3.9. Statistical Analysis of the Experimental Data

All the experimental data, i.e., E, σ_{uts}, %ε, WVP, % water sorption, OP, and % antioxidant activity after 24 h, were statistically treated using the statistical software SPSS, IBM, Chicago, IL, USA, ver. 20. More specifically, for all statistical tests, we assumed a confidence interval (C.I.) of 95%, which is the most common value used for such analyses. Thus, the value of the statistical significance level is $p = 0.05$. The results for mean values and standard deviation of the abovementioned parameters are presented in Tables 2 and 3.

Furthermore, for each one of the abovementioned properties, the hypothesis H0 (mean values could be assumed as equal) was examined for all the combinations of samples. This was performed for supporting the hypothesis that, considering different samples, every property has a statistically different mean value. The normality tests for data sets indicated that some of them cannot be assumed as normal distributions. Also, variance homogeneity tests, according to Levene's criterion, showed that, in some cases of films, we did not have homogeneity of variance. Thus, we could not use the ANOVA method for testing the mean equalities. Instead, we used the nonparametric Kruskal–Wallis method. The results are presented in Table 4. The (*Sig.*) values which are presented in this table were compared with the significance level (p). It is obvious from these comparisons that, in all cases, the (*Sig.*) values are smaller than the significance level (p) values. Thus, in all cases and for all parameters, mean values are statistically different. The smaller the (*Sig.*) value, compared to significance level value (p), the more assured that the mean values are statistically unequal. For *Sig.* = 0, the inequality of means is 100% statistically assured. For (*Sig.*) close to (p) values, the inequality of means is limitedly assured. According to the clarifications mentioned above, we developed the empirical Equation (3) for calculation of an empirical factor we call "inequality assurance" (*IA*). This factor is the percentage of deviation of (*Sig.*) from (p) values toward 0. The "inequality assurance" (*IA*) is calculated as follows:

$$IA = \frac{p - Sig.}{p} * 100 \tag{3}$$

It is obvious from Table 4 that, in all cases, the inequality of mean values is statistically assured strongly (IA $\geq$ 88%).

Table 4. Mean values inequality test of modulus of elasticity (E), tensile strength (σuts), % elongation at break (εb), water vapor permeability WVTR, % water sorption, oxygen permeability (OP), and % antioxidant activity after 24 h of all tested films.

	Sig.	*IA*
E	0.002	96
Σuts	0.005	90
%ε	0.003	94
WVTR	0.005	90
% water sorption	0.005	90
OP	0.004	92
% Antioxidant activity after 24 h	0.001	98

4. Conclusions

Following the spirit of cyclic economy in this work, we developed a fully bio-based and biodegradable active packaging film using the fully biodegradable PLLA polymer in combination with a by-product of the food industry, i.e., chitosan (CS), and a natural antioxidant, i.e., the basil-oil extract (BO). Concluding the previous analysis, we could say that the modification of CS with BO molecules via a "green" adsorption/desorption process leads successfully to a novel hydrophobic CS/BO bioactive blend which could be used as a masterbatch in extrusion molding processes. The development of PLLA/CS/BO active packaging films exhibits enhanced tensile, barrier, and antioxidant properties. The most promising films in this work were the composite PLLA films loaded with a CS/BO blend in compositions from 5%w/w to 10%w/w The last films showed enhanced tensile properties, the lowest % water sorption, enhanced water/oxygen barrier properties, and significant antioxidant activity. According to these results, the composite PLLA/CS/BO5 and PLLA/CS/BO10 films could be novel, promising, active packaging films. Moreover, the novel and promising results of this study could be used as a guide for the encapsulation of other EOs in such PLLA/CS films with more improved characteristics.

Author Contributions: Conceptualization, C.E.S., A.E.G., and M.A.K.; methodology, C.E.S., A.E.G., and M.A.K.; validation, C.E.S., A.E.G., M.A.K., M.B., and A.L.; formal analysis, C.E.S., A.E.G., M.A.K., M.B., and A.L.; investigation, C.E.S., A.E.G., M.A.K., M.B., A.L., and Z.S.; resources, C.E.S., A.E.G., M.A.K., M.B., A.L., and Z.S.; data curation, C.E.S., A.E.G., M.A.K.; M.B., A.L., and Z.S., writing—original draft preparation, C.E.S., A.E.G., and M.A.K.; writing—review and editing, C.E.S., A.E.G., and M.A.K.; visualization, C.E.S., A.E.G., and M.A.K.; supervision, C.E.S., A.E.G., and M.A.K.; and project administration, C.E.S., A.E.G., and M.A.K. All authors have read and agreed to the published version of the manuscript.

Funding: This research received no external funding.

Institutional Review Board Statement: Not applicable.

Informed Consent Statement: Not applicable.

Data Availability Statement: Not applicable.

Conflicts of Interest: The authors declare no conflict of interest.

References

1. John, R.P.; Nampoothiri, K.M.; Pandey, A. Fermentative production of lactic acid from biomass: An overview on process developments and future perspectives. *Appl. Microbiol. Biotechnol.* **2007**, *74*, 524–534. [CrossRef] [PubMed]
2. Connolly, M.; Zhang, Y.; Brown, D.M.; Ortuño, N.; Jordá-Beneyto, M.; Stone, V.; Fernandes, T.F.; Johnston, H.J. Novel polylactic acid (PLA)-organoclay nanocomposite bio-packaging for the cosmetic industry; migration studies and in vitro assessment of the dermal toxicity of migration extracts. *Polym. Degrad. Stab.* **2019**, *168*, 108938. [CrossRef]
3. Rasal, R.M.; Janorkar, A.V.; Hirt, D.E. Poly(lactic acid) modifications. *Prog. Polym. Sci.* **2010**, *35*, 338–356. [CrossRef]
4. Ghorpade, V.M.; Gennadios, A.; Hanna, M.A. Laboratory composting of extruded poly(lactic acid) sheets. *Bioresour. Technol.* **2001**, *76*, 57–61. [CrossRef]
5. Neumann, I.A.; Flores-Sahagun, T.H.S.; Ribeiro, A.M. Biodegradable poly (l-lactic acid) (PLLA) and PLLA-3-arm blend membranes: The use of PLLA-3-arm as a plasticizer. *Polym. Test.* **2017**, *60*, 84–93. [CrossRef]
6. Gerometta, M.; Rocca-Smith, J.R.; Domenek, S.; Karbowiak, T. Physical and Chemical Stability of PLA in Food Packaging. In *Reference Module in Food Science*; Elsevier: Amsterdam, The Netherlands, 2019; ISBN 978-0-08-100596-5.
7. Muzzarelli, R.A.A.; Boudrant, J.; Meyer, D.; Manno, N.; Demarchis, M.; Paoletti, M.G. Current views on fungal chitin / chitosan, human chitinases, food preservation, glucans, pectins and inulin: A tribute to Henri Braconnot, precursor of the carbohydrate polymers science, on the chitin bicentennial. *Carbohydr. Polym.* **2012**, *87*, 995–1012. [CrossRef]
8. Cazón, P.; Vázquez, M. Applications of Chitosan as Food Packaging Materials. In *Sustainable Agriculture Reviews 36: Chitin and Chitosan: Applications in Food, Agriculture, Pharmacy, Medicine and Wastewater Treatment*; Crini, G., Lichtfouse, E., Eds.; Sustainable Agriculture Reviews; Springer International Publishing: Cham, Switzerland, 2019; pp. 81–123, ISBN 978-3-030-16581-9.
9. Bonilla, J.; Fortunati, E.; Vargas, M.; Chiralt, A.; Kenny, J.M. Effects of chitosan on the physicochemical and antimicrobial properties of PLA films. *J. Food Eng.* **2013**, *119*, 236–243. [CrossRef]
10. Stoleru, E.; Dumitriu, R.P.; Munteanu, B.S.; Zaharescu, T.; Tănase, E.E.; Mitelut, A.; Ailiesei, G.-L.; Vasile, C. Novel procedure to enhance PLA surface properties by chitosan irreversible immobilization. *Appl. Surf. Sci.* **2016**, *367*, 407–417. [CrossRef]
11. Râpă, M.; Miteluţ, A.C.; Tănase, E.E.; Grosu, E.; Popescu, P.; Popa, M.E.; Rosnes, J.T.; Sivertsvik, M.; Darie-Niţă, R.N.; Vasile, C. Influence of chitosan on mechanical, thermal, barrier and antimicrobial properties of PLA-biocomposites for food packaging. *Compos. Part. B Eng.* **2016**, *102*, 112–121. [CrossRef]
12. Claro, P.I.C.; Neto, A.R.S.; Bibbo, A.C.C.; Mattoso, L.H.C.; Bastos, M.S.R.; Marconcini, J.M. Biodegradable Blends with Potential Use in Packaging: A Comparison of PLA/Chitosan and PLA/Cellulose Acetate Films. *J. Polym. Environ.* **2016**, *24*, 363–371. [CrossRef]
13. Fathima, P.E.; Panda, S.K.; Ashraf, P.M.; Varghese, T.O.; Bindu, J. Polylactic acid/chitosan films for packaging of Indian white prawn (Fenneropenaeus indicus). *Int. J. Biol. Macromol.* **2018**, *117*, 1002–1010. [CrossRef] [PubMed]
14. Ye, J.; Wang, S.; Lan, W.; Qin, W.; Liu, Y. Preparation and properties of polylactic acid-tea polyphenol-chitosan composite membranes. *Int. J. Biol. Macromol.* **2018**, *117*, 632–639. [CrossRef] [PubMed]
15. Kasirajan, S.; Umapathy, D.; Chandrasekar, C.; Aafrin, V.; Jenitapeter, M.; Udhyasooriyan, L.; Packirisamy, A.S.B.; Muthusamy, S. Preparation of poly(lactic acid) from Prosopis juliflora and incorporation of chitosan for packaging applications. *J. Biosci. Bioeng.* **2019**, *128*, 323–331. [CrossRef] [PubMed]

16. Sánchez-González, L.; Vargas, M.; González-Martínez, C.; Chiralt, A.; Cháfer, M. Use of Essential Oils in Bioactive Edible Coatings: A Review. *Food Eng. Rev.* **2011**, *3*, 1–16. [CrossRef]
17. Sanches-Silva, A.; Costa, D.; Albuquerque, T.G.; Buonocore, G.G.; Ramos, F.; Castilho, M.C.; Machado, A.V.; Costa, H.S. Trends in the use of natural antioxidants in active food packaging: A review. *Food Addit. Contam. Part. A* **2014**, *31*, 374–395. [CrossRef] [PubMed]
18. Salgado, P.R.; Di Giorgio, L.; Musso, Y.S.; Mauri, A.N. *Bioactive Packaging*; Elsevier Inc.: Amsterdam, The Netherlands, 2019; ISBN 978-0-12-814130-4.
19. Arroyo, B.J.; Santos, A.P.; de Melo, E.d.A.; Campos, A.; Lins, L.; Boyano-Orozco, L.C. Chapter 8—Bioactive Compounds and Their Potential Use as Ingredients for Food and Its Application in Food Packaging. In *Bioactive Compounds*; Campos, M.R.S., Ed.; Woodhead Publishing: Sawston, UK, 2019; pp. 143–156, ISBN 978-0-12-814774-0.
20. Giannakas, A.; Tsagkalias, I.; Achilias, D.S.; Ladavos, A. A novel method for the preparation of inorganic and organo-modified montmorillonite essential oil hybrids. *Appl. Clay Sci.* **2017**, *146*, 362–370. [CrossRef]
21. Giannakas, A.; Spanos, C.G.; Kourkoumelis, N.; Vaimakis, T.; Ladavos, A. Preparation, characterization and water barrier properties of PS/organo-montmorillonite nanocomposites. *Eur. Polym. J.* **2008**, *44*, 3915–3921. [CrossRef]
22. Giannakas, A.; Patsaoura, A.; Barkoula, N.-M.; Ladavos, A. A novel solution blending method for using olive oil and corn oil as plasticizers in chitosan based organoclay nanocomposites. *Carbohydr. Polym.* **2017**, *157*, 550–557. [CrossRef]
23. Grigoriadi, K.; Giannakas, A.; Ladavos, A.K.; Barkoula, N.-M. Interplay between processing and performance in chitosan-based clay nanocomposite films. *Polymer Bull.* **2015**, *72*. [CrossRef]
24. Wasanasuk, K.; Tashiro, K.; Hanesaka, M.; Ohhara, T.; Kurihara, K.; Kuroki, R.; Tamada, T.; Ozeki, T.; Kanamoto, T. Crystal Structure Analysis of Poly(l-lactic Acid) α Form On the basis of the 2-Dimensional Wide-Angle Synchrotron X-ray and Neutron Diffraction Measurements. *Macromolecules* **2011**, *44*, 6441–6452. [CrossRef]
25. Hosen, M.S.; Rahaman, M.H.; Gafur, M.A.; Habib, R.; Qadir, M.R. Preparation and Characterization of Poly(L-lactic acid)/Chitosan/Microcrystalline Cellulose Blends. *Chem. Sci. Int. J.* **2017**, 1–10. [CrossRef]
26. Duarte, A.R.C.; Mano, J.F.; Reis, R.L. Novel 3D scaffolds of chitosan–PLLA blends for tissue engineering applications: Preparation and characterization. *J. Supercrit. Fluids* **2010**, *54*, 282–289. [CrossRef]
27. Giannakas, A.; Salmas, C.; Leontiou, A.; Tsimogiannis, D.; Oreopoulou, A.; Braouhli, J. Novel LDPE/Chitosan Rosemary and Melissa Extract Nanostructured Active Packaging Films. *Nanomaterials* **2019**, *9*, 1105. [CrossRef] [PubMed]
28. Sunilkumar, M.; Francis, T.; Thachil, E.T.; Sujith, A. Low density polyethylene–chitosan composites: A study based on biodegradation. *Chem. Eng. J.* **2012**, *204*, 114–124. [CrossRef]
29. Prasanna, K.; Sailaja, R.R.N. Blends of LDPE/chitosan using epoxy-functionalized LDPE as compatibilizer. *J. Appl. Polym. Sci.* **2012**, *124*, 3264–3275. [CrossRef]
30. Rajalakshmi, A.; Krithiga, N.; Jayachitra, A. Antioxidant Activity of the Chitosan Extracted from Shrimp Exoskeleton. *Middle East. J. Sci. Res.* **2013**, *16*, 1446–1451. [CrossRef]

processes

MDPI

Article

Hydroxypropyl Methylcellulose-Based Hydrogel Copolymeric for Controlled Delivery of Galantamine Hydrobromide in Dementia

Sidra Bashir [1], Nadiah Zafar [1,*], Noureddine Lebaz [2], Asif Mahmood [1] and Abdelhamid Elaissari [2,*]

1 Faculty of Pharmacy, The University of Lahore, Punjab 54000, Pakistan; sidrabashir100@gmail.com (S.B.); asif.mahmood@pharm.uol.edu.pk (A.M.)
2 Univ Lyon, Université Claude Bernard Lyon 1, CNRS, LAGEPP UMR 5007, F-69100 Villeurbanne, France; noureddine.lebaz@univ-lyon1.fr
* Correspondence: nadiah.zafar@pharm.uol.edu.pk (N.Z.); abdelhamid.elaissari@univ-lyon1.fr (A.E.)

Received: 20 September 2020; Accepted: 23 October 2020; Published: 25 October 2020

check for updates

Abstract: The study aims to prepare a smart copolymeric for controlled delivery of Galantamine hydrobromide. The synthesis of the hydrogel was executed through free radical polymerization using HPMC (Hydroxypropyl methylcellulose) and pectin as polymers and acrylic acid as monomer. Cross-linking was performed by methylene bisacrylamide (MBA). HPMC-pectin-co-acrylic acid hydrogel was loaded with Galantamine hydrobromide (antidementia drug) as a model drug for treatment of Alzheimer based dementia. Formulated hydrogels (SN1–SN9) were characterized for Fourier transform-infrared spectroscopy, differential scanning calorimetry, thermogravimetric analysis, X-ray diffraction, and energy dispersive X-ray. Drug loading efficiency, gel fraction, measurements of porosity, and tensile strength were reported. Swelling and release studies were performed at pH 1.2 and 7.4. Drug liberation mechanism was evaluated by applying different release kinetic models. Galantamine hydrobromide was released from prepared hydrogels by Fickian release mechanism. Swelling, gel fraction, porosity, and drug release percentages were found to be dependent on hydroxypropyl methylcellulose, pectin, acrylic acid, and methylene bisacrylamide concentrations. By increasing HPMC amount, swelling was increased from 76.7% to 95.9%. Toxicity studies were conducted on albino male rabbits for a period of 14 days. Hematological and histopathological studies were carried out to evaluate safety level of hydrogel. Successfully prepared HPMC-pectin-co-acrylic acid hydrogel showed good swelling and release kinetics, which may help greatly in providing controlled release drug effect leading to enhanced patient compliance for dementia patients.

Keywords: HPMC; galantamine hydrobromide (GH); pectin; hydrogel; methylene bisacrylamide; dementia

1. Introduction

Innovative approaches and advancements in biomedical science help in providing improved drug delivery systems. Formulation of advanced and productive drug delivery systems are a necessity to provide better treatment of different diseases [1]. Oral route is the most acceptable route of drug delivery for the greatest number of patients mainly because of its low cost compared to other dosage forms. Moreover, it is a safe route with an effective outcome and also oral dosage forms are easily ingested. Conventional immediate release oral dosage forms require frequent daily dosing leading to fluctuation of drug level in plasma, which imparts a negative effect on the drug efficacy. Controlled release (CR) tablets are developed not only to surmount the demerits of older oral dosage forms but also to provide other useful advantages [2]. CR systems maintain steady state concentration of drug in plasma and thereby reduce the frequency of daily dose. CR tablets usually face the problem of dose dumping when they undergo disintegration process, which results in severe toxicity. Such demerits are usually overcome by using diverse polymers in controlled release systems. Utilization of polymers has highly impacted the advancement of modern medicine and has led towards increased patient compliance [3,4]. Biocompatible polymers are widely employed in controlled release systems, which can release drug over long intervals of time. Polymeric based drug carriers display lesser side effects and also provide potential for site specific drug delivery [5].

Polymers that display changes in their physical characteristics when external stimuli are applied are termed as stimuli-responsive polymers [6]. Such polymers can be employed to achieve any particular targeted area drug release [7]. They are also employed in hydrogels fabrication [8,9]. Hydrogels are 3D polymeric cross-linked network like structures displaying high hydrophilicity. They are also termed as smart or intelligent systems as they are engineered to be sensitive to environmental stimuli. Hydrogels are widely applicable in diverse biomedical, environmental, and pharmaceutical fields, such as tissue engineering, contact lens, immune-therapy, and most importantly in controlled drug release. Their attractive feature of displaying swelling behavior and releasing entrapped drug molecules upon change in pH and other stimuli such as temperature, light, pressure, and electric field makes them compatible for usage in controlled and targeted drug delivery systems [10]. Free radical polymerization is a widely used method for cross-linking the hydrogels structure [11,12].

Hydroxypropyl methylcellulose (HPMC) is a cellulose derivative, which is water soluble in nature. It is widely employed in many industries like food, plastic, and most importantly, the pharmaceutical industry. Many grades of HPMC are used in formulation of several oral dosage forms. HPMC finds its diverse applications in oral dosages due to its biocompatibility, biodegradability, hydrophilicity, and swelling ability [13].

Pectin is a naturally occurring polysaccharide having a complex structure with hydrophilic characteristic. Pectin is nontoxic, biocompatible, and economic in cost, which is why it has many applications in biotechnology and pharmaceuticals. It has been generally used as stabilizer and gelling agent as well as polymer in hydrogel formation [14].

Galantamine hydrobromide (GH) is an antidementia drug, which is used in treatment of Alzheimer based dementia. Because of its short half-life of 7 h, it is administrated two times a day. Dementia is a neurodegenerative syndrome related to memory loss, mental confusion, and motor dysfunctions. The pharmacological class of Galantamine hydrobromide is cholinesterase inhibitor. It is a reversible competitive inhibitor of acetylcholine esterase. It is metabolized by cytochrome P450, 2D6, and 3A4 enzymes. Drugs that are metabolized through these enzymes affect the pharmacokinetics of Galantamine hydrobromide if they are coadministered [15].

Hydrogels are three dimensional networks that have the ability to absorb a large amount water. These have resemblance to natural tissues because of their soft, elastic nature and low interfacial tension. These are widely employed in the pharmaceutical, agriculture, and medical sector because of their tunable properties, biocompatibility, and gel forming ability. These intelligent networks can be prepared by a number of techniques, but the most widely used approach is aqueous free radical polymerization [16].

In this study, MBA cross-linked HPMC-pectin-co-acrylic acid hydrogels were fabricated by free radical polymerization for controlled delivery of Galantamine hydrobromide. Controlled release oral hydrogels of Galantamine hydrobromide can give a better and more efficient way of administrating the therapeutic amount of drug as a single dose, thereby reducing the frequency of dosing. Such a system shall maintain the steady state drug plasma levels, which would be beneficial for dementia patients as they are much more reluctant to take their medications and also often forget to take their medicines because of their mental condition. In this study, nine hydrogels formulations were prepared by varying the amount of polymers, monomer, and cross-linker. The fabricated hydrogels were evaluated by different characterization tests, and their in vitro drug release profile was studied.

2. Materials and Methods

2.1. Materials

Hydroxypropyl methylcellulose (HPMC E-5), pectin, and acrylic acid were purchased from Sigma Aldrich (Darmstadt, Germany). Methylene bis-acrylamide (MBA), ammonium persulfate (APS), potassium phosphate (monobasic), hydrochloric acid, and ethanol were obtained from Merck-(Darmstadt, Germany). Galantamine hydrobromide was received as a gift from Reko pharmaceutical Multan road Lahore (Pakistan).

2.2. Methods

HPMC-pectin-co-acrylic acid hydrogels were formulated through free radical polymerization. A total of 0.5 g of HPMC was added slowly to already warm 25 mL of distilled water. It was stirred until formation of clear solution A. A total of 0.5 g of pectin was added in small divided portions to 10–15 mL distilled water to get clear solution B. Solution A and B were mixed together under continuous stirring to obtain solution C. At the same time, solutions of ammonium per sulfate (APS) and methylene bisacrylamide (MBA) were also prepared separately. A total of 0.2 g of APS was mixed in approximately 5 mL of distilled water. A total of 0.3 g of MBA was added into water to get a clear solution.

Approximately 15 mL of acrylic acid were added dropwise into already prepared solution C under continuous stirring. Next, addition of APS solution followed by addition of MBA solution into solution C. The prepared solution was kept under stirring for 5–10 min followed by sonication for 3–5 min at 5000 rpm. Sonication was done to eliminate any dissolved oxygen. Test tubes were filled with hydrogel solution, which were sealed with aluminum foil and again sonicated to ensure complete removal of dissolved oxygen. The test tubes were arranged in test tube stand, and this stand was placed in water bath for heating. Firstly, the temperature was set at 45 °C for one hour. Later on, the temperature of the bath was increased in the following sequence: 50 °C for two hours, 55 °C for three hours, 60 °C for four hours, and 65 °C for the next twelve hours. During the above-mentioned time period, free radical polymerization was completed, leading to formation of hydrogels. After twenty-four hours, the test tube stand was removed from bath. Tubes were cooled down at room temperature followed by their careful breakage to recover the prepared hydrogels. Hydrogels were cut by sharp edged blade into discs of specific dimensions. The hydrogel discs were washed with ethanol: water (50:50) solution in order to remove any unreacted species. Discs were transferred into petri dishes and dried at room temperature. Different combinations were prepared by varying quantities of HPMC, pectin, acrylic acid, and MBA (Table 1). Dried hydrogels were stored for further characterization and evaluation process.

Table 1. Composition of Hydroxypropyl methylcellulose (HPMC)-pectin-co-acrylic acid hydrogels formulations (SN1–SN9).

Formulation Codes	HPMC (g)	Pectin (g)	MBA (g)	Acrylic Acid (mL)	APS (g)
SN1	0.5	0.5	0.3	15	0.2
SN2	0.5	0.5	0.5	15	0.2
SN3	0.5	0.5	0.7	15	0.2
SN4	0.5	1	0.3	15	0.2
SN5	0.5	1.5	0.3	15	0.2
SN6	1	0.5	0.3	15	0.2
SN7	1.5	0.5	0.3	15	0.2
SN8	0.5	0.5	0.3	17	0.2
SN9	0.5	0.5	0.3	19	0.2

2.2.1. Drug Loading and Entrapment Efficiency

Drug loading efficiency of all hydrogel formulations was calculated. Drug was loaded into hydrogels through post loading method in which the hydrogel discs were soaked into 1%*w*/*v* drug solution for approximately 1 week. After optimum swelling, discs were removed from the drug solution. Blotting paper was used to remove excess solvent from their surface. Firstly, these swollen discs were dried at room temperature and then dried in an oven at 40–45 °C for one week or complete drying [16]. Drug loading was determined through the equation below:

$$Drug\ Loading\ (\%) = \left(\frac{W_d - W_0}{W_d}\right) \times 100 \quad (1)$$

where W_0 and W_d refer to hydrogel weight before and after soaking, respectively.

2.2.2. Sol-Gel Fraction

Gel fraction of hydrogels was determined. Soxhlet extractor was used in which small dried hydrogel preweighed pieces were placed with addition of deionized water. Temperature was set at 100 °C for 4 h. After extraction of uncross-linked polymer, the hydrogel pieces were removed from Soxhlet apparatus and dried at 40 °C in an oven [16]. Weight of dried hydrogel pieces was noted. Gel percentage was determined by using the following equations:

$$Gel\ fraction\ (\%) = \left(\frac{W_0 - W_i}{W_0}\right) \times 100 \quad (2)$$

$$Sol\ fraction\ (\%) = 100 - gel\ fraction \quad (3)$$

where W_0 is initial weight of hydrogel pieces before extraction and W_i is weight of dried hydrogel pieces after extraction.

2.2.3. Fourier Transform Infrared (FTIR) Spectroscopy

FTIR analysis of all ingredients, i.e., HPMC, pectin, acrylic acid, MBA, Galantamine hydrobromide, and fabricated hydrogels was performed for confirmation of the identity and compatibility of ingredients. Hydrogel discs were crushed and placed on crystal mark of FTIR apparatus. Little force was applied on the sample to make a thin film for FTIR analysis. FTIR spectra of fabricated hydrogels and all employed materials were recorded at 500–4000 cm^{-1} by using FTIR spectrophotometer (Bruker Tensor 27 series, Bremen, Germany) [17].

2.2.4. Differential Scanning Calorimetry (DSC)

DSC test was performed to evaluate the thermal stability of hydrogel formulation over a specific temperature range. Glass transition temperature (T_g) and nature of crystallization were evaluated in DSC, which was performed using TA instrument (West, Sussex, UK). A 10 mg sample was placed on aluminum pans. DSC was performed at heating range of 0–500 °C at heating rate of 20 °C/min using constant nitrogen flow of 20 mL/min [18].

2.2.5. Thermogravimetric Analysis (TGA)

TGA was also carried out to check thermal stability in terms of weight loss percentage of sample by providing heat flow, which reflects the physical and chemical nature of the sample. For TGA analysis, 10mg samples were analyzed using TA instrument SDT Q600 (West, UK). Grounded samples were subjected to a temperature range of 0–500 °C and the heating rate was 10 °C/min under a nitrogen purge (25 mL/min) [19].

2.2.6. Powered X-ray Diffraction (PXRD)

XRD was performed for determining the crystalline or amorphous nature of the samples (Galantamine hydrobromide, HPMC, pectin, and prepared hydrogel) at room temperature under a voltage of 40 kV by using an X-ray diffractometer (Siemens model D500, Cu Kα radiation). Diffractogram was scanned at 2θ angle range of 60–50° at speed of 2°/min [20].

2.2.7. Energy Dispersive X-ray (EDX) Spectroscopy

EDX was performed for identification of elemental composition of prepared hydrogels. Atoms of each element emit specific X-rays, which were used for identification of elements of each sample. An EDX analyzer (Model EX-400; Japan) was used for recording EDX peaks of Galantamine hydrobromide, HPMC-pectin-co-acrylic acid hydrogel and Galantamine hydrobromide loaded HPMC-pectin-co-acrylic acid hydrogel [21].

2.2.8. Porosity Test

To find the percentage of porosity, weight of hydrogel discs was noted and then dipped into absolute ethanol overnight. Weight of disc was noted again after removing excess ethanol from discs. Porosity % test was performed for all hydrogel formulations and calculated as:

$$Porosity\ (\%) = \left(\frac{M_2 - M_1}{\rho V}\right) \times 100 \tag{4}$$

where M_1 and M_2 are hydrogel mass before and after dipping into absolute ethanol respectively, ρ is the absolute ethanol density, and V is the hydrogel disc volume.

2.2.9. Mechanical Strength

Mechanical strength test was performed by using universal testing machine (UTM) (Instron 3367, Norwood, MA, USA). Freshly prepared hydrogel was placed in UTM jaws and Young's modulus and force in Newton (N) was noted at a point where hydrogel was broken [22].

2.2.10. Swelling Studies

Swelling studies were performed at two pH values, i.e., at 1.2 and 7.4. Dried hydrogel discs of all formulations were weighed and dipped into 200 mL of respective pH buffer solutions. After specified time intervals ranging from 0.5 to 72 h, the discs were taken out from buffer solutions and placed onto filter paper for absorption of excessive buffer. Their weight was noted until constant weight was reached. This process was carried out for all hydrogel formulations. Swelling study was conducted for three consecutive days, and percentage was determined through the following formula:

$$\%\ of\ equilibrium\ swelling = \left(\frac{W_s - W_i}{W_s}\right) \times 100 \tag{5}$$

where W_s is the weight of swollen disc at a specific time interval and W_i is its dry weight.

2.2.11. Galantamine Hydrobromide Release Study

Release study of Galantamine hydrobromide from HPMC-pectin-co-acrylic acid hydrogel was performed at pH 1.2 and 7.4. Dissolution apparatus II (USP) was used at 50 rpm, and a temperature of 37 °C was maintained. Drug loaded hydrogel discs were placed in 900 mL of buffer of respective pH, and samples were withdrawn after definite time intervals ranging from 0.5 to 36 h. Absorbance was measured by UV-spectrophotometer (UV-1800, Kyoto, Japan) at 287 nm [16].

2.2.12. Release Kinetics

Data obtained from in vitro drug release process were evaluated to determine the corresponding kinetic model. Dissolution profile data can be interpreted by some mathematical functions, and each function has its own features [23].

Zero order kinetics:

$$Q_t = Q_o - K_o t \quad (6)$$

First order kinetics:

$$LnQ_t = LnQ_o - K_1 t \quad (7)$$

Higuchi Kinetics:

$$Q_t = K_h t^{1/2} \quad (8)$$

Q_o: Initial amount of drug in hydrogel microparticles,
Q_t: Amount of drug after time 't'
K_o, K_1, and K_h are rate constants.
Korsemeyer–Peppas kinetics:

$$\frac{M_t}{M_{inf}} = K_p t^n \quad (9)$$

2.2.13. Acute Oral Toxicity Studies

Toxicity study was conducted as per organization for economic cooperation and development (OECD) guidelines on rabbits for examining the safety profile of fabricated HPMC-pectin-co-acrylic acid hydrogels. All the study protocols were reviewed and approved by Institutional Research Ethics Committee of Faculty of Pharmacy, The University of Lahore notification no. IREC-2019-100. Twelve albino male rabbits were divided into two groups named as controlled and tested groups. Each rabbit in both groups has a weight ranging from 1.5 to 2 kg. Healthy food and water were supplied to both groups on a daily basis. The tested group was given a dose of crushed carrier system according to dose of 2 g/kg. Weight and physical activities were checked on daily basis for 2 weeks. After this time, blood samples were taken from the two groups for different biochemical analysis tests. Then they were sacrificed to remove vital organs, i.e., lungs, brain, heart, liver, small intestine, stomach, and spleen for histopathological examination [24].

3. Results and Discussion

3.1. FTIR Analysis

Pectin IR peaks are illustrated in Figure 1a. The spectrum has prominent and evident peaks at 3363, 1720, 1602, and at 1224 cm^{-1} representing –OH stretch, stretching of ester carbonyl group (>Carbon=Oxygen), Carbon=Carbon, and Carbon–Oxygen stretching vibration, respectively. Pectin spectrum is confirmed by a literature study conducted by Rehmani et al. [25], who developed semi-interpenetrating hydrogel for delivery of active agents in control kinetics. They studied FTIR spectrum of pectin and found characteristic peaks at 3368, 1737, 1227, and at 1605 cm^{-1}.

FTIR spectrum of HPMC E-5 shown in Figure 1b, displays characteristics peaks at 3519, 2867, 1386, and 1045 cm^{-1}, revealing the presence of hydroxyl group (OH) stretching, C–H group, vibration of hydroxyl group (OH), and C–O group stretching, respectively. Spectrum of HPMC is confirmed through a literature study performed by Akhlaq et al. [26]. They observed characteristic peaks at 3421.48, 2891.10, and at 1377 cm^{-1}.

Galantamine hydrobromide FTIR spectrum illustrated in Figure 1c shows a prominent peak at 3561.87 cm^{-1} because of O–H stretching of enol group (C=C–O–H). Another characteristic peak occurring at 1623.769 cm^{-1} represented N–H bending. Peaks at 1141.625 and 600–700 cm^{-1} represent C–N stretch and C–Br bond, respectively. Hanafy et al. [27] have prepared Galantamine hydrobromide loaded chitosan nanoparticles administered through nasal routes for Alzheimer's disease. FTIR analysis showed characteristic peaks at 3559 cm^{-1} for O–H stretching.

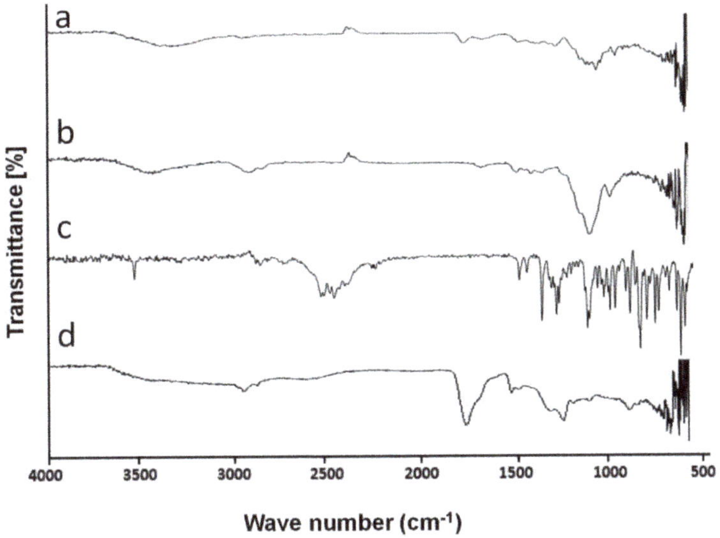

Figure 1. FTIR spectrum of (**a**) Pectin, (**b**) HPMC, (**c**) Galantamine hydrobromide, and (**d**) hydrogel formulation.

Peaks in IR spectrum of acrylic acid appearing at 1634.74 and 1705.45 cm^{-1} represented Carbon=Carbon and C–O vibrations. These results are verified through literature in a study conducted by Feng et al. [28] who prepared pH-sensitive particles based on polymeric systems for controlled delivery of rebar inhibitor. They observed FTIR peaks of acrylic acid appearing at 1626 and 1707 cm^{-1}.

Infrared spectrum of MBA given showed characteristic peaks at 3308.1, 1666.2, 1530.24, 1408.77, 1379.45, and 1220.25 cm^{-1}. These peaks are caused by the presence of Carbon=Oxygen bond and Nitrogen–Hydrogen groups. Similar results are reported by Ibrahim [29] who prepared grafted polyacrylamide/chitosan-based hydrogel. They also found IR peaks of MBA at 3305.05, 1656.84, 1537.12, 1382.45, and 1224.89 cm^{-1}.

IR Spectrum of unloaded formulation of HPMC-pectin-co-acrylic acid hydrogel is shown in Figure 1d. It shows peaks at 786, 1157, 1444, 1693, and 2915 cm^{-1}, respectively. There is an appearance of new peaks, which confirmed the formation of hydrogel structure.

3.2. Swelling Studies

3.2.1. Influence of MBA on Swelling Percentage

Swelling studies of HPMC-pectin-co-acrylic acid hydrogel formulations were conducted at pH 7.4 and pH 1.2 for 72 h, which is the equilibrium stage of swelling for these hydrogels. Figure 2a shows that by increasing MBA concentration from 0.5 to 0.7 g, swelling percentage was decreased from 84.62 to 48.25% at pH 7.4. A higher cross-linker amount resulted in formation of highly interlaced and dense network having high crosslinking density, which displayed low swelling behavior. At pH 1.2, the swelling percentage was only up to 10% because acrylic acid is a pH responsive monomer and it undergoes ionization only at pH 7.4 and remains in unionized state at pH 1.2, thereby no chain relaxation in terms of swelling was observed.

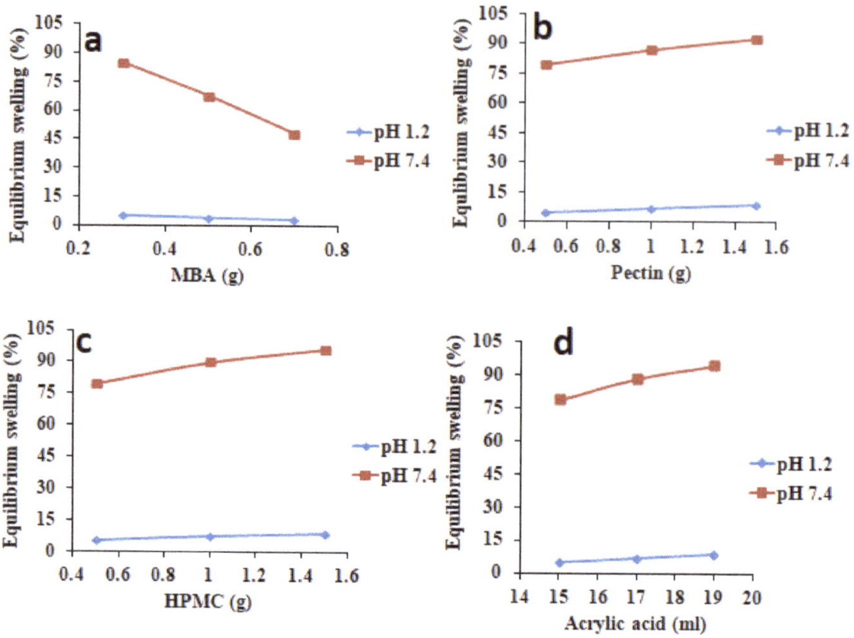

Figure 2. Influence of variable amounts of (**a**) methylene bisacrylamide (MBA), (**b**) pectin, (**c**) HPMC, and (**d**) acrylic acid on swelling percentage of fabricated hydrogels.

3.2.2. Influence of Pectin on Swelling Percentage

Figure 2b shows swelling behavior at pH 7.4 of formulations in which pectin amount was increased from 0.5 to 1.5 g. Results had revealed that if higher pectin quantity was used, swelling percentage was also increased from 79.58 to 92.62% at pH 7.4. There were two main reasons behind this increased swelling pattern. Firstly, pectin contains carboxymethyl groups (–$COOCH_3$), which has greater tendency for ionization at pH 7.4. When (–$COOCH_3$) groups become ionized, electrostatic repulsions occur between them. Such repulsions cause expansion of hydrogel network. Secondly, high porosity was observed by increasing pectin amount at pH 7.4. So, due to high porosity, more solvent influx occurred into the hydrogel resulting in a high swelling percentage. At pH 1.2, it was observed that swelling percentage was up to 10%, which was considered as a negligible result because at this pH, carboxymethyl groups (–$COOCH_3$) do not ionize, resulting in low to negligible swelling.

3.2.3. Influence of HPMC on Swelling Percentage

By increasing HPMC amount from 0.5 to 1.5 g (Figure 2c), swelling percentage was increased from 76.68% to 95.89% at pH 7.4. Swelling was increased because of the hydrophilic nature of the polymer, which in turn showed excellent swelling potentials. HPMC contains numerous hydroxyl group (–OH), which promote swelling at pH 7.4. At pH 1.2, swelling percentage was observed up to 10%, which was considered as negligible. Ionization of hydroxyl group (–OH) does not occur at pH 1.2, leading to negligible swelling.

3.2.4. Influence of Acrylic Acid on Swelling Percentage

Figure 2d shows swelling of formulations in which the amount of acrylic acid was increased gradually from 15 to 19 mL. Swelling percentage range was 78.69–94.38% at pH 7.4. Results had revealed that increased swelling was seen at basic pH 7.4. Ionized carboxylic groups of acrylic acid had triggered network swelling at higher pH, i.e., 7.4. So, by increasing acrylic acid amount, more carboxylic groups were obtained in ionized form at pH 7.4. At pH 1.2, very low swelling percentages were observed, which was considered as negligible, as carboxylic groups do not ionize at acidic pH.

3.3. Drug Loading Efficiency

Drug loading efficiency was established to find out the drug percentage entrapped in hydrogel discs. Impact of MBA, pectin, HPMC, and acrylic acid contents on drug loading efficiency is also discussed below.

3.3.1. Influence of Cross-Linker on Drug Loading Efficiency

It was found that when concentration of cross-linker (MBA) was increased from 0.3 to 0.7 g, entrapment efficiency was decreased from 89.59% to 64.25%, respectively (Figure 3a). The reason behind this low entrapment efficiency was because of the decreasing space availability between polymeric chains as the concentration of MBA increased. This compactness between two polymer chains was due to high cross-linker contents. Extremely cross-linked hydrogels cannot expand much (low swelling). Hence, hydrogels containing higher concentration of cross-linker displayed poor swelling and less Galantamine hydrobromide loading.

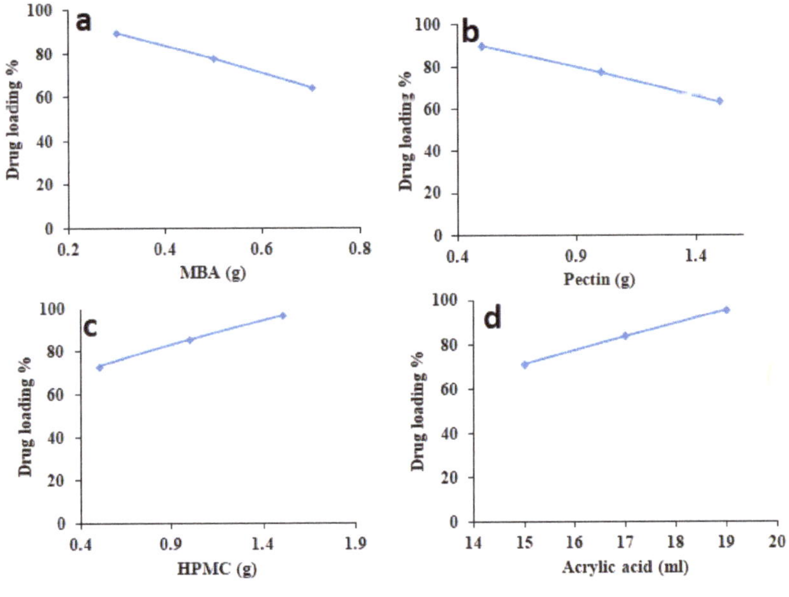

Figure 3. Effect of amount (**a**) MBA, (**b**) pectin, (**c**) HPMC, and (**d**) acrylic acid on drug loading efficiency.

3.3.2. Influence of Pectin on Drug Loading Efficiency

The effect of increasing pectin concentration on drug loading efficiency was also investigated (Figure 3b). It was found that loading efficiency of Galantamine hydrobromide was decreased from 90% to 63.51% as concentration of pectin was increased from 0.5 to 1.5 g. Generally, it is observed that hydrogels displaying maximum/more swelling also display maximum/more drug loading efficiency, but such findings were found to be not applicable in the case of pectin. By increasing pectin concentration, even though swelling was increased, drug loading efficiency was decreased. Higher contents of pectin result in a highly viscous diffusion control film/coat that offers resistance in transport of drug models into the developed network. Moreover, the size of Galantamine hydrobromide molecule is larger than water molecules, hence low drug loading was observed.

3.3.3. Influence of HPMC on Drug Loading Efficiency

When concentration of HMPC was increased from 0.5 to 1.5 g, higher drug loading efficiency of Galantamine hydrobromide was observed (Figure 3c). Drug loading efficiency was increased from 73.25% to 97.15% due to more hydrophilic character and high swelling capability of HPMC.

3.3.4. Influence of Acrylic Acid on Drug Loading Efficiency

Similarly, when the amount of acrylic acid was increased from 15 to 19 mL, drug loading efficiency was increased from 71.26% to 95.64%, as represented in Figure 3d. Such higher entrapment efficiency was because of higher swelling capability of acrylic acid at pH 7.4.

3.4. Sol-Gel Fraction

Influence of variable amounts of MBA, pectin, HPMC, and acrylic acid on sol-gel fraction was investigated, and the results are reported in Figure 4. It was found that increased amount of MBA, pectin, HPMC, and acrylic acid caused increased gel fraction. Figure 4a shows that when MBA amount was increased from 0.3 to 0.7 g, gel fraction was also proportionally increased due to more physical entanglement between polymeric chains leading to eventual more gel formation (83.56–94.25%). When the pectin amount was increased from 0.5 to 1.5 g, gel fraction was also increased from 82.61% to 92.36%, as illustrated in Figure 4b. Gelling ability of pectin is the main reason of this increased gel fraction by increasing pectin concentration. Figure 4c shows that when HPMC amount was increased from 0.5g to 1.5g, percentage of gel increases from 84.89 to 92.86%. Lastly, results also showed that by increasing acrylic acid concentration, gel fraction was increased from 81.97 to 93.36% (Figure 4d) because physical entanglement increased by increasing the monomer (acrylic acid) amount.

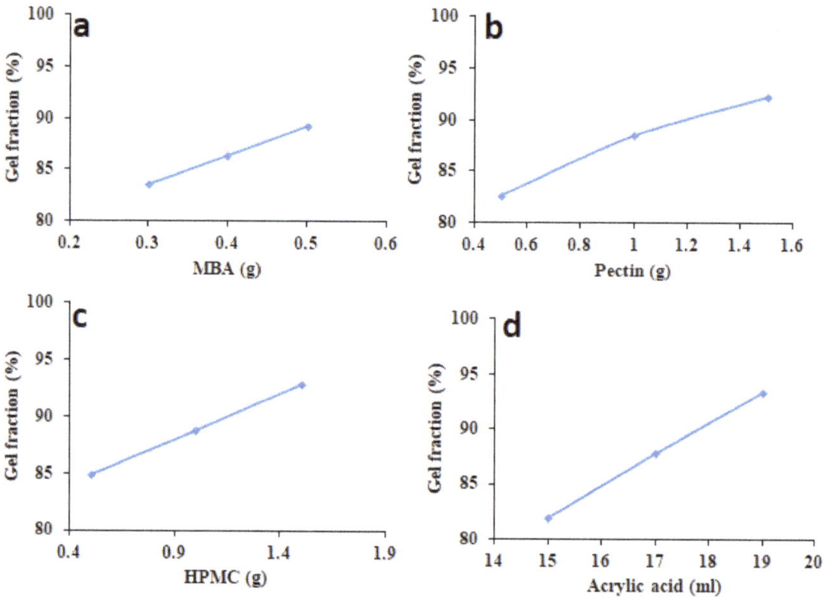

Figure 4. Influence of variable amounts of (**a**) MBA, (**b**) pectin, (**c**) HPMC, and (**d**) acrylic acid on gel fraction of HPMC-pectin-co-acrylic acid hydrogel.

3.5. Thermal Analysis

Thermal analysis is a powerful tool that gives us information about thermal stability of drug, pectin, and prepared hydrogel. DSC and TGA investigations were performed at 0–500 °C range. Figure 5 presents

DSC thermograms of Galantamine hydrobromide, pectin, and fabricated hydrogel. Figure 5a displays a small endothermic peak at 280 °C, which represents the melting point of Galantamine hydrobromide. A sharp exothermic peak was shown at 295 °C due to degradation of drug molecules. DSC thermogram of pectin (Figure 5b) presents first endothermic peak near 110 °C, which is T_g of pectin and is due to moisture content. One more endothermic peak is observed at 185 °C due to its melting point. Exothermic peak at 275 °C was due to complete degradation. Mittal and Kaur [30] reported similar results of thermal analysis of pectin. Thermogram of HPMC-pectin-co-acrylic acid hydrogel is given in Figure 5c. DSC results show that the first endothermic peak occurred at 200 °C and the second one at 285 °C.

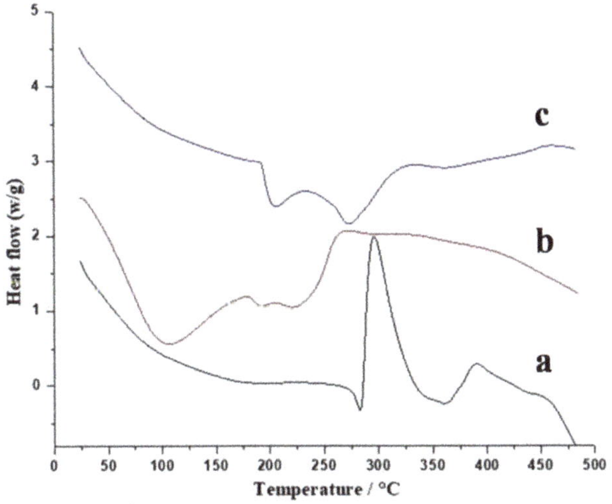

Figure 5. Differential scanning calorimetry (DSC) thermogram of Galantamine hydrobromide (**a**), pectin, (**b**) and formulated hydrogel (**c**).

Figure 6 represents TGA thermograms of Galantamine hydrobromide, pectin, and formulated hydrogel. The TGA thermogram (Figure 6a) shows that Galantamine hydrobromide model drug reveals initial 10% mass loss at 285 °C followed by 50% mass loss at 375 °C. It was observed from the TGA thermogram of pectin (Figure 6b) that weight loss starts above 200 °C. Initial 10% weight loss is observed at 240 °C because of breakage of polymer linkage. At 280 °C, almost 40% weight loss is noted. Above 300 °C, 60% weight loss is noted due to complete breakdown of polymer backbone. Hastuti et al. [31] prepared pectin-carboxymethyl chitosan film. They found similar results in TGA analysis. Initial 10% mass loss occurred at 170 °C, and major mass loss was observed at 285–350 °C. The thermogram of HPMC-pectin-co-acrylic acid hydrogel is given in Figure 6c. The TGA thermogram reveals that the first major weight loss started at 290 °C, and this weight loss continued until 500 °C where 10% weight remained. It could easily be concluded that Galantamine hydrobromide loaded hydrogels were more stable as compared to individual formulation ingredients.

3.6. X-ray Diffraction

XRD investigation of Galantamine hydrobromide, HPMC, pectin, and HPMC-pectin-co-acrylic acid hydrogel formulation was performed, and the corresponding diffractogram is shown in Figure 7. The Galantamine hydrobromide of diffractogram (Figure 7a) exhibits many characteristic sharp peaks such as at 12.84°, 13.56°, 17.56°, and 20.76°. These peaks confirm the crystalline nature of the model drug. Figure 7b shows the pectin diffractogram, which contains many crystalline sharp peaks at 7.08°, 11.72°, 13.16°, 18.84°, 19.48°, 25.16°, 27.72°, and 40.28°. Tan et al. [32] have prepared dual interlinked pulp of sago/pectin hydrogel for drug targeting in colon region. Their XRD pectin diffractogram also represents similar crystalline peaks at 8.8°, 12.30°, 20.2°, 27.8°, and 39.4°. Figure 7c refers to

HPMC diffractogram, which shows only two distinct peaks at 8.12° and 20.2°. Similar results were reported by Wang et al. [33] who had fabricated chitosan/HPMC/glycerol-based hydrogel. They also observed only two peaks at 7.9° and 20.3° in XRD results of HPMC. Less numbers of crystalline peaks confirmed the amorphous nature of HPMC powder. Figure 7d shows HPMC-pectin-co-acrylic acid hydrogel diffractogram. All characteristic peaks exhibited by pure Galantamine hydrobromide, HPMC, and pectin were observed to be missing from hydrogel diffractogram. This has confirmed that crystalline nature of drug as well as other ingredients was transformed into amorphous nature.

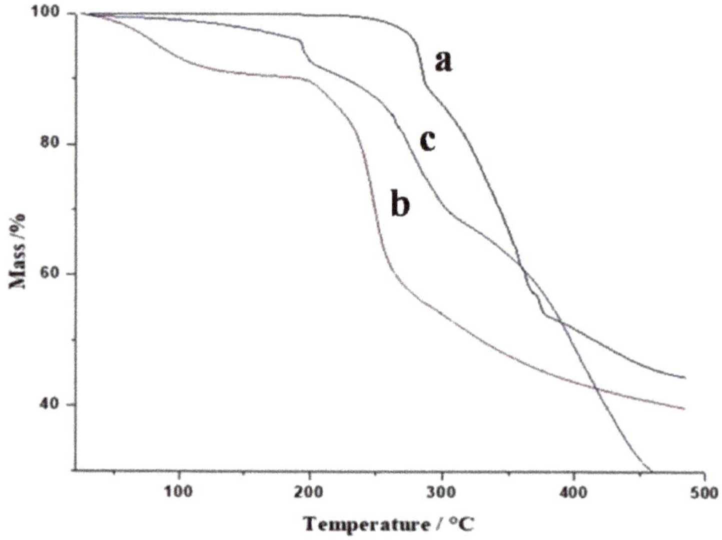

Figure 6. TGA thermogram of Galantamine hydrobromide (**a**), pectin (**b**), and formulated hydrogel (**c**).

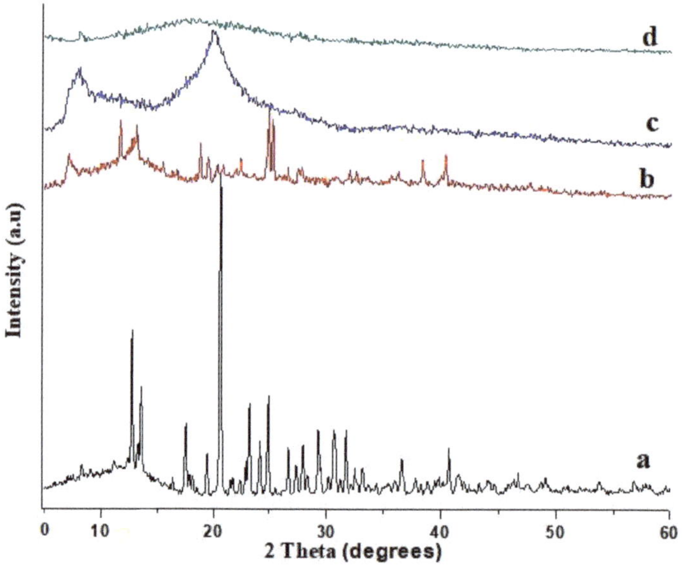

Figure 7. Diffractogram of Galantamine hydrobromide (**a**), pectin (**b**), HPMC (**c**), and formulated hydrogel (**d**).

3.7. Energy Dispersive X-ray Spectrum (EDX) Spectroscopy

EDX was carried out to identify elemental composition of Galantamine hydrobromide, unloaded hydrogel, and hydrogel loaded with Galantamine hydrobromide. Figure 8a shows EDX spectrum of Galantamine hydrobromide having carbon, oxygen, and bromine elements. Generally, EDX spectrum shows not only presence of particular element but also displays % weight of each element present in the compound. It was noted that elements of drug were absent in unloaded hydrogels, but bromine was found in EDX spectrum of Galantamine loaded hydrogels. Presence of bromine in case of Galantamine hydrobromide loaded hydrogels ensured successful loading of drug. Percentage of each element is given in Table 2.

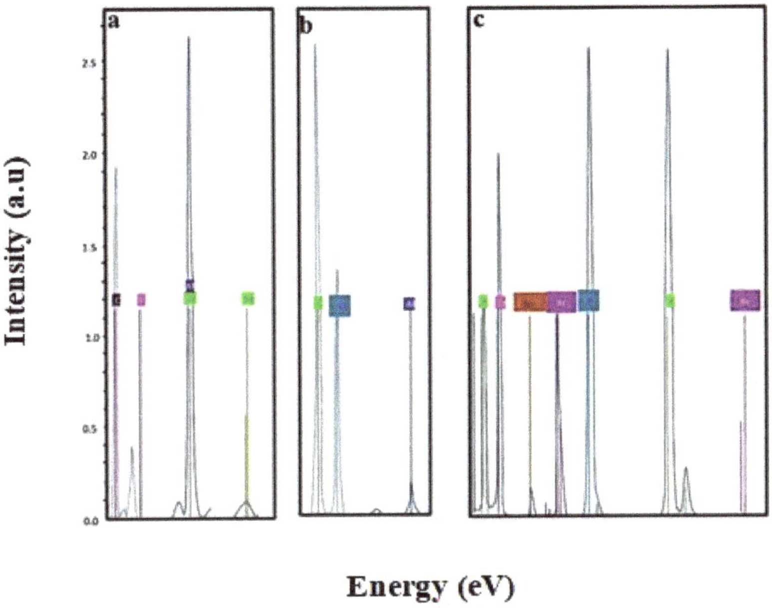

Figure 8. Energy dispersive X-ray (EDX) spectra of Galantamine hydrobromide (**a**), unloaded hydrogel (**b**), and drug loaded hydrogel (**c**).

Table 2. Elemental composition of HPMC-pectin-co-acrylic acid hydrogel.

Type of Material	Elements	% Weight	% Atomic
Galantamine hydrobromide	Carbon	51.96	76.81
	Oxygen	10.83	12.01
	Bromine	25.11	5.58
	Nitrogen	3.64	4.61
Unloaded HPMC-pectin-co-acrylic acid	Carbon	51.77	59.08
	Oxygen	47.09	40.35
Galantamine hydrobromide loaded HPMC-pectin-co-acrylic acid	Carbon	52.83	73.82
	Oxygen	39.99	73.82
	Bromine	2.33	0.86
	Sodium	0.76	0.98
	Potassium	18.73	14.15
	Phosphorous	10.68	10.19

3.8. Porosity Measurements

Porosity of different formulations of HPMC-pectin-co-acrylic acid hydrogel is described in Figure 9. By increasing MBA amount, porosity percentage was decreased from 40.85% to 29.64% (Figure 9a) because of high cross-linking density. It was evident from the fact that total swollen volume decreases with the rise of MBA contents, thus a dense and hard structure is achieved. As density has direct relationship with viscosity, viscosity of the network was greatly improved in response to higher MBA contents. Similar findings in respect of porosity, viscosity, and cross-linker feed have been observed in a study conducted by Yoshinobu et al. [34]. By increasing HPMC amount, porosity percentage was increased from 35.02% to 48.25% (Figure 9c). By increasing pectin amount, porosity increased from 33.54% to 47.51% (Figure 9b). When quantity of acrylic acid was increased, porosity percent increased from 32.89% to 44.89% (Figure 9d). HPMC, pectin, and acrylic acid increase the hydrogel porosity because of increased viscosity of the solution.

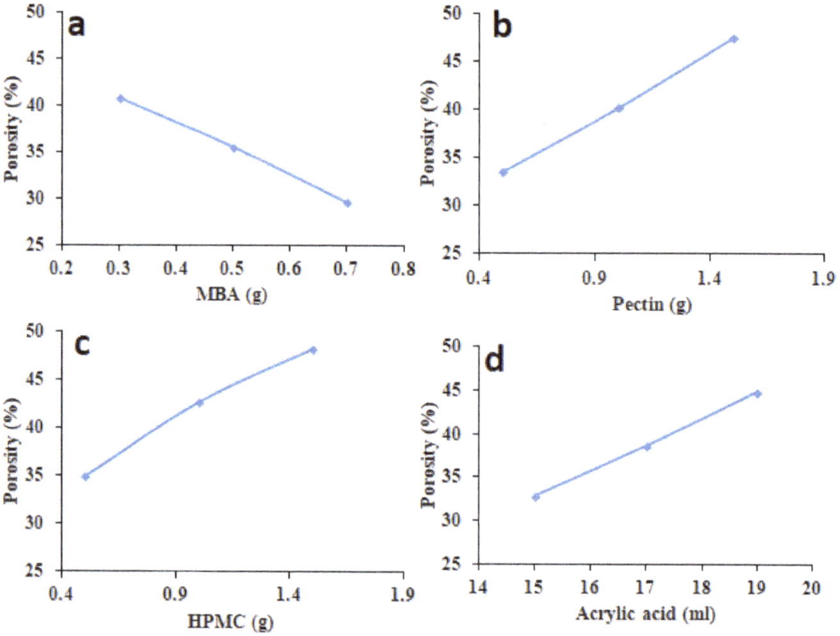

Figure 9. Influence of variable amounts of (**a**) MBA, (**b**) pectin, (**c**) HPMC, and (**d**) acrylic acid on porosity percentage of formulated hydrogels.

3.9. Tensile Strength

Tensile strength of a selected formulation was evaluated by using universal testing machine (UTM) to find out the force at a point where hydrogel breaks. Generally, it is proven from literature that hydrogel formulation, which contains high amount of monomer, would always display the highest amount of tensile strength. Formulation SN9 was selected for testing of tensile strength since it contains the highest amount of acrylic acid (19mL) as compared to other prepared formulations. It is found that this formulation displays 11.043 N/mm^2 of tensile strength (Figure 10). Young's modulus was 62.340 N/mm^2 and percentage of total elongation was 26.464.

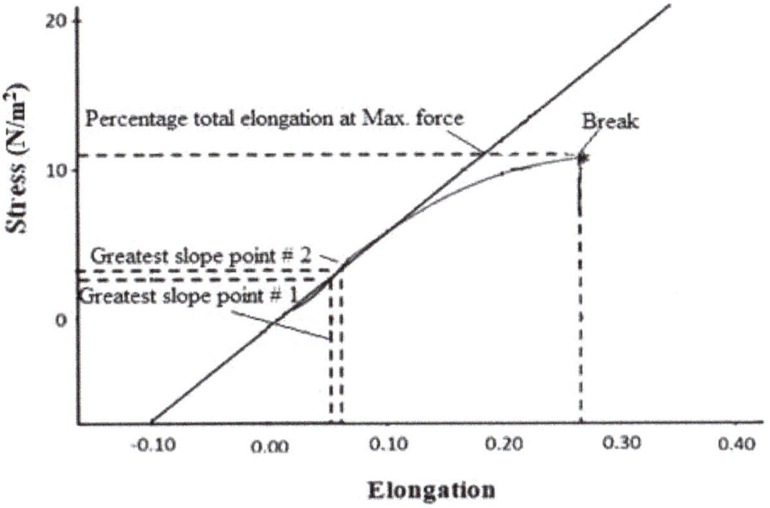

Figure 10. Tensile strength of HPMC-pectin-co-acrylic acid hydrogel.

3.10. In Vitro Release % of Galantamine Hydrobromide from HPMC-Pectin-Co-acrylic Acid Hydrogel

3.10.1. Influence of MBA Amount on Drug Release Percentage

This investigation was carried out on dissolution apparatus II for all formulations of HPMC-pectin-co-acrylic acid hydrogels. Drug loaded hydrogels discs were weighed and then immersed into 900 mL buffers of pH 1.2 and 7.4 individually. Samples were withdrawn after definite time interval from 0 to 36 h and scanned at 287 nm wavelength for absorbance. After applying software program DD solver, release kinetics was calculated. The influence of variable amount of MBA on drug release percentage (SN1–SN3) is represented in Figure 11. Drug release percentage was higher at alkaline pH as compared to acidic. Results showed that when MBA amount was increased, percentage of drug release decreased from 77% to 66% at pH 7.4. Increased cross-linker amount caused more physical entanglement resulting in denser hydrogel structure, which hindered drug release from the polymeric network. At pH 1.2, negligible amount of Galantamine hydrobromide was released as compared to pH 7.4.

3.10.2. Influence of Pectin Amount on Drug Release Percentage

Effect of variable pectin amount on percentage of drug release (SN4–SN5) was observed and the results are displayed in Figure 11. As pectin amount increased from 0.5 to 1.5 g, drug release percentage decreased from 78.09 to 66.35% at pH 7.4. This behavior was due to increased viscosity upon increasing amount of pectin. There was a formation of gelatinous diffusion control layer, which created a barrier for drug release from the hydrogel network. At pH 1.2, negligible amount of Galantamine hydrobromide was released as compared to pH 7.4.

3.10.3. Influence of HPMC and Acrylic Acid Amount on Drug Release Percentage

HPMC (SN6–SN7) and acrylic acid (SN8–SN9) concentration influences the liberation percentage of Galantamine hydrobromide as observed. Hydrogel discs, when subjected to dissolution studies, were solid and having less volume, but with the passage of time, these were swollen as a result of penetration of buffer media. Moreover, geometry of hydrogel discs remained intact even after dissolution experiments. Formulations containing variable amount of HPMC revealed that as the amount of HPMC was gradually increased from 0.5 to 1.5 g, drug release percentage also simultaneously increased from 75.36 to 87.62% at pH 7.4. This is due to higher swell-ability and hydrophilic nature of HPMC at pH 7.4. With increase in concentration of HPMC, hydroxyl groups were increased, which have more tendency to absorb water resulting in greater percentage of drug release. At acidic pH, up to 8.3% Galantamine hydrobromide liberation was observed. Percentage release also increased upon increasing amount of acrylic acid from 15 to 19 mL at pH 7.4 as compared to pH 1.2 because acrylic acid has carboxylic groups, which led to higher ionization at pH 7.4 providing enhanced swelling and drug release.

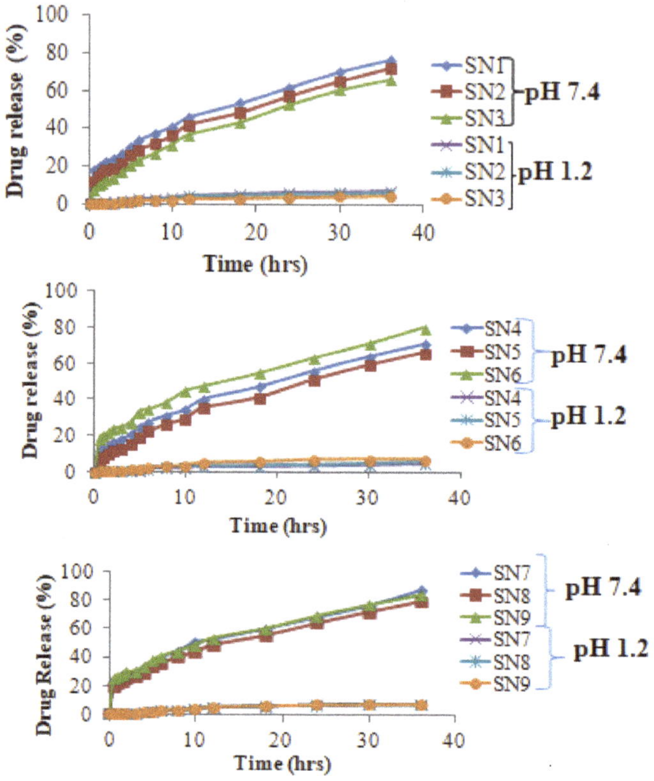

Figure 11. Galantamine hydrobromide release studies from hydrogels (SN1–SN9).

3.11. Assessment of Galantamine Hydrobromide Release through Kinetic Modeling

Kinetic modeling was carried out on Galantamine hydrobromide release data to find out release order from HPMC-pectin-co-acrylic acid hydrogel formulations. Zero order, first order, Higuchi, and Korsemeyer–Peppas models were applied for all formulations. Results are shown in Table 3. It was clear from R^2 values that Galantamine hydrobromide release from HPMC-pectin-co-acrylic acid hydrogel followed zero order kinetics. First order release kinetics was not supported by the release data. But the other two models, i.e., Higuchi and Korsemeyer–Peppas, supported release data of fabricated hydrogels. Based upon value of "n", i.e., 0.439, mechanism of release was Fickian diffusion involving transport of Galantamine hydrobromide solution due to polymer chain relaxation [35].

3.12. Acute Oral Toxicity Study

Toxicity studies were carried out to find out the safety level of HPMC-pectin-co-acrylic acid hydrogel. There was no sign of dermal and ocular toxicity and no mortality was observed on administration of prepared hydrogel to tested group. The observation period was 14 days. During this period, no significant physical change was observed in both tested and control animals. Blood samples were withdrawn from both groups of rabbits for obtaining complete blood count (CBC), uric acid, liver function test (LFT), and renal function test (RFT) profile. Both sets of animals were sacrificed, and their vital organs were removed for histopathological examination. Results of body weight, water, and food intake are reported in Table 4. Hematological (Tables 5 and 6) and histopathological examination (Figure 12) revealed that HPMC-pectin-co-acrylic acid hydrogel was nontoxic, and there was no specific change observed in the controlled and tested group.

Table 3. Kinetic modeling of release data.

Formulation Code	Zero Order	First Order	Higuchi Model	Korsemeyer–Peppas	
R^2	R^2	R^2	R^2	R^2	n
SN1	0.9911	0.7241	0.9715	0.9887	0.431
SN2	0.9907	0.7585	0.9789	0.9902	0.426
SN3	0.9909	0.7909	0.9817	0.9917	0.433
SN4	0.9906	0.7001	0.9630	0.9878	0.440
SN5	0.9908	0.7392	0.9350	0.9848	0.405
SN6	0.9908	0.7921	0.9605	0.9874	0.436
SN7	0.9905	0.7133	0.9165	0.9883	0.389
SN8	0.9901	0.6769	0.9049	0.9852	0.377
SN9	0.9874	0.6679	0.8788	0.9873	0.358

Table 4. Clinical monitoring during oral toxicity studies.

Clinical Monitoring	Control Animal Group (A)	Tested Animal Group (B)
Signs of any illness	None	None
Body Weight (g)		
Before treatment	1653.78 ± 0.40	1738.91 ± 0.40
On Day 1	1651.43 ± 0.60	1735.42 ± 0.60
On Day 7	1651.39 ± 0.50	1733.49 ± 0.40
On Day 14	1650.84 ± 0.20	1733.64 ± 0.50
Food consumption (g)		
Before treatment	75.87 ± 3.045	74.28 ± 2.87
On Day 1	73.49 ± 2.183	77.93 ± 1.06
On Day 7	76.48 ± 4.184	67.85 ± 3.04
On Day 14	68.98 ± 3.789	72.48 ± 3.98
Water intake (mL)		
Before treatment	200.52 ± 2.45	180.62 ± 2.90
On Day 1	190.48 ± 4.21	187.59 ± 1.60
On Day 7	195.82 ± 3.48	204.92 ± 3.10
On Day 14	203.26 ± 2.49	200.65 ± 2.40
Signs of skin allergy	None	None
Signs of ocular toxicity	None	None
Any mortality	None	None

For histopathological assessment, histology slide of each vital organ was prepared, and their micrographs were recorded by single blind assessment method. No change was found in the tested group as compared to the control group (Figure 12). Thus, HPMC-pectin-co-acrylic acid hydrogel was found to be nontoxic and safe to living tissues. Cardiac tissues (control and tested) had normal cardiomyocytes without any hypertrophy and inflammation. Liver section of both control and tested groups showed no sign of inflammation and degradation in hepatic cells. Lung section and alveoli were clear without any sign of inflammation and cellular damage. No accumulation of macrophages and other defender cells was found. Spleen section was normal, and their normal shape was retained in both groups of rabbits. In brain section, cells were observed to be normal and neurons were visible with smooth myelin sheath around them. Kidney section revealed that kidneys were normal in both groups. There was no accumulation of inflammatory cells with absence of cellular damage.

Lastly, histomicrographs of small intestinal section also displayed normal results in both control and tested group.

Table 5. Hematological and biochemical analysis of rabbit's blood.

Finding Parameters	Controlled Animal Group (A)	Tested Animal Group (B)
White blood cells (× 10^3/μL)	4.50 ± 0.34	5.20 ± 0.25
Red blood cells (× 10^6/ μL)	3.98 ± 0.49	4.46 ± 0.51
Hemoglobin (g/dL)	11.93 ± 0.76	12.86 ± 0.69
Platelets (× 10^3/ μL)	42.63 ± 0.87	44.52 ± 0.47
Lymphocytes %	63.10 ± 0.58	68.70 ± 0.81
Monocytes %	3.10 ± 0.35	3.00 ± 0.21
Mean corpuscular volume (fL)	60.90 ± 2.90	61.40 ± 2.32
Mean corpuscular hemoglobin (pg)	19.90 ± 0.50	20.90 ± 0.10
Mean corpuscular hemoglobin concentration (g/dL)	35.80 ± 1.41	33.90 ± 1.84

Table 6. LFT and RFT profile of rabbit blood sample.

Finding Parameters	Controlled Animal Group (A)	Tested Animal Group (B)
ALT(U/L)	52.45 ± 1.42	57.28 ± 1.59
AST (U/L)	126.00 ± 2.33	135.00 ± 2.84
ALP (U/L)	135.00 ± 1.46	148.20 ± 1.68
ALB (g/L)	5.17 ± 0.45	4.90 ± 0.82
Creatinine (mg/dL)	0.79 ± 0.11	0.60 ± 0.12
Uric acid (mg/dL)	2.40 ± 0.25	2.90 ± 0.21
Urea (mg/dL)	16.02 ± 0.15	13.50 ± 0.18

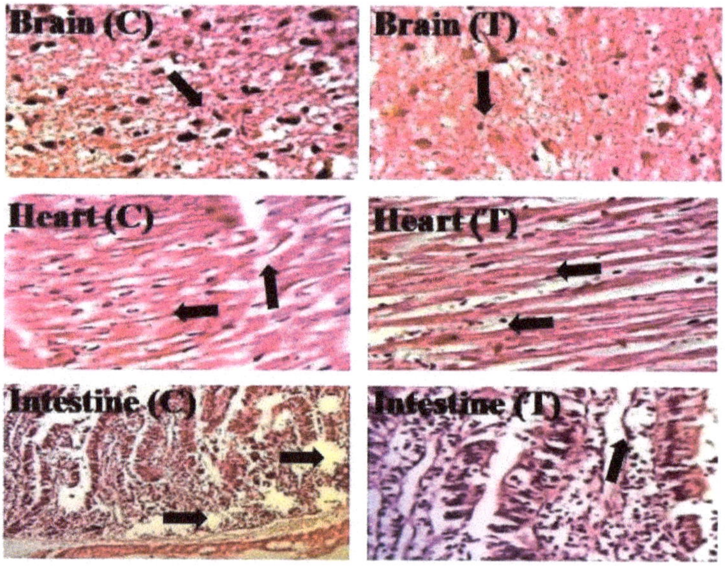

Figure 12. *Cont.*

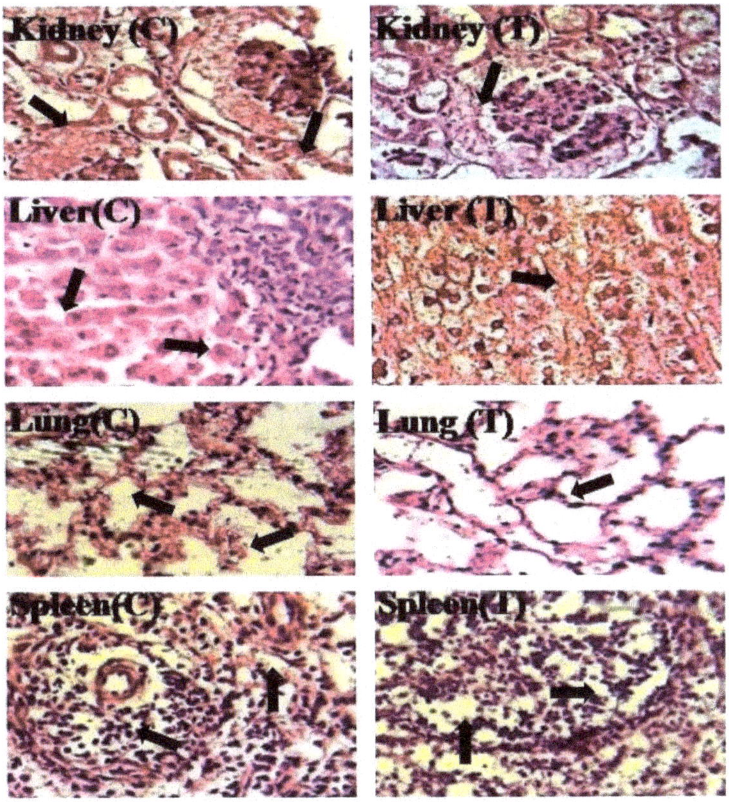

Figure 12. Histopathological examination of vital organs of rabbits.

4. Conclusions

HPMC-pectin-co-acrylic acid hydrogels were successfully fabricated through chemically cross-linked method (free radical polymerization technique). Formulated hydrogels showed pH-dependent swelling and release behavior for controlled delivery of Galantamine hydrobromide. Equilibrium swelling percentage was high, i.e., 96.87% at pH 7.4, whereas, at pH 1.2, up to 10% swelling percentage was observed. Swelling and release results depends on the amount of pectin, acrylic acid, HPMC, and MBA. As HPMC and acrylic acid amount was increased, swelling and release percentages were observed to be simultaneously increased but not in case of pectin. Higher pectin amount provided higher degree of swelling but lower level of drug release. MBA produced an inverse effect on swelling and release kinetics. The high amount of MBA leads to a lower amount of swelling and release. FTIR, DSC, TGA, and XRD confirmed the complex formation of hydrogel, which was more thermally stable as compared to individual ingredients. Gel fraction results were also directly proportional to HPMC, pectin, acrylic acid, and MBA amount. Mechanical strength of HPMC-pectin-co-acrylic acid hydrogel was excellent having capability to hold out against biological stress. EDX results confirmed loading of Galantamine hydrobromide into formulated hydrogel. In toxicity studies, blood samples and histomicrographs showed that fabricated hydrogel was safe, nontoxic, and compatible to living tissues. It was concluded that formulated hydrogels displayed pH dependent behavior and smart response to their environmental conditions. Thus, that is why they can be efficiently used in Alzheimer-related dementia where controlled delivery of Galantamine hydrobromide would be provided. Hence, dosing frequency would be reduced, which would ultimately lead to enhanced patient compliance.

Author Contributions: Conceptualization, N.Z.; methodology, N.L.; software and revision of manuscript, A.M.; validation, S.B., formal analysis, A.E.; investigation, S.B.; resources: S.B., N.Z.; data curation, S.B.; writing original draft, N.Z.; visualization, A.E.; review & editing, N.L.; supervision, N.Z. All authors have read and agreed to the published version of the manuscript.

Funding: This research received no external funding.

Conflicts of Interest: The authors declare no conflict of interest.

References

1. Moghanjoughi, A.A.; Khoshnevis, D.; Zarrabi, A. A concise review on smart polymers for controlled drug release. *Drug Deliv. Transl. Res.* **2016**, *6*, 333–340. [CrossRef] [PubMed]
2. Gupta, P.K.; Robinson, J.R. Oral controlled-release delivery. In *Treatise on Controlled Drug Delivery*; Taylor & Francis group, CRC Press: London, UK, 2017; pp. 255–313.
3. Arslan, M.; Tasdelen, M.A. Click chemistry in macromolecular design: Complex architectures from functional polymers. *Chem. Afr.* **2019**, *2*, 195–214. [CrossRef]
4. Robertis, S.D.; Bonferoni, M.C.; Elviri, L.; Sandri, G.; Caramella, C.; Bettini, R. Advances in oral controlled drug delivery: The role of drug-polymer and interpolymer non-covalent interactions. *Expert Opin. Drug Deliv.* **2014**, *12*, 441–453. [CrossRef] [PubMed]
5. Moulton, S.E.; Wallace, G.G. 3-dimensional (3D) fabricated polymer based drug delivery systems. *J. Controlled Release* **2014**, *193*, 27–34. [CrossRef] [PubMed]
6. Ofridam, F.; Lebaz, N.; Gagnière, É.; Mangin, D.; Elaissari, A. Effect of secondary polymer on self-precipitation of pH-sensitive polymethylmethacrylate derivatives Eudragit E100 and Eudragit L100. *Polym. Adv. Technol.* **2020**, *31*, 1270–1279. [CrossRef]
7. Ranucci, E.; Manfredi, A. Polyamidoamines: Versatile bioactive polymers with potential for biotechnological applications. *Chem. Afr.* **2019**, *2*, 167–193. [CrossRef]
8. Deen, G.; Loh, X. Stimuli-responsive cationic hydrogels in drug delivery applications. *Gels* **2018**, *4*, 13. [CrossRef]
9. Levourch, G.; Lebaz, N.; Elaissari, A. Hydrophilic submicron nanogel particles for specific recombinant proteins extraction and purification. *Polymers* **2020**, *12*, 1413. [CrossRef]
10. Grande, D.; Rohman, G. Oligoester-derivatized (semi-)interpenetrating polymer networks as nanostructured precursors to porous materials with tunable porosity. *Chem. Afr.* **2019**, *2*, 253–265. [CrossRef]
11. Parhi, R. Cross-linked hydrogel for pharmaceutical applications: A review. *Adv. Pharm. Bull.* **2017**, *7*, 515–530. [CrossRef]
12. Gupta, P.; Vermani, K.; Garg, S. Hydrogels: From controlled release to pH-responsive drug delivery. *Drug Discov. Today* **2002**, *7*, 569–579. [CrossRef]
13. Moussa, E.; Siepmann, F.; Flament, M.P.; Benzine, Y.; Penz, F.; Siepmann, J.; Karrout, Y. Controlled release tablets based on HPMC:lactose blends. *J. Drug Deliv. Sci. Technol.* **2019**, *52*, 607–617. [CrossRef]
14. Raj, A.A.S. A review on pectin: Chemistry due to general properties of pectin and its pharmaceutical uses. *Sci. Rep.* **2012**, *1*, 550.
15. Yen, W.F.; Basri, M.; Ahmad, M.; Ismail, M. Formulation and evaluation of galantamine gel as drug reservoir in transdermal patch delivery system. *Sci. World J.* **2015**, *2015*, 1–7.
16. Mahmood, A.; Ahmad, M.; Sarfraz, R.M.; Minhas, M.U. β-CD based hydrogel microparticulate system to improve the solubility of acyclovir: Optimization through in-vitro, in-vivo and toxicological evaluation. *J. Drug Deliv. Sci. Technol.* **2016**, *36*, 75–88. [CrossRef]
17. Khanum, H.; Ullah, K.; Murtaza, G.; Khan, S.A. Fabrication and in vitro characterization of HPMC-g-poly(AMPS) hydrogels loaded with loxoprofen sodium. *Int. J. Biol. Macromol.* **2018**, *120*, 1624–1631. [CrossRef]
18. Khan, Z.; Minhas, M.U.; Ahmad, M.; Khan, K.U.; Sohail, M.; Khalid, I. Functionalized pectin hydrogels by cross-linking with monomer: Synthesis, characterization, drug release and pectinase degradation studies. *Polym. Bull.* **2019**, *77*, 339–356. [CrossRef]
19. Vega-Chacón, J.; Tarhini, M.; Lebaz, N.; Jafelicci, M.; Zine, N.; Errachid, A.; Elaissari, A. Protein-silica hybrid submicron particles: Preparation and characterization. *Chem. Afr.* **2020**, *3*, 1–9. [CrossRef]
20. Ouanji, F.; Ellouzi, I.; Kacimi, M.; Ziyad, M. Ca-hydroxyzincate: Synthesis and enhanced photocatalytic activity for the degradation of methylene blue under uv-light irradiation. *Chem. Afr.* **2019**, *2*, 395–400. [CrossRef]
21. Hebeish, A.; Hashem, M.; El-Hady, M.M.A.; Sharaf, S. Development of CMC hydrogels loaded with silver nano-particles for medical applications. *Carbohydr. Polym.* **2013**, *92*, 407–413. [CrossRef]

22. Shin, M.-S.; Kim, S.J.; Kim, I.Y.; Kim, N.G.; Song, C.G.; Kim, S.I. Swollen behavior of crosslinked network hydrogels based on poly(vinyl alcohol) and polydimethylsiloxane. *J. Appl. Polym. Sci.* **2002**, *85*, 957–964. [CrossRef]
23. Saqib, M.; Bhatti, A.S.A.; Ahmad, N.M.; Ahmed, N.; Shahnaz, G.; Lebaz, N.; Elaissari, A. Amphotericin B loaded polymeric nanoparticles for treatment of leishmania infections. *Nanomaterials* **2020**, *10*, 1152. [CrossRef] [PubMed]
24. Mahmood, A.; Sharif, A.; Muhammad, F.; Sarfraz, R.M.; Abrar, M.A.; Qaisar, M.N.; Anwer, N.; Amjad, M.W.; Zaman, M. Development and in vitro evaluation of (β-cyclodextrin-g-methacrylic acid)/Na+-montmorillonite nanocomposite hydrogels for controlled delivery of lovastatin. *Int. J. Nanomedicine* **2019**, *14*, 5397–5413. [CrossRef] [PubMed]
25. Rehmani, S.; Ahmad, M.; Minhas, M.U.; Anwar, H.; Zangi, M.I.; Sohail, M. Development of natural and synthetic polymer-based semi-interpenetrating polymer network for controlled drug delivery: Optimization and in vitro evaluation studies. *Polym. Bull.* **2016**, *74*, 737–761. [CrossRef]
26. Akhlaq, M.; Maryam, F.; Elaissari, A.; Ullah, H.; Adeel, M.; Hussain, A.; Ramzan, M.; Ullah, O.; Danish, M.Z.; Iftikhar, S.; et al. Pharmacokinetic evaluation of quetiapine fumarate controlled release hybrid hydrogel: A healthier treatment of schizophrenia. *Drug Deliv.* **2018**, *25*, 916–927. [CrossRef]
27. Hanafy, A.S.; Farid, R.M.; ElGamal, S.S. Complexation as an approach to entrap cationic drugs into cationic nanoparticles administered intranasally for Alzheimer\textquotesingles disease management: Preparation and detection in rat brain. *Drug Dev. Ind. Pharm.* **2015**, *41*, 2055–2068. [CrossRef]
28. Feng, L.; Yang, H.; Dong, X.; Lei, H.; Chen, D. pH-sensitive polymeric particles as smart carriers for rebar inhibitors delivery in alkaline condition. *J. Appl. Polym. Sci.* **2017**, *135*. [CrossRef]
29. Ibrahim, A.G. Synthesis of poly(acrylamide-graft-chitosan) hydrogel: Optimization of the grafting parameters and swelling studies. *Am. J. Polym. Sci. Technol.* **2019**, *5*, 55–62. [CrossRef]
30. Mittal, N.; Kaur, G. In situ gelling ophthalmic drug delivery system: Formulation and evaluation. *J. Appl. Polym. Sci.* **2013**, *131*. [CrossRef]
31. Hastuti, B.; Mudasir, M.; Siswanta, D.; Triyono, T. Preparation and Pb(II) adsorption properties of crosslinked pectin-carboxymethyl chitosan film. *Indones. J. Chem.* **2015**, *15*, 248–255. [CrossRef]
32. Tan, H.L.; Tan, L.S.; Wong, Y.Y.; Muniyandy, S.; Hashim, K.; Pushpamalar, J. Dual crosslinked carboxymethyl sago pulp/pectin hydrogel beads as potential carrier for colon-targeted drug delivery. *J. Appl. Polym. Sci.* **2016**, *133*. [CrossRef]
33. Wang, T.; Chen, L.; Shen, T.; Wu, D. Preparation and properties of a novel thermo-sensitive hydrogel based on chitosan/hydroxypropyl methylcellulose/glycerol. *Int. J. Biol. Macromol.* **2016**, *93*, 775–782. [CrossRef] [PubMed]
34. Yoshinobu, M.; Morita, M.; Sakata, I. Porous structure and rheological properties of hydrogels of highly water-absorptive cellulose graft copolymers. *J. App. Polym. Sci.* **1992**, *45*, 805–812. [CrossRef]
35. Ouyang, L.; Highley, C.B.; Rodell, C.B.; Sun, W.; Burdick, J.A. 3D printing of shear-thinning hyaluronic acid hydrogels with secondary cross-linking. *ACS Biomat. Sci. Eng.* **2016**, *2*, 1743–1751. [CrossRef]

Publisher's Note: MDPI stays neutral with regard to jurisdictional claims in published maps and institutional affiliations.

processes

Article

Process Transferability of Friction Riveting of AA2024-T351/Polyetherimide (PEI) Joints Using Hand-Driven, Low-Cost Drilling Equipment

Anamaria Feier [1], Andrei Becheru [1], Mihai Brîndușoiu [1] and Lucian Blaga [2,*]

[1] Department of Materials and Manufacturing Engineering, Mechanical Faculty, Politehnica University Timișoara, Bl. Mihai Viteazu No. 1, 300222 Timișoara, Romania; ana-maria.feier@upt.ro (A.F.); becheru.andrei@yahoo.ro (A.B.); mihaibrindusoiu@gmail.com (M.B.)

[2] Department Solid State Materials Processing, Institute for Materials Mechanics, Helmholtz Zentrum Hereon, 21502 Geesthacht, Germany

* Correspondence: lucian.blaga@hereon.de; Tel.: +49-(0)4152-87-2055

Abstract: The present work deals with the transferability of Friction Riveting joining technology from laboratory equipment to adapted in-house, low-cost machinery. A G13 drilling machine was modified for the requirements of the selected joining technique, and joints were performed using polyethermide plates and AA2024 aluminum alloy rivets of 6 mm diameter. This diameter was not previously reported for Friction Riveting. The produced joints were mechanically tested under tensile loading (pullout tests) with ultimate tensile forces of 9500 ± 900 N. All tested specimens failed through full-rivet pullout, which is the weakest reported joint in Friction Riveting. In order to understand this behavior, FE models were created and analyzed. The models produced were in agreement with the experimental results, with failure initiated within the polymer under stress concentrations in the polymeric material above the deformed metallic anchor at an ultimate value of the stress of 878 MPa at the surface of the joint. Stresses decreased to less than half of the maximum value around the anchoring zone while the rivet was removed and towards the surface. The paper thus demonstrates the potential ease of applying and reproducing Friction Riveting with simple machinery, while contributing to an understanding of the mechanical behavior (initialization of failure) of joints.

Keywords: Friction Riveting; metal-polymer hybrid joints; friction-based multi-material connections; anchoring FE modelling; rivet failure modes

Citation: Feier, A.; Becheru, A.; Brîndușoiu, M.; Blaga, L. Process Transferability of Friction Riveting of AA2024-T351/Polyetherimide (PEI) Joints Using Hand-Driven, Low-Cost Drilling Equipment. *Processes* **2021**, *9*, 1376. https://doi.org/10.3390/pr9081376

Academic Editors: Shaghayegh Hamzehlou, Eduardo Vivaldo-Lima and M. Ali Aboudzadeh

Received: 8 June 2021
Accepted: 31 July 2021
Published: 6 August 2021

Publisher's Note: MDPI stays neutral with regard to jurisdictional claims in published maps and institutional affiliations.

1. Introduction

The continuously increasing demand for cost reduction, together with high performance product requirements, is leading to substantial research and engineering developments in new materials and tailored joining technologies [1]. The outcomes of these efforts are mixed and hybrid structures in which the properties and performance of products are improved through combining the properties and behaviors of each specific material [2]. Hybrid polymer–metal structures are used in such way in a several engineering applications, as exemplified in Figure 1.

Due to strong dissimilarities in physical-chemical properties, hybrid joints between metals and polymers are challenging, more so because of geometrical and design considerations [2]. To overcome some of the limitations of the current state of the art in hybrid joining, Friction Riveting (FricRiveting) has been developed at the Helmholtz-Zentrum Geesthacht (now Helmholtz Zentrum Hereon) in Germany as a process for joining metallic bolts (rivets) with polymeric plates [3]. This research studies the material combination of aluminum AA204 with polyetherimide (PEI) to be joined via Friction Riveting. The feasibility of this combination has already been proven, and the joints have been characterized and optimized in a series of studies using the equipment available at the institute that holds the patent for Friction Riveting [4–6].

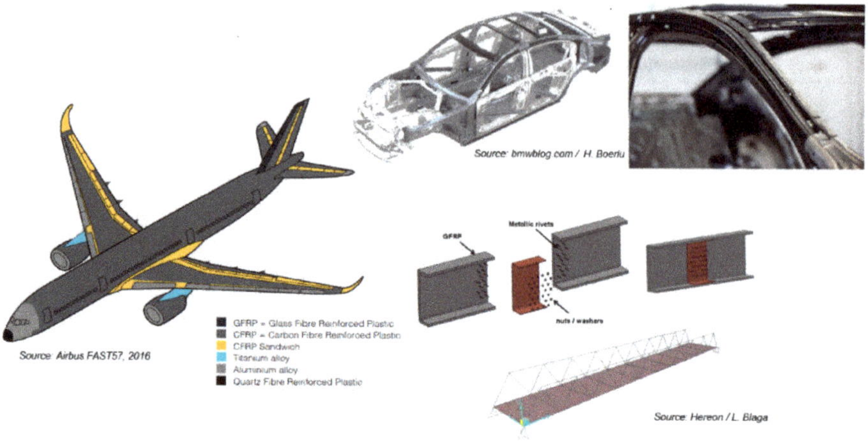

Figure 1. Application examples of hybrid polymer–metal joints.

Friction Riveted metallic-insert/point-on-plate joints for polymer–metal hybrid structures could be used in transportation industries or civil engineering, as well as in lower-scale electronics, in clips, stinger-skin connections, sandwich panels, or additional safety connections to welded or adhesively bonded structures. Such joints can be applied both to axial loads, whereby the rivet anchoring efficiency is of uttermost importance, and to shear loads, in which the joint configuration is single-, or double-lapped [2]. The scope of this research paper is to validate the previous experiments for Friction Riveting under axial loading (anchoring efficiency) by using and partially adapting an existing drilling machine, contributing thus to the industrial transferability of the technique. Furthermore, the paper intends to generate new knowledge regarding the failure mode and fracture initiation within the polymeric part resisting the removal of the anchored rivet.

2. Friction Riveting

In its basic process variant involving the so-called point-on-plate or metallic-insert joint (Figure 2), Friction Riveting consists of a rotating cylindrical metallic rivet being plastically deformed and subsequently anchored within a polymer or polymeric composite plate. The joining occurs mainly through the mechanical interlocking and anchoring of the plastically deformed tip of the metallic rivet inside the polymer part. The rotating rivet widens its diameter inside the polymer because of the increasing heat due to the friction, corroborated by heat concentration in the joint formation area due to the insulating properties of polymers [7]. Adhesion between the partially softened or molten polymer after its reconsolidation might also contribute to the joining mechanisms, as well as to the micromechanical interlocking at the rivet–polymer interface. The softened/molten polymeric material from the joining area, along the path of the inserting rivet, is expelled during the process as flash material, which can be removed during or after the process by cutting [8].

Despite significant studies and characterizations of different material combinations in Friction Riveting, the technique was not applied outside of laboratory environments and industrial machinery. The scope of this research paper is to validate the previous experiments by using and partially adapting an existing drilling machine, contributing thus to the industrial transferability of the technique and concomitantly to the modelling of joint mechanical behavior for an improved understanding of the failure of such joints under axial loading.

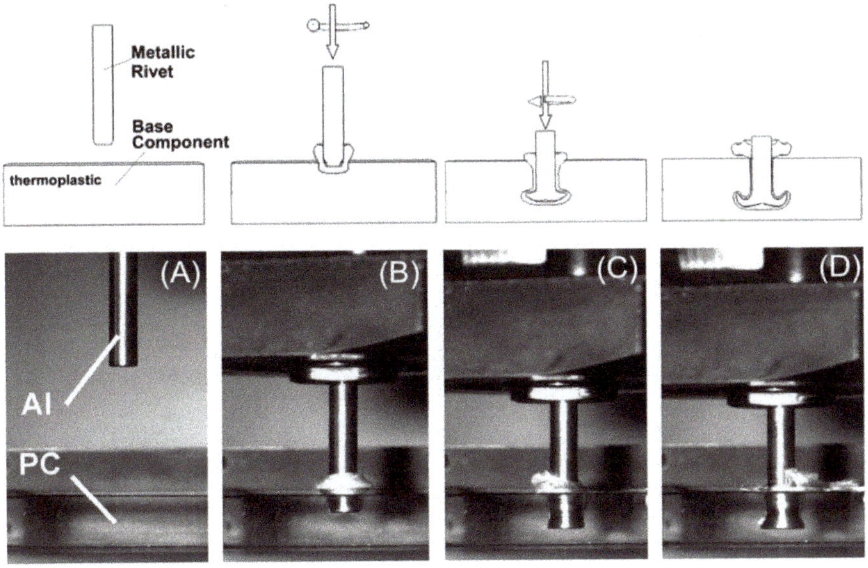

Figure 2. Friction Riveting process steps shown on an aluminum–polycarbonate joint: (**A**) Clamping/positioning of joining partners, (**B**) Rivet rotation and insertion, (**C**) Rivet plastic deformation and forging, (**D**) Joint consolidation/anchoring [7]. Reproduced with permission from [7].

3. Materials and Methods

3.1. Base Materials

Aluminum AA2024-T351 extruded rivets were joined with polyetherimide (PEI) plates, the exact material combination and specifications as in [5,6]. Plain, featureless rivets were used, with a diameter of 6 mm and 60 mm length. The choice of materials was based on the process development of Friction Riveting, as this was the initial material combination used for the process development and validation [2].

PEI is a high-performance thermoplastic amorphous polymer characterized by high specific strength, increased rigidity under high temperatures, thermal stability, and good electrical properties. It is used widely in electronics and medical devices as well as the chemical and petroleum industries [9]. The AA2024-T3521 alloy is widely used in the transportation industries due to its relatively high strength under both quasi-static and dynamic loading, while its corrosion resistance is rather moderate.

3.2. Joining Equipment and Procedure

The equipment used in this study was a G13 drilling machine (producer, country), presented in Figure 3. The G13 is a fixed-tool drilling machine for the small-series drilling of steel up to Ø 13 mm, weighing a total of 162 kg.

As the maximum rotational speed of the G13 is 4000 rpm, it was by far not matching the equipment used for previous work in FricRiveting, where rotational speeds of up to 23,400 rpm were applied to join the same combination of materials, with a reduced rivet diameter indeed, of only 5 mm. Therefore, the equipment was modified by replacing the original motor with a commercial, off-the-shelf motor capable of achieving 10,000 rpm rotational speed. A hand-driven force of 1950 N was applied, constant for all trials. The displacement of the rivet (penetration rate) was manually controlled via an indicator on the drilling machine. The successfully anchored joints tested and reproduced were performed using the maximum configuration of this adapted equipment, with any lower setting not capable of obtaining sufficient deformation and anchoring.

Figure 3. G13 drilling machine used for Friction Riveting.

3.3. Joint Mechanical Performance

Five specimens were produced via the described method, and their mechanical performance was tested in terms of ultimate pull-out force using a Testwell/UTS universal testing machine with a 260 kN load cell. Tests were performed at room temperature at a testing speed of 1 mm/min. The testing configuration is shown in Figure 4.

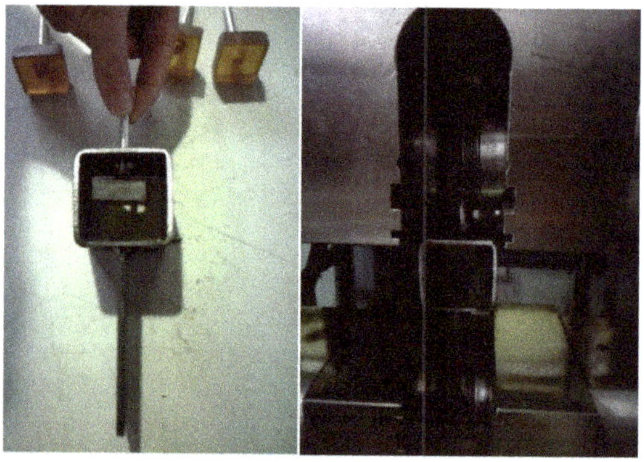

Figure 4. Pullout tests of friction-riveted joints.

3.4. Numerical Simulation of Pullout Tests

The joints were modelled numerically via ANSYS v19.1 FEM software. First, the joining partners were discretized and the material properties defined. Figure 5 shows the cross section of a joint used for FE modelling in this study.

For the discretization of the model, tetrahedral elements were applied in the meshing. In the anchoring zone/joining area, a coarse mesh was used for an improved evaluation of the results. The mechanical properties in Table 1 were used as material inputs for the FE model.

Table 1. Mechanical properties of materials used in FE model.

Component	Material	Density (g/cm^3)	E Modulus (MPa)	ν (-)	R_m (MPa)	$R_{p0.2}$ (MPa)
Rivet	AA 2024-T351	2.78	76,000	0.33	470	324
Base plate	PEI	1.27	3500	0.44	129	129

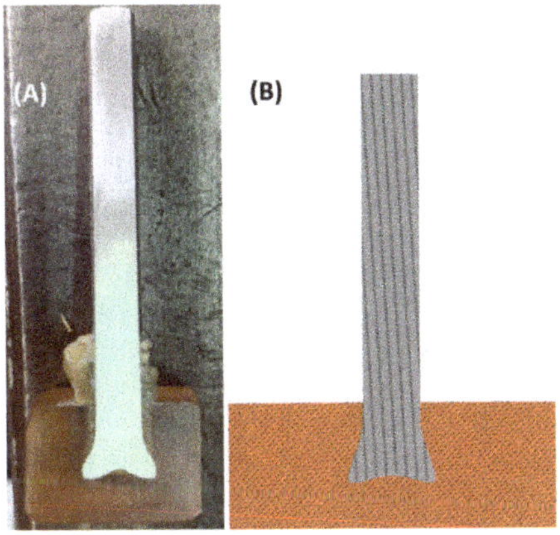

Figure 5. Friction riveted joint (**A**) cross section and (**B**) FE model.

4. Results and Discussion

As described in Section 3.2, AA2024-T351 rivets were successfully deformed and anchored within PEI plates. The final joining parameters were a rotational speed of 10,000 rpm, a joining force of 1950 N, and displacement at friction manually controlled as described in Section 3.2. Lower rotational speeds were tested but led to insufficient deformation and thus no presence of rivet anchoring, while increased loads produced either rivet buckling or full drilling of the PEI plate.

Four specimens were mechanically tested, as described in Section 3.3, leading to an ultimate pullout force of 9500 ± 900 N. All tested samples failed by full rivet pullout, as shown in Figure 6.

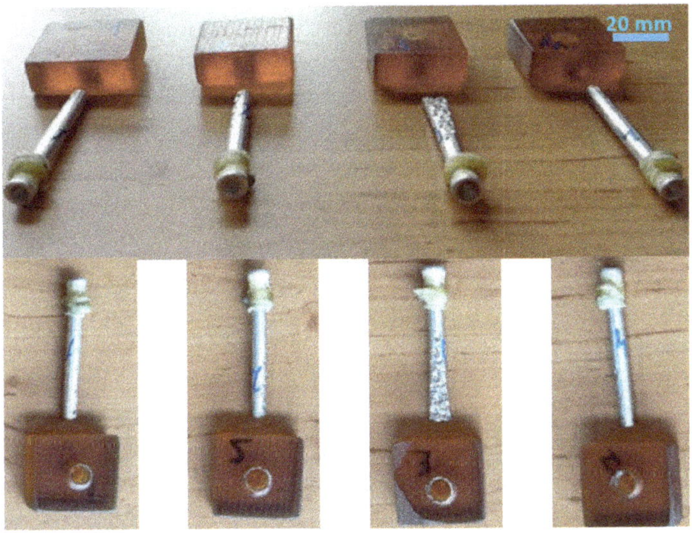

Figure 6. Tested specimens displaying full rivet pullout failure.

One can observe that a certain volume of polymeric material is adhered to the rivet. This consists of both the flash expelled during the friction riveting process, as well as a narrow layer of polymer displaced from the composite plate's surface, which corresponds to the full rivet pullout failure observed by Pina Cipriano et al. [6] and first identified as Type III failure of Friction Riveting joints by Rodriguez et al. [7].

For a better comprehension of the mechanical behavior of the anchored rivet under tensile loading, the results of the mechanical tests were compared with the FE models. Figure 7 presents a numerical model of the development of the deformations and Misses stresses during the rivet displacement.

The ultimate failure (Figure 8) occurs in the experiments via complete removal of the anchored rivet, without additional cracks in the polymeric material. The ultimate value of the stress was 878 MPa, present only at the surface of the joint, indicating the weakening of the polymeric material above the deformed rivet, leading to a lower resistance to the removal of the metallic anchor. Deeper in the joining area, close to the actual anchoring, the stresses decrease to less than a half of the surface value; therefore, the rivet does not achieve the ultimate yield stress and is instead removed and shifts toward the surface. Furthermore, the initialization of the removal of the rivet from the anchoring area can be observed. Figure 9 compares the stress–stress curves of both the FE model and the experimental data, showing their agreement. Pina Cipriano et al. previously investigated the correlation of the anchoring deformation, ultimate tensile force, and subsequent joint failure for the same material combination as in the present paper [5,6]. By addressing the anchoring efficiency in terms of volumetric ratio and via statistical analysis, they showed that the most significant process parameter is the joining force (friction force in their work). It has to be noted that the work of Pina Cipriano et al. was performed on dedicated laboratory equipment, specifically designed for Friction Riveting [5,6]. As also mentioned in their research report, the joining force is of highest importance for joint formation, given sufficient heat input provided by the parameters of rotational speed and the rotational speed–joining force interaction. The current study shows to some extent that the rotational speed required can be as low as 10,000 rpm for deformation and anchoring to occur. Indeed, increased mechanical performance and an improved failure type still has to be accomplished when transferring the technology to simpler devices. The recommendation would be to increase the joining force and future research will have to confirm this. Nonetheless, factors such as rivet buckling and oversized deformation (leading to weak anchoring) will have to be considered in that case, as well as joining-energy efficiency.

Borges et al. analyzed previous FE models of PEI/AA2024 Friction Riveting joints [10], which cannot be applied directly to the current work, as the joining conditions, equipment, and most important, the mechanical behavior differ. In the work of Borges et al., optimized joints were modelled with a final fracture within the shaft of the metallic rivet. The disagreement between models and experiments of around 10% was supposedly due to geometrical simplifications [10]. Moreover, the Johnson–Cook failure model was applied to the rivet, which could not be the case in the present investigation, as the failure initiates and finalizes within the polymeric material only (metallic rivet completely removed with some adhesion and fracture at the PEI-AA2024 interface).

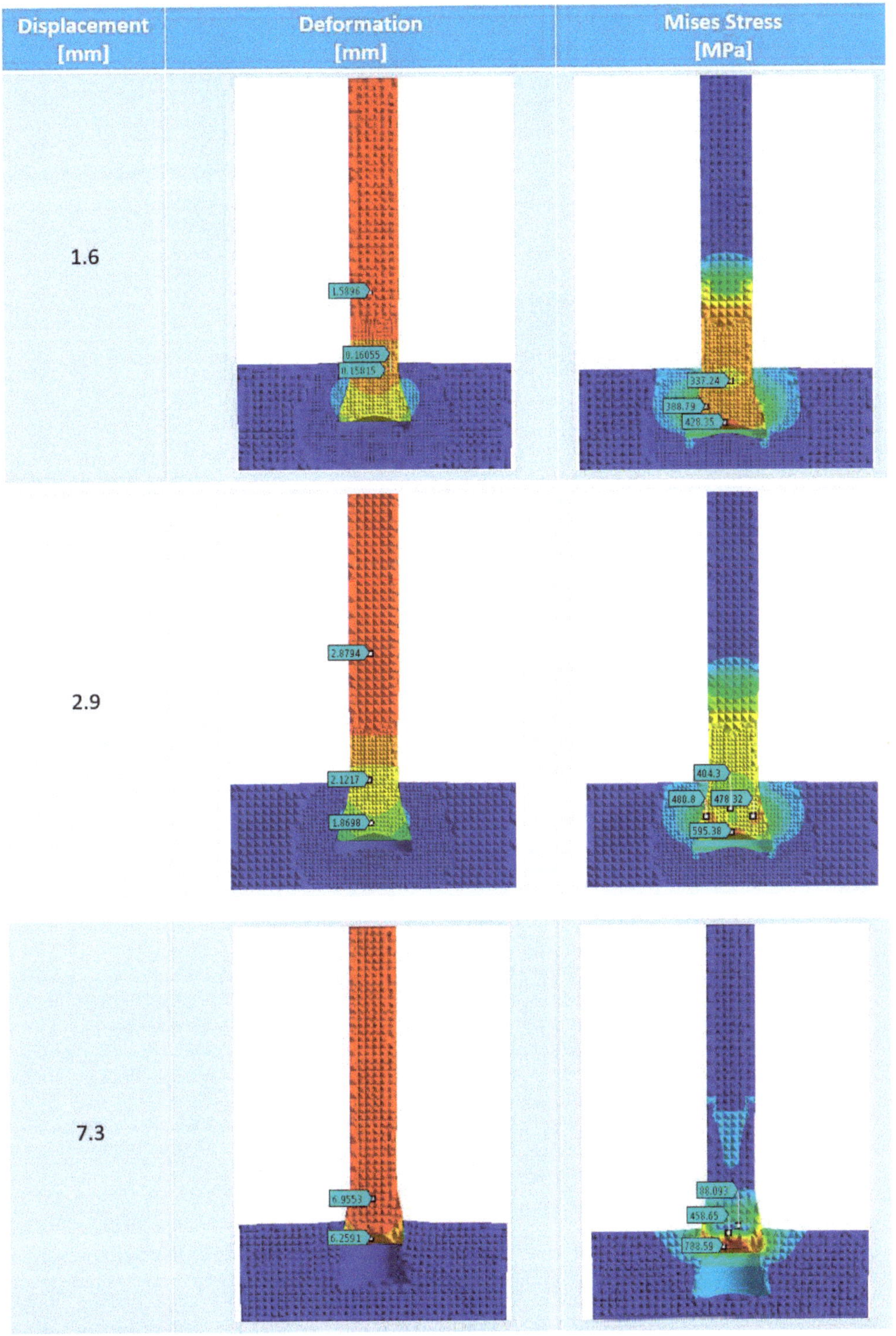

Figure 7. FE model displaying the development of Misses stresses of the friction riveted joints under tensile loading.

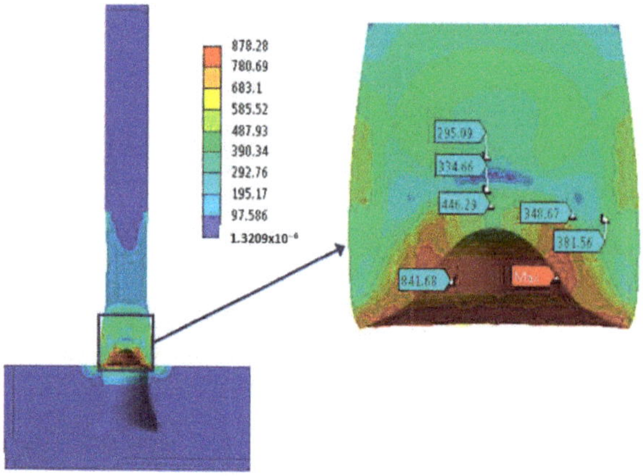

Figure 8. FE model of joint failure.

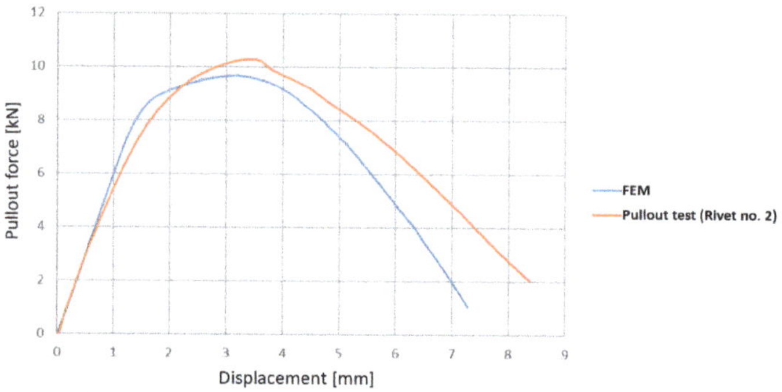

Figure 9. Comparison between experimental values and FE model of pullout tests.

A limited number of researchers have successfully attempted Friction Riveting. Hynes et al. performed low-speed Friction Riveting of AA1100 threaded rivets with polymethyl methacrylate (PMMA) using in-house adapted Friction Riveting equipment with a spindle capable of up to 3000 rpm [11]. The authors also reported the Type III failure for some of their joints, with SEM observations linking this to the rupture of the polymeric material due to the applied tensile load, similar to the current work. This was indicated by serrations on the surface of the molten PMMA [11]. Future research will investigate changes in physical-chemical properties within the PEI-AA2024 interface as well as the expelled flash material. Furthermore, the influence of threads on both joint formation and mechanical performance will have to be analyzed and correlated with the microstructural changes within the materials. Gagliardi et al. used an adapted milling machine for Friction Riveting of titanium grade two-holed cylindrical rivets with both pure and GF-reinforced polyamide 6 (PA6), showing that the spindle speed (referred as rotational speed in other publications on Friction Riveting) is the main process parameter affecting the mechanical performance/anchoring force [12].

5. Conclusions

The present paper demonstrated the transferability of Friction Riveting from dedicated laboratory equipment to modified simple drilling machines using a commercially available off-the-shelf motor for achieving relatively high rotational speeds and thus sufficient energy for rivet tip deformation and anchoring. This investigation shows that Friction Riveting can be performed with easy-to-achieve modification of low-cost machinery. FE models were in agreement with the experimentally obtained data, with rivet failure at 9500 $\pm$ 900 N by full-rivet pullout. This failure type was previously reported for Friction Riveting as a rather weak joint and is related to the ductile character of the material or insufficient rivet tip deformation/anchoring. Further work on the optimization and characterization of the rivet–polymer interface will clarify the nature of the failure and the possibilities for strength increase. Increasing the heat input might improve the deformation and anchoring of the rivet. This could be achieved by increasing the joining force while keeping the relatively low rotational speed of the current study, as compared with previous published work. To the best knowledge of the authors, this is the largest rivet diameter (Ø 6 mm) successfully reported in Friction Riveting.

Author Contributions: The authors contributed to this research work as follows: conceptualization, L.B. and A.F.; methodology, L.B. and M.B.; modelling, A.B.; experiments and validation, A.F. and M.B.; resources, A.F.; data curation, all equally; writing—original draft preparation, A.B.; writing—review and editing, L.B.; supervision, A.F.; project administration, L.B. and A.F.; funding acquisition, A.F. All authors have read and agreed to the published version of the manuscript.

Funding: This research was funded by the Romanian Ministry for Research, Innovation and Digitalization through national grant number PN-III-P11.1-MCT-2018-0032.

Institutional Review Board Statement: Not applicable.

Informed Consent Statement: Not applicable.

Data Availability Statement: The data presented in this study are available on request from the corresponding author.

Conflicts of Interest: The authors declare no conflict of interest. The funders had no role in the design of the study; in the collection, analyses, or interpretation of data; in the writing of the manuscript; or in the decision to publish the results.

References

1. Modi, S.; Stevens, M.; Chess, M. Mixed Material Joining Advancements and Challenges. *Cent. Automot. Res.* 2017. Available online: http://www.cargroup.org/wp-content/uploads/2017/05/Joining-Whitepaper-Final_May16.pdf (accessed on 1 May 2021).
2. Filho, S.T.A.; Blaga, L.-A. *Joining of Polymer-Metal Hybrid Structures: Principles and Applications*; Wiley: Hoboken, NJ, USA, 2018.
3. Amancio-Filho, S.T.; Beyer, M.; Santos, J.F. Method of Connecting a Metallic Bolt to a Plastic Workpiece. U.S. Patent No. 7,575,149, 18 August 2009.
4. Amancio-Filho, S.T. *Friction Riveting: Development and Analysis of a New Joining Technique for Polymer-Metal Multi-Materials Structures*; Technische Universität Hamburg-Harburg: Hamburg, Germany, 2007.
5. Cipriano, G.P.; Blaga, L.A.; Dos Santos, J.F.; Vilaça, P.; Amancio-Filho, S.T. Fundamentals of Force-Controlled Friction Riveting: Part I—Joint Formation and Heat Development. *Materials* **2018**, *11*, 2294. [CrossRef] [PubMed]
6. Cipriano, G.P.; Blaga, L.A.; Dos Santos, J.F.; Vilaça, P.; Amancio-Filho, S.T. Fundamentals of Force-Controlled Friction Riveting: Part II—Joint Global Mechanical Performance and Energy Efficiency. *Materials* **2018**, *11*, 2489. [CrossRef] [PubMed]
7. Rodrigues, C.; Blaga, L.; dos Santos, J.; Canto, L.; Hage, E.; Amancio-Filho, S. FricRiveting of aluminum 2024-T351 and polycarbonate: Temperature evolution, microstructure and mechanical performance. *J. Mater. Process. Technol.* **2014**, *214*, 2029–2039. [CrossRef]
8. De Proença, B.C.; Blaga, L.; Dos Santos, J.F.; Canto, L.B.; Amancio-Filho, S.T. Friction riveting ('FricRiveting') of 6056 T6 aluminium alloy and polyamide 6: Influence of rotational speed on the formation of the anchoring zone and on mechanical performance. *Weld. Int.* **2017**, *31*, 509–518. [CrossRef]
9. Johnson, R.O.; Burlhis, H.S. Polyetherimide: A new high-performance thermoplastic resin. *J. Polym. Sci. Polym. Symp.* **2007**, *70*, 129–143. [CrossRef]
10. Borges, M.; Amancio-Filho, S.; dos Santos, J.; Strohaecker, T.; Mazzaferro, J. Development of computational models to predict the mechanical behavior of Friction Riveting joints. *Comput. Mater. Sci.* **2012**, *54*, 7–15. [CrossRef]
11. Hynes, N.R.J.; Vignesh, N.J.; Velu, P.S. Low-speed friction riveting: A new method for joining polymer/metal hybrid structures for aerospace applications. *J. Braz. Soc. Mech. Sci. Eng.* **2020**, *42*, 1–16. [CrossRef]
12. Gagliardi, F.; Conte, R.; Ciancio, C.; Simeoli, G.; Pagliarulo, V.; Ambrogio, G.; Russo, P. Joining of thermoplastic structures by Friction Riveting: A mechanical and a microstructural investigation on pure and glass reinforced polyamide sheets. *Compos. Struct.* **2018**, *204*, 268–275. [CrossRef]

processes

Article

Simulation of Glass Fiber Reinforced Polypropylene Nanocomposites for Small Wind Turbine Blades

Yasser Elhenawy [1], Yasser Fouad [2,*], Haykel Marouani [2] and Mohamed Bassyouni [3,4]

[1] Department of Mechanical Power Engineering, Faculty of Engineering, Port Said University, Port Said 42526, Egypt; dr_yasser@eng.psu.edu.eg
[2] Faculty of Applied Engineering, Muzahimiyah Branch, King Saud University, Riyadh 11451, Saudi Arabia; hmarouani@ksu.edu.sa
[3] Department of Chemical Engineering, Faculty of Engineering, Port Said University, Port Said 42526, Egypt; m.bassyouni@eng.psu.edu.eg or mbassyouni@zewailcity.edu.eg
[4] Materials Science Program, University of Science and Technology, Zewail City of Science and Technology, October Gardens, 6th of October 12578, Egypt
* Correspondence: yfouad@ksu.edu.sa; Tel.: +966-059-681-1898

Abstract: This study aims to evaluate the effect of functionalized multi-walled carbon nanotubes (MWCNTs) on the performance of glass fiber (GF)-reinforced polypropylene (PP) for wind turbine blades. Support for theoretical blade movement of horizontal axis wind turbines (HAWTs), simulation, and analysis were performed with the Ansys computer package to gain insight into the durability of polypropylene-chopped E-glass for application in turbine blades under aerodynamic, gravitational, and centrifugal loads. Typically, polymer nanocomposites are used for small-scale wind turbine systems, such as for residential applications. Mechanical and physical properties of material composites including tensile and melt flow indices were determined. Surface morphology of polypropylene-chopped E-glass fiber and functionalized MWCNTs nanocomposites showed good distribution of dispersed phase. The effect of fiber loading on the mechanical properties of the PP nanocomposites was investigated in order to obtain the optimum composite composition and processing conditions for manufacturing wind turbine blades. The results show that adding MWCNTs to glass fiber-reinforced PP composites has a substantial influence on deflection reduction and adding them to chopped-polypropylene E-glass has a significant effect on reducing the bias estimated by finite element analysis.

Keywords: polymer nanocomposite; MWCNTs; horizontal axis wind turbine; finite element analysis; Ansys

Citation: Elhenawy, Y.; Fouad, Y.; Marouani, H.; Bassyouni, M. Simulation of Glass Fiber Reinforced Polypropylene Nanocomposites for Small Wind Turbine Blades. *Processes* **2021**, *9*, 622. https://doi.org/10.3390/pr9040622

Academic Editors: Shaghayegh Hamzehlou, Anil K. Bhowmick and M. Ali Aboudzadeh

Received: 19 January 2021
Accepted: 31 March 2021
Published: 1 April 2021

1. Introduction

Carbon dioxide and other greenhouse gases that contribute to global warming are not released by renewable energy sources such as sun and wind [1]. In the power field, composites are of benefit in the wind sector, especially in manufacturing of turbine blades and the use of the finite element method. Finite element analysis is a feasible tool for simulating/predicting how wind, heat, and solar radiation, centrifugal force, and gravitational loading will affect the blades and the ideal geometry for turbine optimization [2–5]. A wind turbine blade with a class 1 kW horizontal axis was constructed and investigated by Park [6]. The wind turbine blade was fabricated using a natural flax fiber composite. The design outcomes of areas of the flax/epoxy composite blade were compared with the layout outcomes of glass/epoxy composite blade. To calculate the structural design of the composite blade, a static finite element analysis was performed. Structural analysis confirms that structural protection, blade tip deflection, structural integrity, resonance, and weight are suitable for the engineered natural fiber reinforced polymer composites.

A selection process for an airfoil and the aerodynamic design of a rotor blade were addressed by Habali and Saleh [7]. To build the rotor blade, they used glass fiber reinforced plastic (GFRP) and performed a static resistance test to evaluate its load-bearing potential. Sharma and Shukla [8] studied the impact of carbon nanotubes dispersion and functionalization on the relevant properties of multigrade carbon epoxy composites. With the addition of carbon nanotubes (CNTs) in plastic, Young's modulus, interlayer shear strength, and flexural modulus improved by 51.46, 39.62, and 38.04%, respectively.

Using finite element analysis, the structural response of a small composite wind turbine blade subjected to gyroscopic load was studied by Costa et al. [9]. As a case study, a 5 kW small wind turbine blade was used for aerogenesis. A finite element model of the 2.5 m long composite blade was constructed, and the accuracy of its forecasts against experimental data was verified. The predicted results indicated that gyroscopic loading caused the blade to deflect in any direction of the fin and forward sheet, resulting in relatively excessive stresses within the structure of the composite blade.

Ullah et al. [10] used a comprehensive finite element modeling to simulate structural integrity and failure in composite blades using Ansys software. The outcomes confirmed that the blade was subjected to high compressive force, causing local displacement of the skin. In addition, the analysis outcomes indicated skin–spar debonding due to warping on the adhesive surface. This is the preliminary failure mode that can lead to gradual deterioration of the blade structure.

In terms of the mechanical properties of polymer composites, fibers play a crucial role. They have qualities that are reinforcing [11–13]. In each case, however, there are certain properties to be considered. Glass fibers (GFs), carbon fibers (CFs), and aramid fibers (AFs) are the most widely used in the production of wind turbine blades [14–19]. Other examples are nylon, polyester (PE), polytetrafluoroethylene (PTFE), jute, flax, and steel fibers, which are used for specific purposes [4].

In contrast to carbon fibers, glass fibers have a less complex modulus and are usually less costly. They have the advantage of being electrically non-conductive and hence the metallic-based construction frame of the turbine blades, will not produce a galvanic cell over time. In the fiber phase in the matrix, diverse characteristics include combinations of different materials will be durable and able to withstand a range of influences the blade material is exposed to.

In contrast to thermoset matrices, composites with thermoplastic matrices show greater resistance to impact loads [20]. Thermoplastic composites exhibit ductile modes of failure rather than brittle failure [21]. Unlike ruptures in thermosets, thermoplastic composites undergo plastic deformation [22]. The strength of composites strengthened by glass fibers depends on the interface between the glass fibers and the polymers. Due to their high specific strength, impact resistance, infinite shelf life, high strain before failure, and short processing time, glass fiber-reinforced thermoplastic composites are used in many applications [23]. The physical and chemical properties depend upon the composition. High compatibility makes GF material effective. The compatibility of GF with polymer matrices can be increased by modifying the glass fiber surface or introducing additives in a polymer matrix. However, application of this method is highly cost-dependent. E-glass fibers are preferred in the plastic industry, as they are cheaper than available grades of GFs [24].

With the application of heat, thermoplastics flow readily [25–27]. They are perfect for manufacturing by extrusion or molding injection. It is a complex art to impregnate thermoplastics with fibers. The term "semipreg" is used to explain how thermoplastics are impregnated with fibers. This is because rather than impregnating, the resin acts more like a coating.

Carbon nanotubes can be used to enhance the mechanical properties of wind blades. However, when processing composites, CNTs tend to aggregate in bundles. The main properties of CNTs are high modulus, on the order of 100 GPa; high tensile strength of approximately 50,000 MPa; and density of 1.33 g/cm^3. Other properties include high electrical conductivity, thermal conductivity, and insensitivity to thermal degradation [28].

Many factors are involved in the extrusion process to keep reinforcement undamaged, e.g., the loading of reinforcing fibers, the dimensions of the extruder with respect to the barrel, and the heating and opening of the die. High barrel pressure can reduce the tensile strength; however, the orientation of fibers improve with higher injection speed, and thus improve the mechanical properties [29–32]. In injection molding, fiber orientation is crucial, as shear forces are present. Usually, orientation is ignored hence strength gained in one direction is not attainable in another direction. Short-chopped fiber, particle whiskers, and particles are common reinforcement tools used in injection molding. In general, the length of reinforcement is not greater than the size of mold gate. The typical length of fibers used in injection molding can be up to 25.4 mm. Fiber length is an important factor in the mechanical properties of a part obtained from injection molding. The average length of short fibers is 0.33 mm, whereas long fibers average 2.9 mm. The higher aspect ratio of long fibers provides superior mechanical properties compared to short fibers [20].

In a study done by Crowson et al. [33], neat polypropylene (PP) was compared with 20 wt.% glass fiber-filled PP. Entrance pressure drop was found to increase six times in the presence of glass fibers [34]. During injection molding processing, fibers are subjected to high stresses in the viscous melt. This causes them to break and deform. By studying fiber behavior under different flow conditions, adjustments can be made to maintain a high aspect ratio. This will enhance the reinforcing capacity of fibers. There is a direct relation between the loading level of glass fiber and the mechanical properties of GF/PP composite. In one study, PP was loaded by 10, 20, 28, and 35 wt.% glass fiber in an injection molding process. It was found that a higher percentage of glass fiber loading resulted in lower reinforcement efficiency [35].

During polymer melting, fiber is subjected to high shear and bending forces. This makes the bending load more than the failure load, and breakage occurs. Once the critical stress value is reached, the glass fibers break, and natural fibers break by repeated deformation cycles. Initially, reprocessing leads to increased fiber dispersal within the matrix, but repeated reprocessing will cause a loss of mechanical properties.

Excessive reprocessing causes mechanical as well as thermal degradation of matrix and reinforcement. The matrix is degraded by random chain scissions in the polymer chain, which is likely to decrease the molecular weight of the polymer and cause a loss in mechanical properties. In a study on the effect of molecular weight on interfacial properties of GF/PP composites, it was found that recycled PP with lower molecular weight has low viscosity that enables it to penetrate easily. This leads to higher interfacial strength between GF and PP, resulting in more breakage of GF during tensile tests [36].

In this work, PP was chosen as the thermoplastic matrix material for a small wind turbine blade used in residential applications because (i) it is relatively cheap, (ii) readily available, and (iii) commercial PP blades are available for comparative analysis using Ansys software. Fabrication of wind turbine rotor blades with thermoplastics is rarely covered in the literature, but some examples are available. In this research, multi-walled carbon nanotubes (MWCNTs) are mixed with polypropylene composite to improve the mechanical properties. Moreover, an intensive study was conducted on the interfacial force between PP and glass fiber in the presence of functionalized MWCNTs. This study is aimed at evaluating the suitability of glass fiber (GF)-reinforced polypropylene (PP) composites for wind turbine blades. The rotor blade length is in the range of 0.5 to 1.3 m, typically used for small-scale wind turbine systems such as for residential applications. To our knowledge, most researchers have studied fiber-reinforced thermoset (epoxy and vinyl ester) nanocomposites for wind turbine blade fabrication, and no detailed study has been conducted on glass fiber-reinforced thermoplastic-functionalized nanocomposites for blade manufacture. Desirable properties of functional CNTs, including excellent electrical, thermal conductivity, and mechanical properties, will help in enhancing the performance of thermoplastic composites for wind blade applications.

2. Materials and Methods

Glass fiber-reinforced polypropylene samples were fabricated in the presence of MWCNTs. Different chopped glass fiber and MWCNT loadings were applied and tested.

2.1. Material Characterization

2.1.1. Glass Fiber

Glass fiber properties are provided according to the technical data sheet from the supplier as listed in Table 1.

Table 1. Physical and mechanical properties of glass fiber.

Parameter	Value
Tensile strength at break (MPa)	Warp: 1555; weft: 1387
Moisture content	0.11%
Density (kg/m^3)	2.55×10^3
Thermal conductivity (W/m °C)	1.3
Electrical resistivity (Ω·m)	1.0 × 1014
Length (mm)	2.5
Diameter (μm)	13
Density (kg/m^3)	2600
Coupling agent	Polyvinyl acetate with silane

2.1.2. Multi-Walled Carbon Nanotubes (MWCNTs)

The MWCNTs were received from Grafen (Ankara, Turkey). Surface morphology and manufacturer's specifications are listed in Table 2. It is found that the aspect ratio (ratio of length to diameter) of MWCNTs is 170. The catalytic chemical vapor deposition technique was used to produce CNTs. This process gives more than 90% carbon nanotubes and less than 10% metal oxides [37].

Table 2. Specifications of multi-walled carbon nanotubes (MWCNTs).

Multi-Walled Carbon Nanotubes		SEM _MWCNTs
Surface area	275 m^2/g	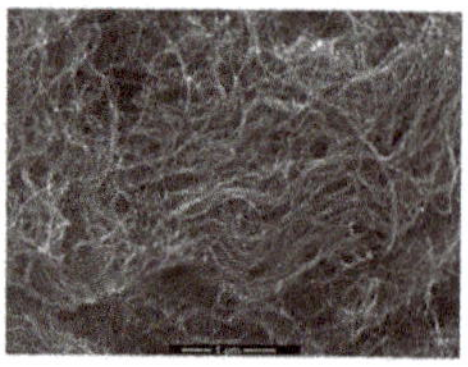
Carbon purity	> 90%	
Diameter	~10 nm	
Length	~1.7 μm	
Metal oxides	9%	

2.1.3. Synthesis of Glass Fiber-Reinforced Polypropylene-Carbon Nanotube Composites

Molding grade polypropylene (PP) was selected as the matrix material. For excellent mold filling, good shine, high stiffness, and antistatic properties, a homopolymer (density 0.9 g/cm^3) was applied. E-glass (UK) chopped strand with an alkali content of less than 1% and polyvinyl acetate coated with a silane binding agent was used. The density of the glass fiber was 2.56 g/cm^3, the mean length of the fiber was 2.5 mm, and the diameter was 10 μm, with an aspect ratio of 250.

Glass fiber-reinforced PP MWCNT composites were synthesized and manufactured using extrusion and injection molding processes. The weight percentages of MWCNTs varied between 0 and 8 wt./wt.%. Physical and mechanical properties were measured using tensile tests. The effect of MWCNTs on the melt flow index (MFI) and mechanical properties of PP-GF composites was studied. The fracture surface of PP composites was examined using SEM analysis.

2.1.4. PP-Nanocomposite Processing

Glass fiber (~10 wt.%) was mixed in a co-rotating twin extruder at 200 °C and 270 rpm with PP-MWCNTs as a master batch (1–15%) and pristine PP. The composite was then shredded and processed at 200 °C and 500 bar using injection molding (Allrounder, Arburg, Germany) to produce tensile and sheet sample shapes. Specimens for determination of tensile properties according to ISO 527-4 189 (multidirectional and fiber-reinforced materials) were fabricated as rectangles with dimensions of 25 mm width, 250 mm length, and 3 mm thickness. A tensile evaluation was carried out using a Zwick B066550 universal testing machine (Germany). A 10 kN load cell was employed. The rate of tension at room temperature was 2 mm/s. To investigate the effect of MWCNTs on the PP flow property, the melt flow index was determined.

2.2. Model Wind Turbine

Based on an optimization analysis, the HAWT geometry and its operating conditions were chosen. More information on optimizing wind turbine performance is given below. The HAWT variable is a three-bladed HAWT with a diameter of 1.26 to 2.83 m. Each blade followed the profile of the NACA20015 airfoil and had a blade length between 0.5 and 1.3 m. This gave an aspect ratio of 3, 4, 5, 6, or 7 based on the ratio between the diameter of the turbine and the length of the blade. The turbine rotates at 34.26 rad/s under optimum conditions when the wind speed is 7 m/s. Table 3 shows the key parameters of the optimized turbine. The 3D geometry and main dimensions of the optimized HAWT are shown in Figure 1. The shape of the blade cross-section, comprising two shear webs, is shown in Figure 2.

Table 3. Main parameters of optimized horizontal axis wind turbine (HAWT).

Parameters	Values	Unit
No. of blades	3	
Radius of rotor (R)	r + L	m
Radius of hub (r)	0.13	m
Length of blade (L)	0.5, 0.7, 0.9, 1.1, 1.3	m
Rated speed of wind (C_w)	7	m/s
Rated speed of the rotor (ω)	34.26	rad/s
Airfoil	NACA0015	

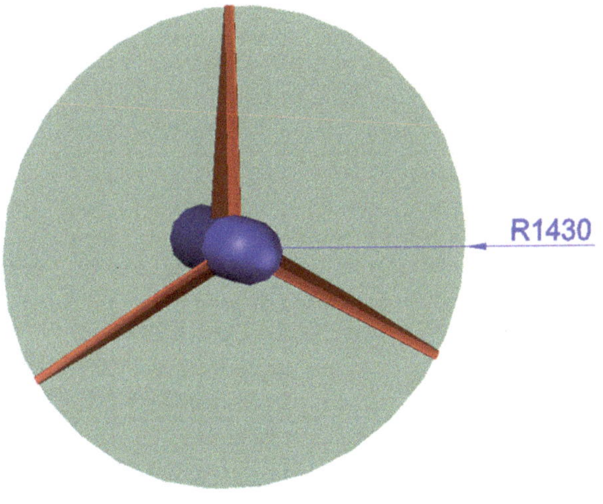

Figure 1. 3D model of geometry of optimized HAWT.

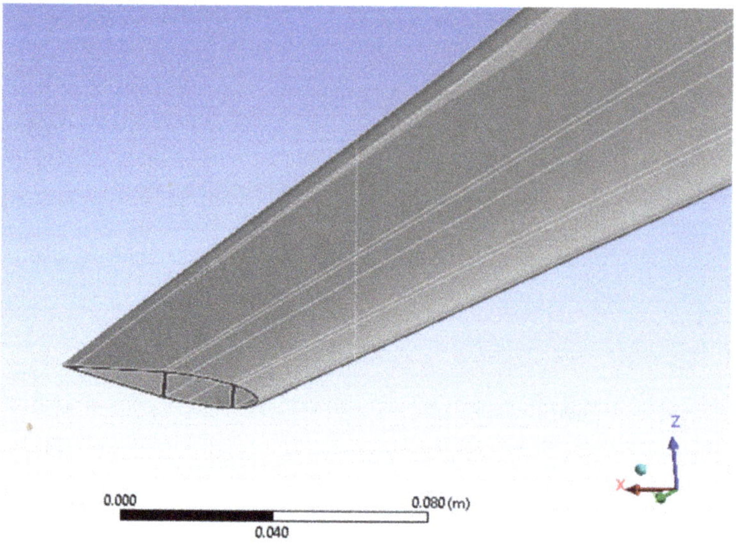

Figure 2. 3D model of geometry of optimized HAWT blade.

The tip speed ratio (TSR), which determines the relative wind speed and the blade velocity, is shown in Equation (1):

$$TSR = \frac{\omega R}{C_W} \quad (1)$$

where ω is rotor speed (rad/s), R is radius (m), and C_W is wind speed (m/s).

3. Results and Discussion

3.1. Tensile Properties of GF-PP-MWCNT Composites

The glass fiber-reinforced PP-MWCNT system showed superior strength compared to PP/GF composites. A substantial improvement (up to 36%) in the strength of GF-reinforced PP was obtained in the presence of MWCNTs (2% wt./wt.) compared to neat polypropylene (31 MPa). At higher CNT loading, the yield strength of GF-PP nanocomposites did not show remarkable improvement, as shown in Figure 3. This can be attributed to the high affinity of CNTs as they create tiny clusters/agglomerates and impact the composite structure. The modulus of elasticity of PP-MWCNTs improved remarkably from 1.5 to 2.5 GPa in the presence of MWCNTs (2% wt./wt), as observed in Figure 4. The loading of MWCNTs showed significant improvement in modulus of elasticity behavior from 1 to 8% wt./wt. It was shown that stiffness did not primarily depend on the filler–matrix interface, but more on the absolute filler content in the direction of tensile loading, as the elastic modulus is determined at low strain values (0.05–0.25%) as a tangent modulus, where no interfacial de-bonding is considered to occur even in the event of weak adhesion [20]. Therefore, the expected enhancement of the bonding of fillers/matrix does not affect the composite stiffness/modulus of elasticity. Around 49% reduction in elongation in the presence of 2% wt./wt. of MWCNTs was recorded, as shown in Figure 5.

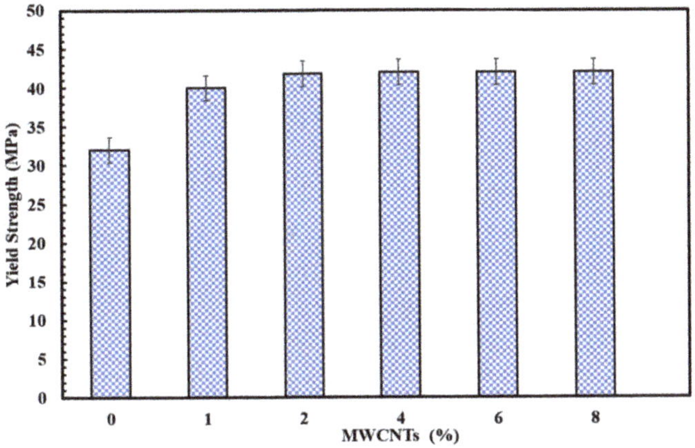

Figure 3. Effect of MWCNTs on the yield strength of PP-GF composites.

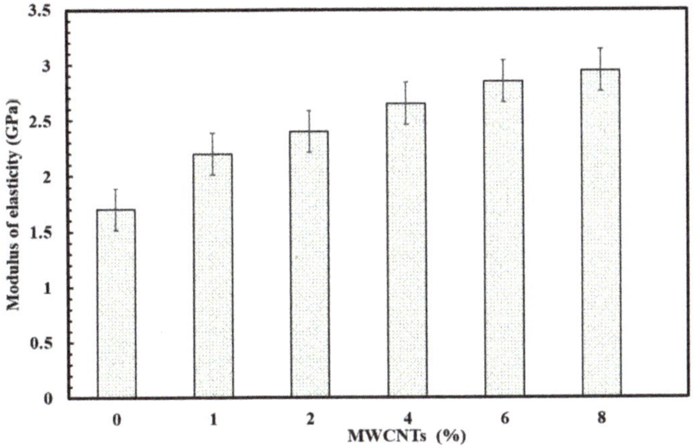

Figure 4. Effect of MWCNTs on modulus of elasticity of PP-GF composites.

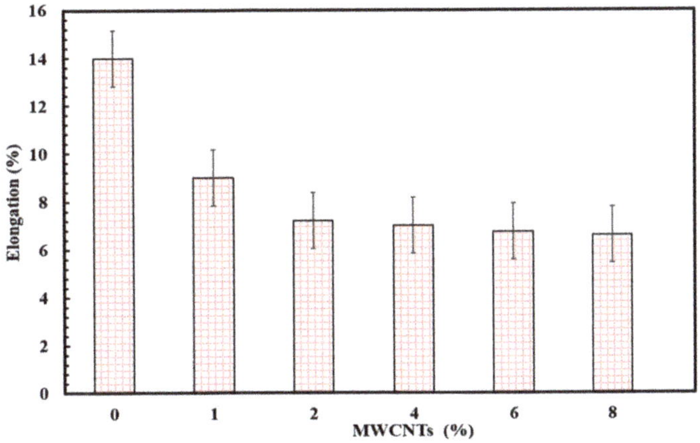

Figure 5. Effect of MWCNTs on elongation of PP-GF nanocomposites.

3.2. Melt Flow Index of PP-Chopped E-Glass Fiber Composites

Since the PP composites were manufactured using injection molding, the melt flow index (MFI) has a significant influence on their quality. The melt flow index is a single point measurement that gives an indication of the ability of the polymer to flow. To this end, MFI was measured according to ISO 1133 at a temperature of 230 °C and a weight of 2.16 kg. The average of 10 MFI values was considered.

The MFI of PP-MWCNT composites decreased with increased MWCNT %, as shown in Table 4. Higher-density PP composites are obtained as MWCNTs are added. The MFI of the neat PP is around 11 g/10 min. The PP-MWCNT composite samples had much lower MFI values than the neat PP at higher concentrations. Higher MFI correlates with lower viscosity at that fixed shear rate.

Table 4. Effect of MWCNTs on melt flow index (MFI) of polypropylene.

MWCNTs wt.%	0	1	2	4	6	8
MFI	11.0	8.4	6.20	3.30	0.47	0.03

Results are in agreement with previous studies that showed lower MFI in the presence of fillers [38].

3.3. Surface Analysis of PP-Glass Fiber Composites

Figure 6 shows the direction and length of the glass fibers using digital microscopy (Keyence VHX-500F, Osaka, Japan). When fiber length is greater than critical length, the tension is expected to rise linearly. It is observed that the average fiber length was reduced to 0.63 mm due to extrusion and injection processes. If the fiber length is less than the critical length, fiber pull-out or matrix–fiber interfacial bond failure can occur. It was reported that PP-glass fiber composite stiffness reached the maximum value with a fiber length of 1 mm [39].

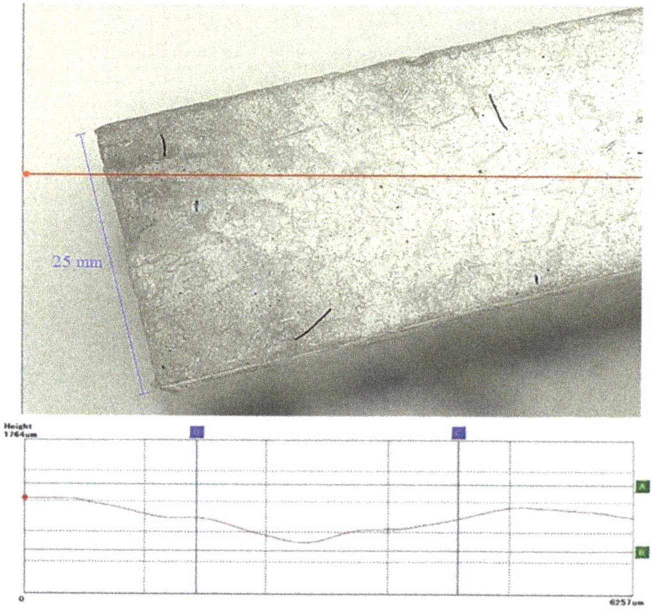

Figure 6. Glass fiber length and distribution in GF-PP composites using 3-D surface images.

3.4. Surface Morphology

It is possible to visually detect the presence of interfacial adhesion due to chemical coupling. The fracture surface can be investigated by scanning electron microscopy (SEM). Surface morphology analysis can discriminate between uncoupled and chemically bound composites. If the fiber surface is smooth, this means the surface did not adhere to the matrix. There is also confirmation at the base of the glass fiber that de-bonding has taken place.

The tensile fracture morphology of the PP–glass fiber interface was studied using SEM at different chopped glass fiber and MWCNTs loadings (Figure 7). In the case of PP-GF (Figure 7a,b), many cavities were observed. Some glass chopped fibers were observed where the fiber pulled out. This indicates poor interfacial bonding between the matrix and the dispersing phase. As the glass fiber content increases, the poor interfacial bonding and the brittleness of the filler affect the tensile property (increasing the modulus). However, the mechanical properties are significantly modified in the presence of modified MWCNTs. Figure 7c,d shows fewer pores at the fracture surface, good glass fiber dispersibility, and matrix layers covering the fiber surface, which reflects good adhesion between polypropylene, glass fiber, and MWCNTs.

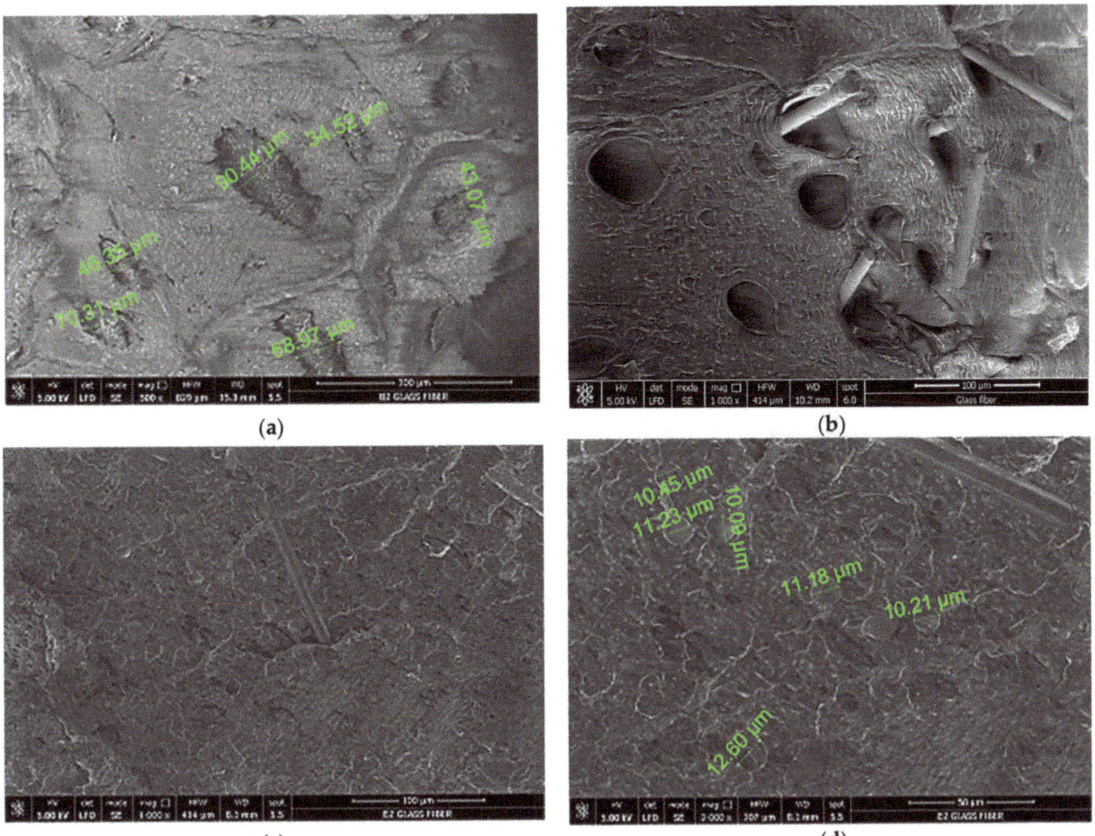

Figure 7. SEM PP composites: (**a**) cavities, (**b**) glass fiber surface, (**c**) PP-GF- modified MWCNTs, (**d**) fiber distribution in PP GF–modified MWCNTs.

3.5. Finite Element Method Modelling

The finite element method (FEM) simulation of the optimized HAWT with composite blades was carried out using the Ansys Static Structural module, a widely used FEA modeling program. To optimize the HAWT geometry, the FEM modeling procedure was applied. A comprehensive overview of the geometry, meshing, materials, and boundary conditions in the FEM model are included in this section.

3.5.1. Geometry

A shaft and three blades form the geometry of the built HAWT as stipulated. The geometry in the SolidWorks software is generated. Figure 8 shows the shape of the HAWT geometry. Table 5 lists the properties of the thermoplastic (PP) and two different composites formulated based on it the FEM mode of the proposed HAWT.

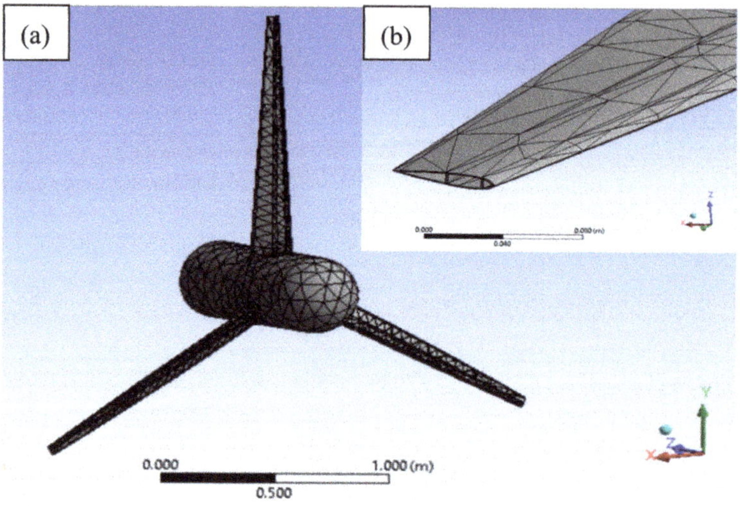

Figure 8. Finite element analysis of mesh: (**a**) blade; (**b**) close view of blade tip.

Table 5. Various properties of glass fiber-reinforced thermoplastic composites.

Property	PP	PP/GF (Short Fiber Composite) at 10% wt./wt.	PP/GF/MWCNTs
Density (kg/m^3)	900	1020	1030
Thermal expansion coefficient (μstrain/°C)	27.1	10.4	9.7
Thermal conductivity (Wm^{-1}k^{-1})	0.88	0.267	0.307
Specific heat capacity (JKg^{-1}k^{-1})	1932	1650	1700
Tensile strength (MPa)	32	34.5	42
Poisson's ratio	0.43	0.32	0.30
Young's modulus (GPa)	1.34	1.72	2.5

3.5.2. FEA Mesh

For meshing of the HAWT, the standardized mesh topology was used. The baseline mesh for the entire turbine is shown in Figure 8a, and the standardized mesh for the

turbine blade is shown in Figure 8b. In order to select a suitable mesh that would provide good accuracy and have a fair computing cost, a mesh sensitivity analysis was carried out. Three meshes were developed for this analysis: coarse mesh with 33,489 nodes, baseline mesh with 96,700 nodes, and fine mesh with 137,467 nodes. Further details are available in Table 6. Figure 9 shows the effect of mesh size using PP/GF material at the expected maximum deflection for the highest rotational velocity of 26.22 rad/s. It is noted that the baseline and fine mesh predictions were in strong agreement, while in the coarse mesh it is significantly underestimated. The baseline mesh with 96,700 nodes and 20,100 elements was therefore considered to be the appropriate mesh and was used for further research.

Table 6. Information on various meshes used in mesh sensitivity analysis.

	Coarse Mesh	Baseline Mesh	Fine Mesh
Nodes	33,489	96,700	137,467
Elements	7210	20,100	27,523
Maximum deflection (mm)	3.652	6.322	6.721

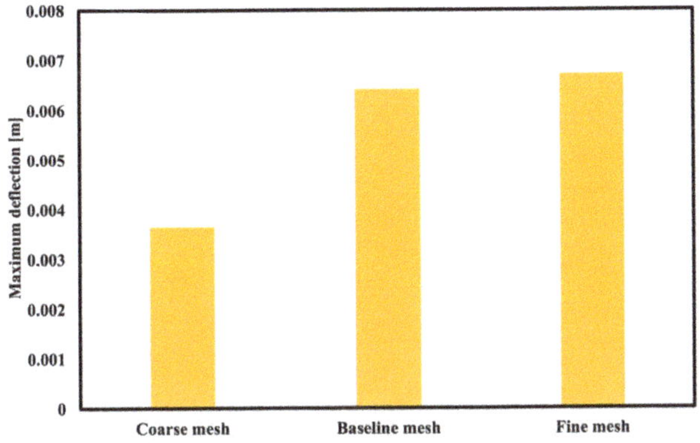

Figure 9. Effect of mesh size using PP/GF material on expected maximum deflection for highest rotational velocity of 26.22 rad/s.

3.5.3. Boundary Conditions

Horizontal axis wind turbines (HAWTs) are subjected to several loads, including aerodynamic, gravitational, and centrifugal loads. The current research focuses primarily on the effects of the various structural materials on the blade deflection at different centrifugal loads. Centrifugal loads correspond to the tip speed ratio (TSR). By applying the relevant rotational velocity, these centrifugal loads are applied to the HAWT rotor in the Ansys Static Structural program.

3.5.4. Solution and Post-Process Results

For the solution and post-processing of observations, the Ansys Static Structural program was used. This software, including static and modal analysis, is capable of numerous structural analyses. Several findings, such as blade deformation and stress distribution, can be derived. In the present analysis, the blade deformation is derived from simulations of the static structure.

The FEM structure analysis of an optimized HAWT was performed based on the method described in Section 2. Graphical representations of the deformation of turbine blades made with PP, glass fiber-reinforced PP, and glass fiber-reinforced PP-MWCNT materials are shown in Figures 10–12, respectively, at an ideal TSR of 3. It was noted that maximum deflections occurred at the blade ends, while strong deflections were observed at the blade midspan. In order to study the impact of applying different materials to maximum deflection at different centrifugal loads, various experiments were carried out. The maximum deflection of the blades at TSR of 7 for different materials is shown in Figure 13.

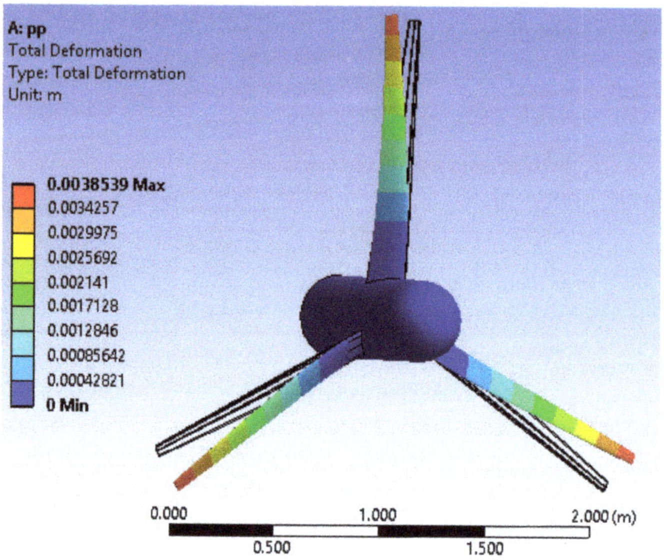

Figure 10. Graphical representation of maximum turbine blade deformation with PP material at TSR = 3.

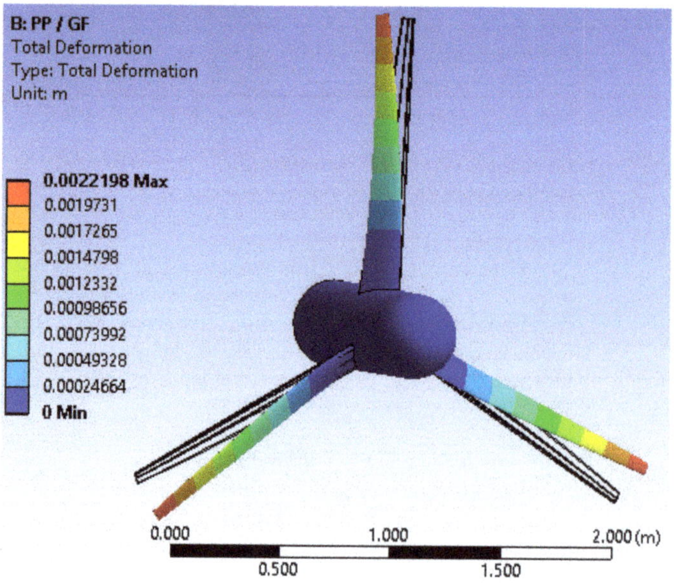

Figure 11. Graphical representation of maximum turbine blade deformation with PP/GF material at TSR = 3.

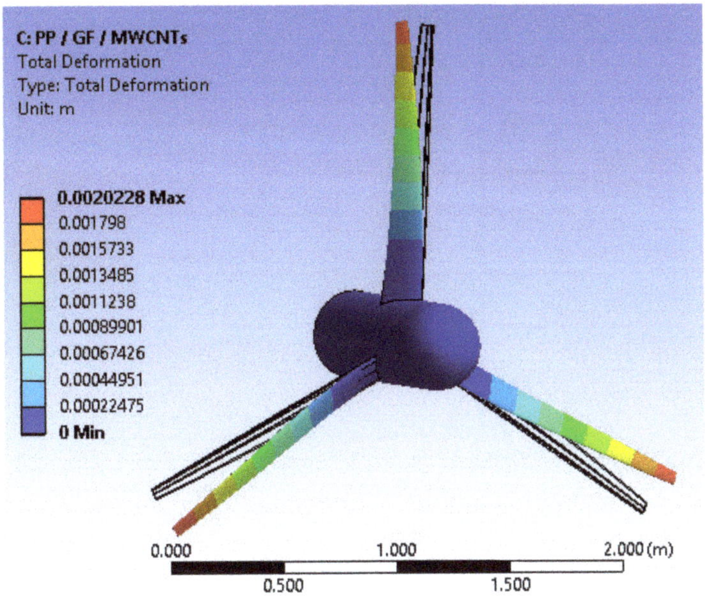

Figure 12. Graphical representation of maximum turbine blade deformation with PP/GF/MWCNT content at TSR = 3.

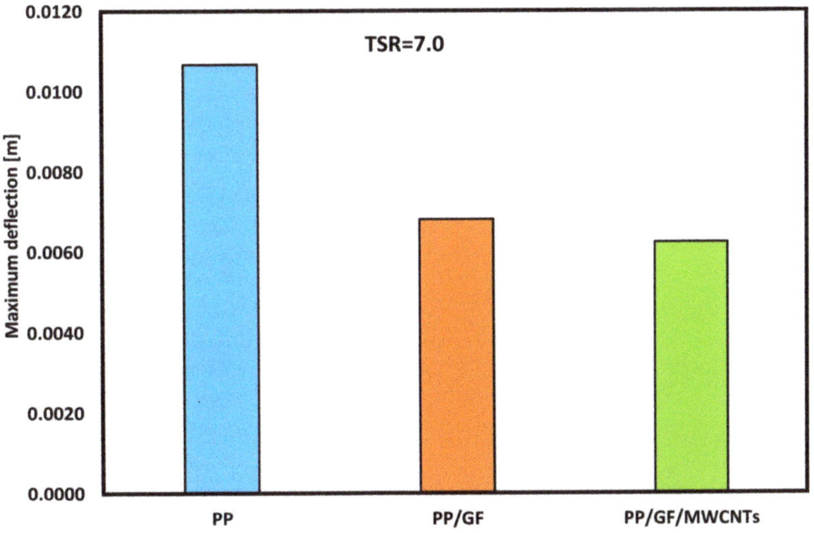

Figure 13. Maximum deflection of blades with various materials at TSR = 7.

Figure 14 shows the maximum deflections found using the Ansys Static Structural program. It can be seen that adding MWCNTs of 2% to glass fiber-reinforced PP composite material leads a reduction in deflection by 10% compared with GF-reinforced PP.

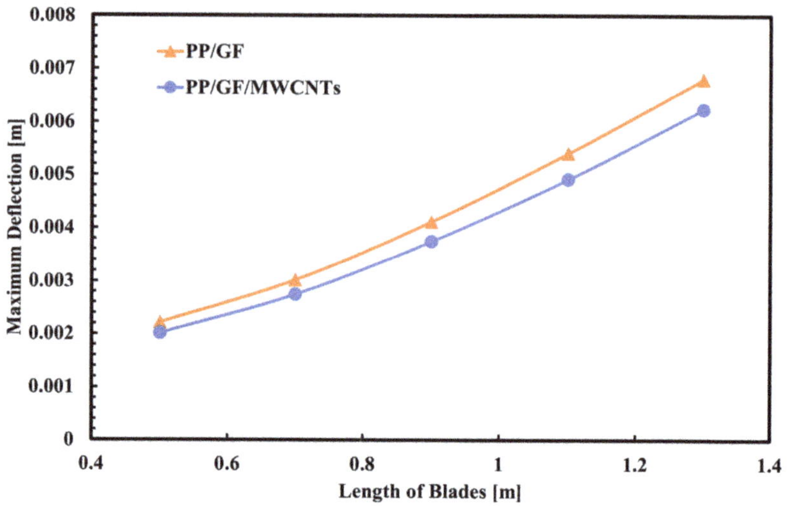

Figure 14. Maximum deflection of different blade material.

4. Conclusions

Glass fiber-reinforced polypropylene nanocomposites were fabricated using the injection molding process. Addition of 2 wt./wt.% MWCNTs and 10 wt./wt.% GF to polypropylene yielded remarkable improvement in the tensile strength of polypropylene. Ansys analysis showed enhancement of blade deflection using MWCNTs in wind blades of different lengths. Compared with GF-reinforced PP, the addition of 2 wt./wt.% MWCNTs to glass fiber-reinforced PP composite material contributed to a 10% reduction in deflection. It is found that short glass fiber-reinforced PP nanocomposites can be effectively used to fabricate small/micro wind turbine blades. For larger wind blades, GF-reinforced PP nanocomposites can be used with continuous glass fiber to synthesize hybrid materials with superior mechanical properties.

Author Contributions: Conceptualization, Y.F. and Y.E.; methodology, Y.F.; software, Y.E.; validation, H.M., M.B. and Y.F.; formal analysis, Y.F.; investigation, Y.F. and M.B.; resources, Y.F. and H.M.; data curation, M.B.; writing—original draft preparation, Y.F.; writing—review and editing, M.B.; visualization, Y.E.; supervision, M.B.; project administration, Y.F.; funding acquisition, Y.F. All authors have read and agreed to the published version of the manuscript.

Funding: The authors extend their appreciation to the Deputyship for Research & Innovation, Ministry of Education in Saudi Arabia for funding this research work through the project number (DRI-KSU-1410).

Institutional Review Board Statement: Not applicable.

Informed Consent Statement: Not applicable.

Data Availability Statement: The data presented in this study are available on request from the corresponding author.

Conflicts of Interest: The authors declare no conflict of interest.

References

1. Zoromba, M.S.; Tashkandi, M.; Alshehri, A.; Abdel-Aziz, M.; Bassyouni, M.; Mahmoud, S.; Ben Slimane, A.; Al-Hossainy, A. Polymer solar cell based on doped o-anthranilic acid and o-aminophenol copolymer. *Opt. Mater.* **2020**, *104*, 109947. [CrossRef]
2. Ashby, M.; Bréchet, Y.; Cebon, D.; Salvo, L. Selection strategies for materials and processes. *Mater. Des.* **2004**, *25*, 51–67. [CrossRef]
3. Elhenawy, Y.; Hafez, G.; Abdel-Hamid, S.; Elbany, M. Prediction and assessment of automated lifting system per-formance for multi-storey parking lots powered by solar energy. *J. Clean. Prod.* **2020**, *266*, 121859. [CrossRef]
4. Marsh, G. Reinforced plastics transform 'small wind' market. *Reinf. Plast.* **2004**, *48*, 22–26. [CrossRef]

5. Bassyouni, M.; Gutub, S. Materials selection strategy and surface treatment of polymer composites for wind turbine baldes fabrication. *Polym. Polym. Compos.* **2013**, *21*, 463–471.
6. Park, H. A study on structural design and analysis of small wind turbine blade with natural fibre (flax) composite. *Adv. Compos. Mater.* **2015**, *25*, 1–18. [CrossRef]
7. Habali, S.; Saleh, I. Local design, testing and manufacturing of small mixed airfoil wind turbine blades of glass fibre reinforced plastics: Part I: Design of the blade and root. *Energy Convers. Manag.* **2000**, *41*, 249–280. [CrossRef]
8. Sharma, K.; Shukla, M. Three-phase carbon fiber amine functionalized carbon nanotubes epoxy composite: Pro-cessing, characterisation, and multiscale modeling. *J. Nanomater.* **2014**, *2014*. [CrossRef]
9. Da Costa, M.S.P.; Clausen, P.D. Structural Analysis of a Small Wind Turbine Blade Subjected to Gyro-scopic Load. *J. Phys. Conf. Ser.* **2020**, *1618*, 042006. [CrossRef]
10. Ullah, H.; Ullah, B.; Silberschmidt, V.V. Structural integrity analysis and damage assessment of a long composite wind turbine blade under extreme loading. *Compos. Struct.* **2020**, *246*, 112426. [CrossRef]
11. Junaedi, H.; Baig, M.; Dawood, A.; Albahkali, E.; Almajid, A. Mechanical and Physical Properties of Short Carbon Fiber and Nanofiller-Reinforced Polypropylene Hybrid Nanocomposites. *Polymers* **2020**, *12*, 2851. [CrossRef] [PubMed]
12. Fanaradelli, T.D.; Rousakis, T.C. 3D Finite Element Pseudodynamic Analysis of Deficient RC Rectan-gular Columns Confined with Fiber Reinforced Polymers under Axial Compression. *Polymers* **2020**, *12*, 2546. [CrossRef] [PubMed]
13. Sherif, G.; Chukov, D.I.; Tcherdyntsev, V.V.; Torokhov, V.G.; Zherebtsov, D.D. Effect of Glass Fi-bers Thermal Treatment on the Mechanical and Thermal Behavior of Polysulfone Based Composites. *Polymers* **2020**, *12*, 902. [CrossRef] [PubMed]
14. Sommer, V.; Stockschläder, J.; Walther, G. Estimation of glass and carbon fiber reinforced plastic waste from end-of-life rotor blades of wind power plants within the European Union. *Waste Manag.* **2020**, *115*, 83–94. [CrossRef] [PubMed]
15. Joustra, J.; Flipsen, B.; Balkenende, R. Structural reuse of high end composite products: A design case study on wind turbine blades. *Resour. Conserv. Recycl.* **2021**, *167*, 105393. [CrossRef]
16. Boyano, A.; Lopez-Guede, J.; Torre-Tojal, L.; Fernandez-Gamiz, U.; Zulueta, E.; Mujika, F. Delamination Fracture Behavior of Unidirectional Carbon Reinforced Composites Applied to Wind Turbine Blades. *Materials* **2021**, *14*, 593. [CrossRef]
17. Rizk, P.; Al Saleh, N.; Younes, R.; Ilinca, A.; Khoder, J. Hyperspectral imaging applied for the detection of wind turbine blade damage and icing. *Remote. Sens. Appl. Soc. Environ.* **2020**, *18*, 100291. [CrossRef]
18. Elhenawy, Y.; Fouad, Y.; Marouani, H.; Bassyouni, M. Performance Analysis of Reinforced Epoxy Functionalized Carbon Nanotubes Composites for Vertical Axis Wind Turbine Blade. *Polymers* **2021**, *13*, 422. [CrossRef]
19. Rodrigues, S.; Marta, A. On addressing wind turbine noise with after-market shape blade add-ons. *Renew. Energy* **2019**, *140*, 602–614. [CrossRef]
20. Bassyouni, M.; Taha, I.; Abdel-Hamid, S.M.-S.; Steuernagel, L. Physico-mechanical properties of chemically treated polypropylene rice straw bio-composites. *J. Reinf. Plast. Compos.* **2012**, *31*, 303–312. [CrossRef]
21. Bassyouni, M.; Sherif, S.A.; Sadek, M.; Ashour, F. Synthesis and characterization of polyurethane–Treated waste milled light bulbs composites. *Compos. Part B Eng.* **2012**, *43*, 1439–1444. [CrossRef]
22. Cosco, F.; Serratore, G.; Gagliardi, F.; Filice, L.; Mundo, D. Experimental Characteri-zation of the Torsional Damping in CFRP Disks by Impact Hammer Modal Testing. *Polymers* **2020**, *12*, 493. [CrossRef] [PubMed]
23. Feih, S.; Wei, J.; Kingshott, P.; Sørensen, B.F. The influence of fibre sizing on the strength and fracture toughness of glass fibre composites. *Compos. Part A Appl. Sci. Manuf.* **2005**, *36*, 245–255. [CrossRef]
24. Thomason, J.; Jenkins, P.; Yang, L. Glass Fibre Strength—A Review with Relation to Composite Recycling. *Fibers* **2016**, *4*, 18. [CrossRef]
25. Rehman, S. Wind energy resources assessment for Yanbo, Saudi Arabia. *Energy Convers. Manag.* **2004**, *45*, 2019–2032. [CrossRef]
26. Biron, M. *Future Prospects for Thermoplastics and Thermoplastic Composites Technical Information for Plastic Users*; Elsevier: Oxford, UK, 2007.
27. Chandler, H. *Wind Energy, the Facts an Analysis of Wind Energy in the EU-25*; European Wind Energy Association: Brussels, Belgium, 2007.
28. Theodore, M.; Hosur, M.; Thomas, J.; Jeelani, S. Influence of functionalization on properties of MWCNT–epoxy nanocomposites. *Mater. Sci. Eng. A* **2011**, *528*, 1192–1200. [CrossRef]
29. Sathishkumar, T.P.; Satheeshkumar, S.; Naveen, J. Glass fiber-reinforced polymer composites–A review. *J. Reinf. Plast. Compos.* **2014**, *33*, 1258–1275. [CrossRef]
30. Jarukumjorn, K.; Suppakarn, N. Effect of glass fiber hybridization on properties of sisal fiber–polypropylene composites. *Compos. Part B Eng.* **2009**, *40*, 623–627. [CrossRef]
31. Arbelaiz, A.; Fernández, B.; Cantero, G.; Llano-Ponte, R.; Valea, A.; Mondragon, I. Mechanical properties of flax fibre/polypropylene composites. Influence of fibre/matrix modification and glass fibre hybridization. *Compos. Part A Appl. Sci. Manuf.* **2005**, *36*, 1637–1644. [CrossRef]
32. Huda, M.S.; Drzal, L.T.; Mohanty, A.K.; Misra, M. Chopped glass and recycled newspaper as reinforcement fibers in injection molded poly(lactic acid) (PLA) composites: A comparative study. *Compos. Sci. Technol.* **2006**, *66*, 1813–1824. [CrossRef]
33. Wharton, S.; Lundquis, J. Atmospheric stability affects wind turbine power collection. *Environmental Res. Lett.* **2012**, *7*, 1–9. [CrossRef]
34. Crowson, R.; Folkes, M.; Bright, P. Rheology of short glass fibre-reinforced thermoplastics and its application to injec-tion moulding I. Fibre motion and viscosity measurement. *Polym. Eng. Sci.* **2004**, *20*, 925–933. [CrossRef]

35. Thomason, J. The influence of fibre length and concentration on the properties of glass fibre reinforced polypropylene. 6. The properties of injection moulded long fibre PP at high fibre content. *Compos. Part A Appl. Sci. Manuf.* **2005**, *36*, 995–1003. [CrossRef]
36. Kam, T.; Jiang, J.; Yang, H.; Chang, R.; Lai, F.; Tseng, Y. Fabrication and Testing of Composite Sandwich Blades for a Small Wind Power System. In Proceedings of the PEA-AIT International Conference on Energy and Sustainable Development: Issues and Strategies. Asian Institute of Technology, Klong Luang, Thailand, 2–4 June 2010.
37. Wang, X.-D.; Vinodgopal, K.; Dai, G.-P. Synthesis of Carbon Nanotubes by Catalytic Chemical Vapor Deposition. *Perspect. Carbon Nanotub.* **2019**. [CrossRef]
38. Nurul, M.; Mariatti, M. Effect of thermal conductive fillers on the properties of polypropylene composites. *J. Thermoplast. Compos. Mater.* **2011**, *26*, 627–639. [CrossRef]
39. Thomason, J. The influence of fibre length and concentration on the properties of glass fibre reinforced polypropylene: 5. Injection moulded long and short fibre PP. *Compos. Part A Appl. Sci. Manuf.* **2002**, *33*, 1641–1652. [CrossRef]

processes

Article

Characterizations of Polypropylene/Single-Walled Carbon Nanotube Nanocomposites Prepared by the Novel Melt Processing Technique with a Controlled Residence Time

Dongho Kang [1,†], Sungwook Hwang [2,†], Bichnam Jung [1,3] and Jinkie Shim [1,*]

1 Korea Packaging Center, Korea Institute of Industrial Technology, Bucheon 14449, Korea; kangppp@kitech.re.kr (D.K.); jbn5666@kitech.re.kr (B.J.)
2 Department of Chemical Engineering, Keimyung University, Daegu 42601, Korea; swhwang@kmu.ac.kr
3 Department of Chemical and Biological Engineering, Korea University, Seoul 02841, Korea
* Correspondence: jkshim@kitech.re.kr
† This author contributes equally to this work.

Abstract: Melt processing is considered one of the favored techniques to produce polymer nanocomposites with various inorganic fillers such as graphene and carbon nanotubes (CNTs). Due to their superior conductivity and tensile properties, among others, CNTs have been applied in broad applications. When a low filler fraction is desired, a high degree of dispersion is required in order to benefit from the intrinsic properties of CNTs. However, due to their high cohesive energy, dispersing CNTs in polymer melts is a difficult task. This study employed the melt mixing technique with a controlled residence time of 20 min to disperse single-walled carbon nanotubes (SWNTs) into a polypropylene matrix. The composites were prepared by using a corotating twin-screw extruder equipped with a back-conveying element with varying amounts of SWNTs from 0.29 to 6.56 wt.%. Mechanical, electrical, morphological, and rheological properties were evaluated. Due to the filler effect, storage, loss modulus, and complex viscosity increased with the SWNT content. Based on the van Gurp–Palmen plot, 0.29 wt.% SWNTs was the rheological percolation threshold, and the electrical property measurement revealed a 1.4 wt.% SWNT electrical percolation threshold based on the statistical percolation theory. Relatively large agglomerates were found when the SWNT content increased more than 1.28 wt.%.

Keywords: carbon nanotube; controlled residence time; melt mixing; polymer composites; percolation network

Citation: Kang, D.; Hwang, S.; Jung, B.; Shim, J. Characterizations of Polypropylene/Single-Walled Carbon Nanotube Nanocomposites Prepared by the Novel Melt Processing Technique with a Controlled Residence Time. *Processes* **2021**, *9*, 1395. https://doi.org/10.3390/pr9081395

Academic Editors: Shaghayegh Hamzehlou and M. Ali Aboudzadeh

Received: 7 July 2021
Accepted: 9 August 2021
Published: 12 August 2021

1. Introduction

The development of new functional polymer materials is often accomplished by several strategies, such as the mixing of a conventional matrix polymer with other polymer components (polymer blends) and/or solid fillers (composites). The advantage of this method and such composites has resulted in better conditions for market launch and higher customer satisfaction compared to homopolymers [1].

Carbon nanotubes (CNTs) are prominent fillers. Following their first detailed description in the scientific literature in 1991, these "coaxial tubes of graphitic sheets" [2] attracted the attention of the wider scientific community and found practical application due to their outstanding performance, especially their electrical [3–8], thermal [9–12], and mechanical properties [12–19].

Because of these excellent properties, CNTs are utilized in a variety of technologically important applications, such as automotive and aerospace engineering, defense, electronics, energy, and sporting goods [20]. With their popularity, the global production of CNTs has already exceeded the kiloton level and is expected to more than double in the next 4 years due to the increasing production of existing manufacturers and the emergence of new manufacturers [21].

As CNTs tend to form strong agglomerates caused by physical interactions such as van der Waals forces, several methods have been applied to disperse and distribute these primary agglomerates in order to utilize the benefits of the properties of individual CNTs. Depending on the nature of the polymer, various processing techniques, such as high-shear mixing [22], electrospinning [23], surfactant-assisted processing [24], in situ polymerization [25], latex fabrication [26], and melt processing [1,27–29], can be found in the literature. Among these, melt processing is the preferred method for the processing of thermoplastic-based composites and is highly relevant for industrial applications.

For industrial applications of the melt mixing process, two-step masterbatch (MB) dilution is considered a favorable option over direct nanotube incorporation. In order to create electrically percolated networks within CNT composites, the secondary agglomeration of individualized CNTs is important and has been explained in many studies [27–31]. Secondary agglomeration, however, requires well-dispersed CNT masterbatches, which should be generated from the masterbatch production step.

When it comes to the melt mixing process of masterbatches, processing conditions need to be considered, namely rotation speed, residence time, and temperature profile. Among these factors, it is often reported that rotation speed (shear stress) is highly significant for better dispersion of CNTs in polymer matrices [1,22,27,32]. Kasaliwal et al. [27] applied the response surface methodology to investigate the effect of temperature and rotation speed on the dispersion of multiwalled CNTs (MWNTs) in polycarbonate (PC). It was found that the dispersion of MWNTs improved linearly with increasing rotation speed, whereas temperature was not significant. Villmow et al. [1] discussed the influence of extrusion conditions, screw configuration, throughput, and rotation speed. It was revealed that rotation speed has a high influence on the macrodispersion of MWNTs in polycaprolactone (PCL). Verma et al. [33] applied the melt recirculation approach to prepare MCNT/PPC composites containing up to 15 wt.% of two different aspect ratios of MCNTs. It was found that these composites display a very low percolation threshold and an improvement in tensile strength.

Polypropylene (PP) homopolymer is a nonpolar partial crystalline polymer that has a lower density, higher melting temperature, higher heat resistance, and higher stiffness than low- and high-density polyethylene (LDPE and HDPE, respectively) [34]. Thus, PP has been popularly used as a matrix to reinforce both single-walled CNTs (SWNTs) [35–37] and MWNTs [38–40], whereas, due to their high interfacial energy with CNTs ($\approx$10 mJ/m^2) compared to polar polymers such as polycarbonate ($\approx$2.4 mJ/m^2) [41], CNTs are difficult to disperse in PP, resulting in the further physical or chemical modification of CNTs or PP. Chemical or physical modifications of CNTs and PP, however, are not suitable for industrial applications.

According to the aforementioned studies, it is well known that the most important factors for achieving good dispersion of CNTs are high mixing energy and long residence time, resulting in two dominant dispersion mechanisms: rupture and erosion [42,43]. In particular, the long residence time induces an erosion mechanism toward the agglomeration of CNTs. In order to utilize the benefits of the long residence time, in this study, a specially designed corotating twin-screw compounder with a back-conveying element and a closed plate is used. With a controlled residence time of 20 min, PP/SWNT composites with varying amounts of SWNTs (0.29, 0.43, 0.62, 1.28, 2.94, 3.66, 4.29, 4.90, and 6.56 wt.%) were prepared. To evaluate the dispersion state of SWNTs, rheological, morphological, electrical, and mechanical properties were measured.

2. Materials and Methods

2.1. Materials

PP (HF 429, Samsung Total, Daesan, Korea) with a melt flow index of 8 g/10 min at 240 °C and a melting temperature of 239.4 ± 0.23 °C was used as the matrix. Nonfunctionalized, commercially available single-walled carbon nanotubes (TUBALL^{TM}; OCSiAl Korea, Incheon, Korea) were used without further purification. According to the specifications provided by the manufacturer, TUBALL™ has more than 85% carbon purity. The outer mean nanotube diameter is 1.8 ± 0.4 nm and longer than the 5 μm length of the nanotube.

2.2. Sample Preparation

A schematic diagram of the sample preparation is illustrated in Figure 1. A Brabender microcompounder TSC 42/6 (Brabender®, Duisburg, Germany), designed with a small-scale, conical, counter-rotating twin-screw compounder with a barrel capacity of 50 g and a screw diameter of 42 mm (L/D = 6), was used. Based on the PP specifications provided by the manufacturer and preliminary experiments, PP/SWNT composites were produced under a barrel temperature of 250 °C, a rotation speed of 50 rpm, and a residence time of 20 min. The concentrations of SWNTs in the PP matrix were 0.3, 0.5, 0.8, 1.2, 3, 3.5, 4, 5, and 7 wt.%.

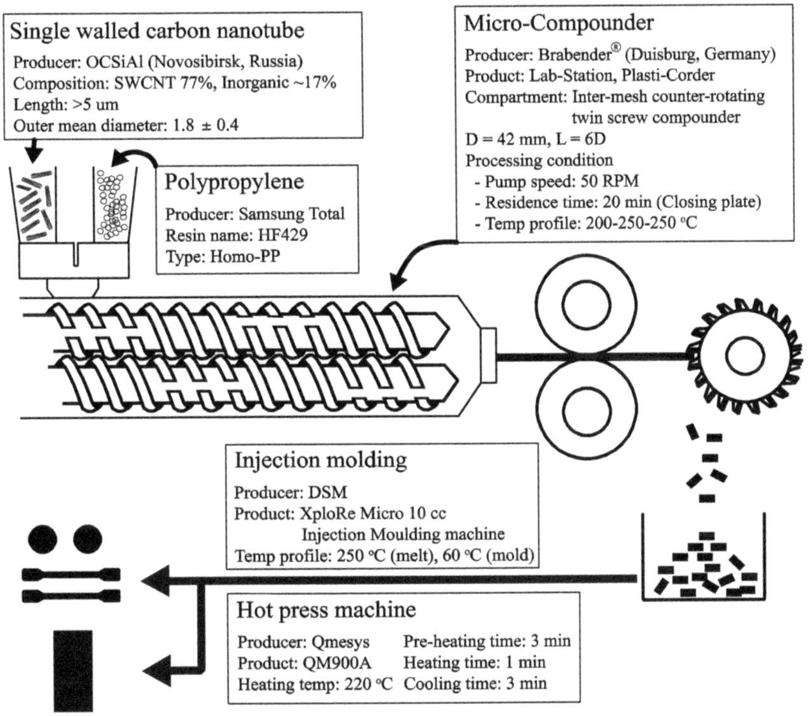

Figure 1. Schematic diagram of the sample preparation of the PP/SWNT composite for this study.

In order to estimate the shear stress of the composite melt inside the barrel, the specific mechanical energy (SME) was calculated using Equation (1), where the drive torque τ (kJ), rotation speed N (min^{-1}), and throughput $\dot{m}$ (kg/min) are involved. This parameter is considered to explain the good characterization of the extrusion process.

$$\text{SME} = \frac{\tau \cdot N}{\dot{m}} (kJ/kg) \tag{1}$$

In this study, drive torque τ was obtained from the compounder, throughput $\dot{m}$ was averaged at 0.0017 ± 0.0002 kg/min, and rotation speed N was fixed at 50 min^{-1}.

After pelletizing, the amount of CNTs in the composite granules was analyzed using a thermogravimetric analyzer (TGA) (Model: TGA Q500, TA Instrument, New Castle, DE, USA)). Pelletized granules were cryofractured to observe the dispersion state of SWNTs.

After that, the composite granules were compression molded to measure morphological and electrical properties and injection molded to measure mechanical and rheological properties. Compression molding was performed using a QMESYS hot press (Model: QM900A, Quality & Measurement System, Uiwang, Korea). Molding was carried out at 220 °C with a preheating time of 3 min, a pressing time of 1 min, and a cooling time of 3 min. For accurate characteristic measurement, it was necessary to prepare specimens with a consistent shape and size. Thus, the composite granules were also injection molded at 250 °C using the Xplore Micro 10cc Injection Molding Machine (Xplore Instruments, Geleen, The Netherlands) into test samples with a disk shape diameter of 25.4 mm, a 1 mm thickness, and a "dogbone" shape with size in accordance with ASTM D638.

2.3. *Characterization*

2.3.1. Rheological Measurements

Oscillatory shear measurements and stress relaxation in the linear viscoelastic (LVE) region were carried out on an Anton Paar oscillation rheometer (Model: Physica MCR 302, Anton Paar GmbH, Graz, Austria) using a parallel plate geometry (diameter = 25 mm, gap = 1 mm) at 200 °C on samples previously compression molded into a disk shape. Frequency sweeps were performed after approximately 3 min of temperature equilibration with decreasing frequency from 100 to 0.1 rad/s and strains at 1%. According to Kasaliwal et al. [44], it was observed that increasing the matrix molecular weight produced larger undispersed agglomerates. Thus, different residence times were tested to lower the molecular weight of PP, which was evaluated using the stress relaxation spectrum. The conversion of the stress relaxation spectrum to MWD was performed using RheoCompass^{TM} software (v1.30.1064, Anton Paar GmbH, Graz, Austria). Specifically, it used one of the modulus models, which can be expressed in terms of the relaxation modulus $G(t)$. This model converts the relaxation spectrum from the time domain to the molecular weight domain to recover the MWD curve using a regularized integral inversion, as shown in Equation (2).

$$\frac{G(t)}{G_N^0} = \left(\int_{\ln(M_e)}^{\infty} F^{1/\beta}(t, M) w(M) d(\ln M) \right)^{\beta} \tag{2}$$

where G_N^0 is the plateau modulus to normalize $G(t)$, $F(t,M)$ is a kernel function describing the relaxation behavior of molecular weight M, and $w(M)$ is the weight fraction of the MWD function. The exponent β is a parameter that corresponds to the mixing behavior of the chains. M_e is the average molecular weight between entanglements. Based on material and calculation parameters incorporated in the software, kernel functions β and M_e of PP were single exponentials, 2.5 and 3500 g/mol, respectively.

2.3.2. Morphological Measurements

The state of the macrodispersion of the compression-molded sample was analyzed by optical microscopy. Light transmission microscopy (Model: KYENCE VK-X200K, KEYENCE, Itasca, IL, USA) was performed using a 10× objective magnification and an optical laser with a 408 nm violet laser. Particle analysis using the incorporated software developed by KYENCE was performed in order to quantify the area fraction A_A of undispersed primary SWNT agglomerates according to Equation (3) [1].

$$A_A(\%) = \frac{A_{CNT}}{A_0} \times 100 \tag{3}$$

where A_{CNT} is the area of primary SWNT agglomerates, and A_0 is the overall micrograph area. According to ISO 18553, agglomerates with diameters smaller than 5 µm were neglected. In order to increase the analyzed size of the micrograph area, 3 different areas of each sample were captured.

Additionally, field emission scanning electron microscopy (Model: HITACHI-SU8020, Hitachi, Ltd., Tokyo, Japan) was used for the direct observation of PP/SWNT composites. Palletized composite granules were cryofractured after freezing in liquid nitrogen. A thin layer of carbon was sputter deposited onto the sample. Measurement was performed under high vacuum with an acceleration voltage of 2 kV.

2.3.3. Mechanical Measurements

Dog-bone-shaped samples were injection molded (50 mm in length, 4 mm in width, and 2 mm in thickness) and were tested on an INSTRON 3367 (INSTRON, Norwood, MA, USA) equipped with a 1 kN cell force, measuring force over displacement for each sample at 10 mm/min. At least 3 samples were tested for each composite and tensile stress and strain at yield point obtained from the raw data. One-way ANOVA and Tukey's honestly significant difference (HSD) test were performed to conduct the multiple comparisons of each experimental data set ($\alpha = 0.05$) with R software (version 3.0.2).

2.3.4. Electrical Measurements

In-plane electrical conductivity measurements were made on compression-molded square plates in accordance with ASTM D4496-13. At least 10 measurements were made on samples to obtain the geometric mean value with the associated standard deviation of resistivity. The AiT resistivity measurement system (Model: CMT-SR1000N, Advanced Instrument Technology, Suwon, Korea), equipped with a 4-point probe with a sheet resistance measurement range between 1 m Ω/square and 2 M Ω/square and pin spacing of 20–50 mm, was used. The statistical percolation theory predicts the relationship between the electrical resistivity of the composite and the filler concentration using Equation (4) [45]:

$$\sigma = \sigma_0 \cdot (\Phi - \Phi_C)^t \tag{4}$$

where σ is the composite volume resistivity, σ_0 is the volume resistivity of the filler, and Φ, Φ_C, and t are the fraction of filler, percolation threshold, and critical exponent, respectively. The critical exponent t is expected to depend on the system dimensionality with calculated values of $t \approx 1.33$ in two and $t \approx 2$ in three dimensions [46,47]. By fitting the experimental results in the linear regression of $\log \sigma$ and $\log (\Phi - \Phi_C)$, the percolation threshold and the critical exponent were calculated. In this study, the final critical exponent t was calculated as 2.05, representing the three dimensions of the SWNT percolation network.

3. Results and Discussion

3.1. Processing Properties

Figure 2 explains the specific mechanical energy applied during the melt mixing process to produce composite materials with different SWNT content. Due to the residue of SWNTs in the compounder, after TGA analysis, the final amount of SWNTs in the PP/SWNT nanocomposites was 0.29, 0.43, 0.62, 1.28, 2.94, 3.66, 4.29, 4.80, and 6.56 wt.%.

As the SWNT content increases in this composite system, the SME increases, possibly due to the dominance of nanotube–nanotube interactions, which may affect the dispersion of SWNTs in the composites. Adding more SWNTs results in higher shear stress for the composites during melt mixing. This trend can be also found in the rheological properties of the composites.

3.2. Rheological Properties of the Composites

As discussed earlier, polymers with a high molecular weight result in large agglomerates of composites. It is also known that a long residence time during the melt mixing process may cause a polymer backbone scission resulting in reduced molecular weight. As shown in Figure 3a, the molecular weight (MW) of PP decreased from a residence time to 10 min from 20 min. The weight-average MW, which is more sensitive to molecular size than the number-average MW, reached a low level at a residence time of 20 min, indicating a low polydispersity index. In other words, not only is the length of the molecule reduced, but it can also be evenly distributed at a residence time of 20 min. Thus, the polymer melt infiltration into the primary agglomerates can be enhanced with lower viscosity and molecular weight [42]. Figure 3b shows the storage and loss modulus of PP as a function of residence time at 200 °C.

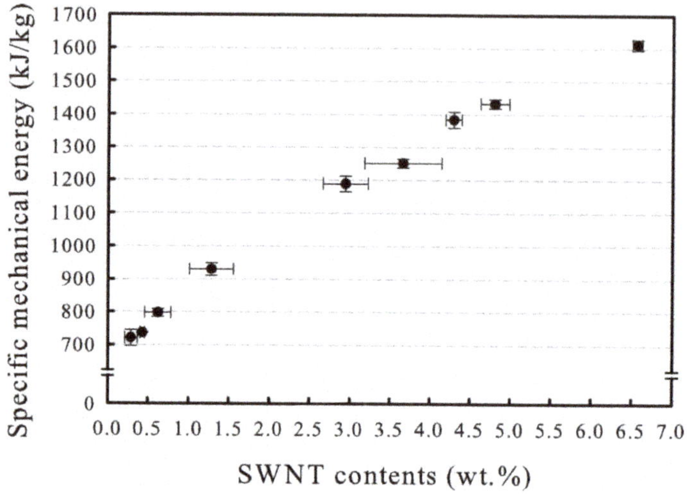

Figure 2. Specific mechanical energy applied to produce composites with different SWNT content.

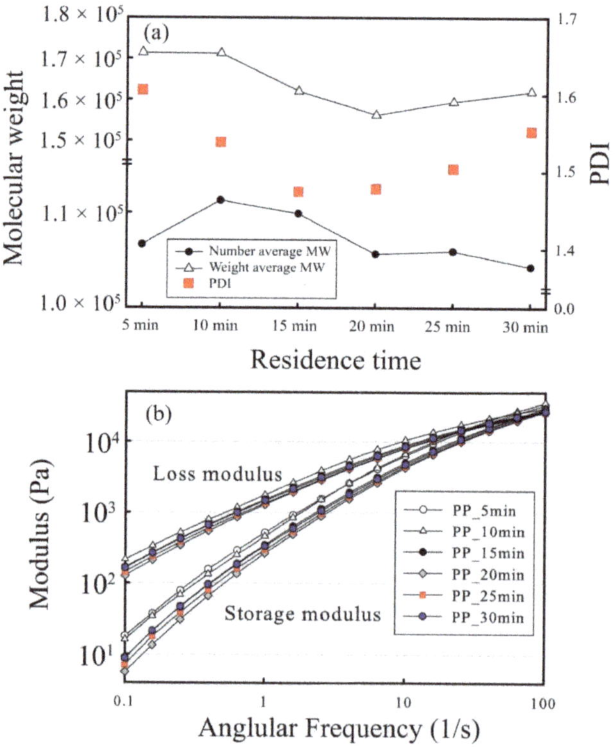

Figure 3. (**a**) Number, weight-average molecular weight, and polydispersity index (PDI) of PP as a function of residence time; (**b**) storage modulus G' and loss modulus G'' of PP as a function of residence time at 200 °C.

At low frequencies, PP without SWNT loading exhibits typical homopolymer-like terminal behavior by full relaxation of the PP chain. PP chains have sufficient time to free themselves from the constraints of entanglement. However, when the nanotube loading reaches 0.29 wt.%, the terminal behavior becomes weak, and the dependence of G' and G'' on ω is limited. As shown in Figure 4a, G' begins to show a plateau at low frequencies of 0.29 wt.% SWNT loading, which indicates a transition from liquid-like to solid-like viscoelastic behavior [48]. As the SWNT content increases in the composite, G' at low frequencies is increased and becomes independent of frequency. This behavior can be explained by the formation of interconnected nanotube networks in the polymer. While the low frequency dependency of G'' shows a similar trend, as shown in Figure 4b, the corresponding increase in the loss modulus G'' is lower than the storage modulus G'. According to Potschke et al. [49], this limited increase in the loss modulus as the SWNT content increases is due to the insensitivity of the interfacial energy or compatibility between the polymer and the SWNTs. Increasing the SWNT content also leads to an increase in the complex viscosity. From 0.29 wt.%, a transition from liquid-like to solid-like behavior appeared, which can be considered as the rheological percolation threshold, as supported by the van Gurp–Palmen plot (Figure 4d). As shown in Figure 4d, the curves approach a phase angle of 90° at 0 wt.%, indicating a viscous behavior, whereas from 0.29 wt.%, the rheological behavior changes from a viscous fluid to an elastic solid, indicating a percolation threshold [50].

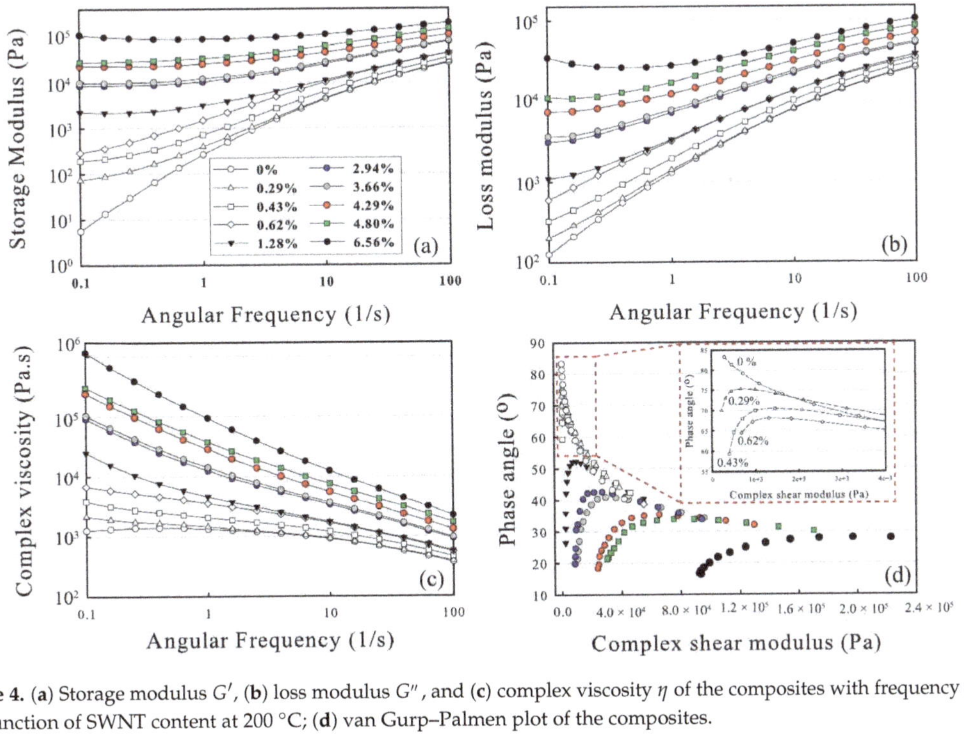

Figure 4. (**a**) Storage modulus G', (**b**) loss modulus G'', and (**c**) complex viscosity η of the composites with frequency sweep as a function of SWNT content at 200 °C; (**d**) van Gurp–Palmen plot of the composites.

3.3. Morphological Properties of the Composites

Images of macrodispersion, size distribution, and area ratio of undispersed SWNT agglomerates (A_A) evaluated by light microscope for composites with varying amounts of SWNTs are described in Figure 5. When it comes to the dispersion of any nanofillers, there are two main mechanisms: erosion and rupture. The erosion mechanism is dominated under low shear stresses with long mixing times, whereas SWNT agglomerates undergo dispersion mainly by the rupture mechanism at high shear stresses. Generally, longer mixing times lead to better dispersion, but they may induce polymer degradation, resulting in reduced viscosity or deteriorated properties of the matrix or the SWNTs. On the other hand, high rotation speed for high shear stresses may cause undesired breakage of SWNTs, reducing the aspect ratio and leading to higher electrical percolation thresholds and higher resistivity values at a given nanotube concentration. Thus, it is important to balance the shear stresses and mixing times for an optimal erosion and rupture mechanism.

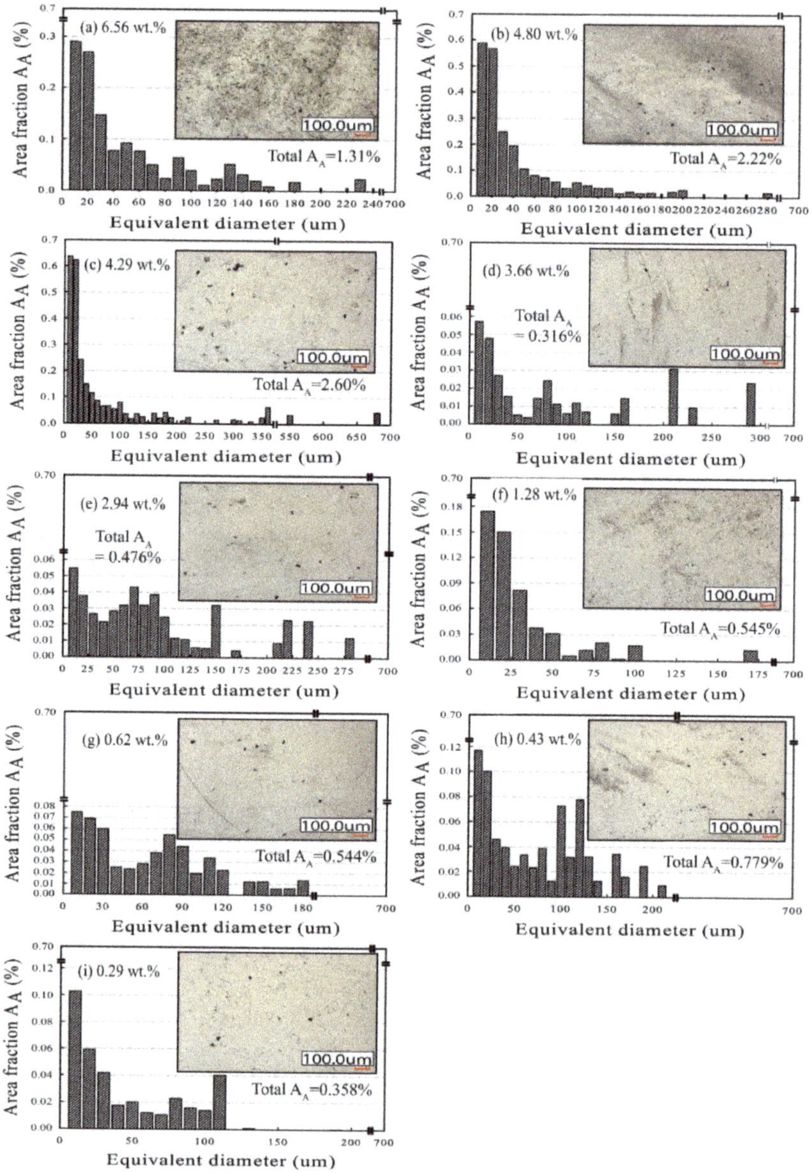

Figure 5. Size distribution of undispersed SWNT agglomerates observed in composites; SWNT content and total agglomerate area (A_A) are described in each figure. (**a**) 6.56 wt.%; (**b**) 4.80 wt.%; (**c**) 4.29 wt.%; (**d**) 3.66 wt.%; (**e**) 2.94 wt.%; (**f**) 1.28 wt.%; (**g**) 0.62 wt.%; (**h**) 0.43 wt.%; (**i**) 0.29 wt.%.

In Figure 5, all composites show that the total area of undispersed SWNT agglomerates is less than 3%, which may imply that both mechanisms of dispersion are applied during the method employed in this study. Due to the amount of SWNTs inserted, the already-dispersed nanotubes enhance the matrix viscosity, and thus, the shear stresses act on the remaining initial agglomerates, which is also shown in Figure 2. In other words, as the SWNT content increases, enough high shear stress is generated to rupture the primary agglomerates into smaller agglomerates of different sizes. Given a controlled mixing time of 20 min, an erosion mechanism may take place to erode the free tube, small fragments, and agglomerates from the surface of the larger agglomerates.

It is generally observed that in composites prepared with high loading of SWNT content of more than 1.28 wt.%, relatively large agglomerates (more than 200 μm) are presented. As expected, the number of large agglomerates decreases with decreasing SWNT content. The decrease in the number of agglomerates from the larger diameter classes may indicate that this method may be effective in rupturing the large agglomerates for composites up to approximately 1.28 wt.% SWNT content.

In order to observe the dispersion state of SWNTs in composites and their network formation, FE-SEM was used to examine the composites. FE-SEM images of the composites with the addition of 0.29 wt.%, 1.28 wt.%, and 6.56 wt.% SWNTs are shown in Figure 6. FE-SEM images of the remaining composites are illustrated in Figures A1–A6. At low SWNT content (Figure 6a), most of the SWNTs are dispersed individually in the PP matrix, and some are formed as small aggregates. As confirmed from the results of the rheological property measurement, despite the small amount of SWNTs introduced, the dispersion state of SWNTs was good enough to transfer the mechanical momentum, resulting in a rheological percolation threshold at 0.29 wt.% SWNTs. As the SWNT content increases (Figure 6b), individual SWNTs gradually grow, and they start to form SWNT networks, resulting in small interparticle distances. Figure 6c illustrates the SEM image of the highest PP/SWNT composite in this study. At this high loading of SWNTs, the SWNT network dominates the properties of the PP/SWNT composite, leading to solid-like behavior at low frequencies of the rheological measurement in Figure 4.

3.4. Mechanical Properties of the Composites

Due to their enormous modulus and high aspect ratio, it is reasonable to expect that the incorporation of SWNTs into polymers will be effective for the reinforcement of polymers, i.e., mechanical properties. The nanometer dimension of the interfacial regions surrounding the SWNTs allows the applied load to be easily transferred from the matrix to the SWNTs.

It has been reported that the formation of large agglomerates as the nanotube content increases acts as a mechanical failure concentrator. Thus, tensile strength and modulus are increased with increasing nanotube content until a certain level, but these properties are decreased when the nanotube content reaches a level where large agglomerates begin to form [51,52]. However, as shown in the upper part of Figure 7, the tensile strength increases almost linearly as the SWNT content increases with the coefficient of determination (R^2) of 0.9. Even though the morphological analysis (Figure 5) reveals that relatively large agglomerates are found when the SWNT content is more than 1.28 wt.%, it can be speculated that there are no agglomerates large enough to induce stress cracking with high loading of SWNTs in composites.

The Young's modulus of the composites is described at the bottom of Figure 7. It continuously increases with increasing SWNT content, revealing the reinforcement effect of SWNTs.

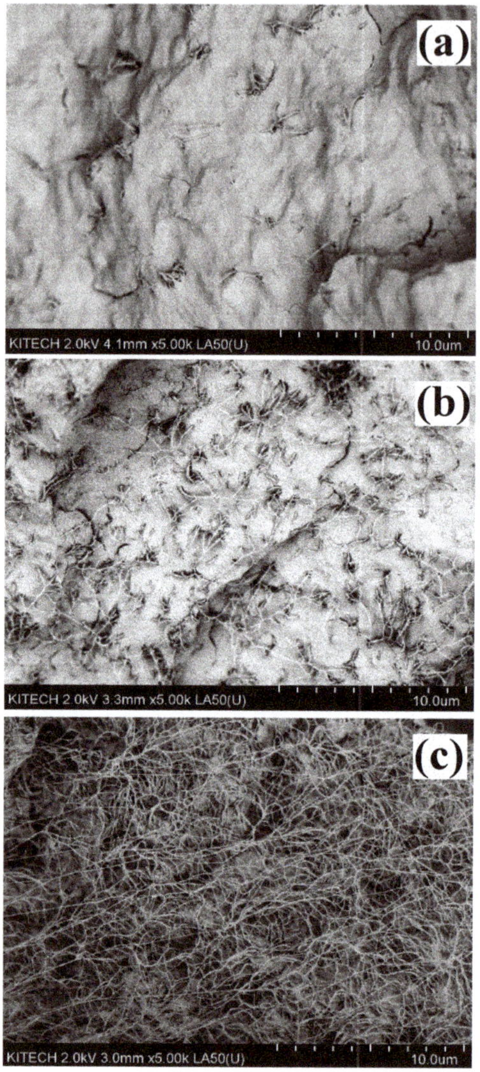

Figure 6. SEM images of (**a**) 0.29 wt.%, (**b**) 1.28 wt.%, and (**c**) 6.56 wt.% SWNT in the composite.

3.5. Electrical Properties of the Composites

One of the main advantages of CNTs is their electrical conductivity, and thus, their inclusion into their PP matrix can construct conductive pathways and increase the electrical conductivity of PP. Electrically conductive polymers based on CNTs and PP have been widely studied [48,53–57]. Electrical percolation thresholds range widely from 0.07 to 2.62 wt.% depending on many factors, such as the processing method and type of CNT (SWNT or MWNT). At these electrical percolation thresholds, the electrical conductivity for PP/CNT composites varies from 10^{-8} to 10^{-2} S/m. Compared to this range, the electrical conductivity in this study shows a profound improvement indicating that better microstructure in the PP/SWNT composite can be achieved with the novel processing technology without any additional additives and chemical treatments of PP or SWNTs.

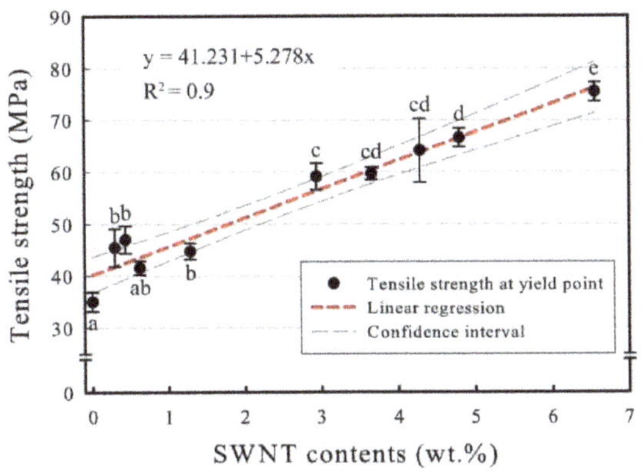

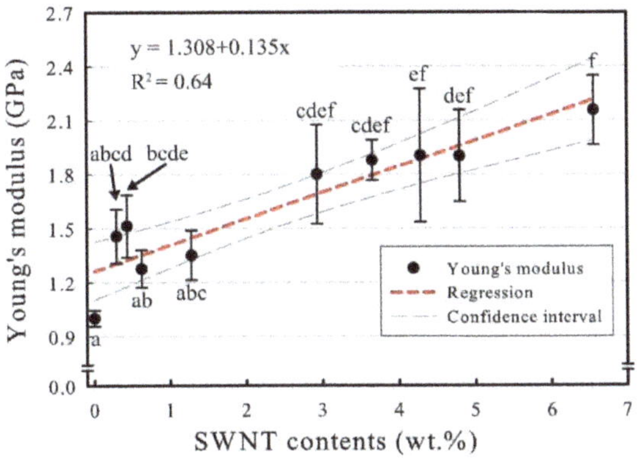

Figure 7. Tensile strength (**top**) and Young's modulus (**bottom**) of the composites as a function of SWNT content; different letters on the top or bottom of the error bar indicate a statistically significant difference at the 95% confidence level.

As visible in Figure 8, the results of the statistical percolation theory show that the electrical percolation threshold is 1.4 wt.% with the critical exponent t = 2.05. This result is in parallel with the analysis of the FE-SEM image (Figure 6b), where the 1.28 wt.% SWNT composite shows a well-grown SWNT network, resulting in efficient electron transport through the network. On the other hand, the weight fraction of the rheological percolation threshold (0.29 wt.%) is much lower than the electrical percolation threshold. This different behavior may come from the difference between charge transfer and mechanical momentum transfer. The electrical percolation threshold is mainly determined by the CNT network and the connecting contacts, while the rheological percolation threshold is predominantly related equally with the CNT network, the polymer matrix, and the CNT–polymer network [42].

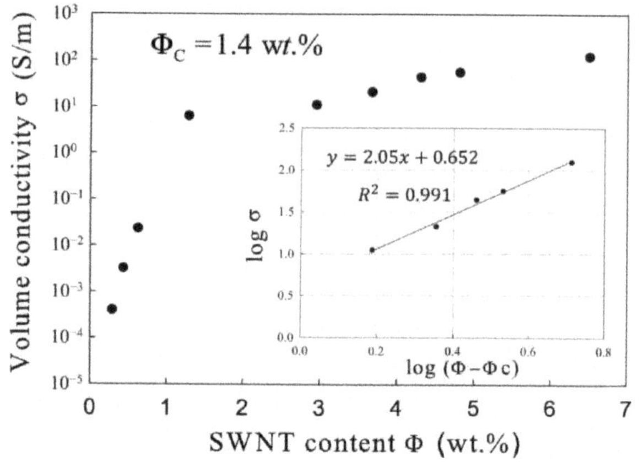

Figure 8. Electrical conductivity as a function of SWNT content.

4. Conclusions

This study discussed the mechanical, electrical, morphological, and rheological properties of SWNT/PP composites prepared by the melt mixing method with a controlled residence time of 20 min. The composites were produced with varying SWNT content, and thermal gravimetric analysis confirmed the SWNT content as 0.29, 0.43, 0.62, 1.28, 2.94, 3.66, 4.29, 4.80, and 6.56 wt.%.

In the rheological measurement, it was found that a residence time of 20 min showed the lowest number-average MW and a low weight-average MW, resulting in a low polydispersity index. Therefore, based on the knowledge that polymer melt infiltration is enhanced with low melt viscosity and molecular weight, a residence time of 20 min for PP was used in this study. The storage modulus began to plateau at low frequencies at 0.29 wt.% loading of SWNTs. As the SWNT content increased, the storage modulus also increased, indicating the formation of interconnected nanotube networks. Based on the van Gurp–Palmen plot, a weight fraction of 0.29 wt.% SWNTs showed the rheological percolation threshold.

Morphological properties were assessed by optical microscopy and FE-SEM. All composites had less than 3% of their total area containing undispersed SWNT agglomerates. Relatively large agglomerates greater than 200 μm were found for composites containing more than 1.28 wt.% SWNTs. Additionally, FE-SEM analysis revealed that individual SWNTs and their small agglomerates were found in the 0.29 wt.% SWNT composite. As SWNT content increased in the PP matrix, the SWNT network also increased, forming an efficient electrical percolation network at 1.28 wt.%.

Despite the large agglomerates found in the morphological measurement at high loading of the SWNT composite, tensile strength and Young's modulus increased linearly with increasing SWNT content, which may imply that relatively large agglomerates were not enough to induce stress cracking.

The result of the statistical percolation theory revealed that the electrical percolation threshold was 1.4 wt.% with a critical exponent t of 2.05. Volume conductivity increased with increasing SWNT content in the composites.

In summary, due to the controlled residence time and excellent dispersion state, an improvement in the electrical conductivity of the SWNTs in the PP matrix was achieved with the novel melt mixing process without further modifications, such as compatibilizers or chemical treatment of CNTs. Rheological, morphological, and electrical measurements supported that an optimized dispersion and electrical percolation network of SWNTs was formed with around 1.28 wt.% SWNTs in PP. However, given the fact that SWNTs have better properties than MWNTs, an electrical percolation threshold of around 1.28 wt.% loading of SWNTs may not meet expectations. This may be because a possible filler breakage, reducing the aspect ratio of SWNTs, occurred during the melt mixing process with a residence time of 20 min. This breakage, however, may help to form an efficient electrical percolation network, generating high electrical conductivity. Further analysis of SWNT length reduction by the melt mixing process and its effects will be needed to confirm this statement.

Author Contributions: Main author, D.K.; main coauthor, S.H.; sample preparation, B.J.; corresponding author, J.S. All authors have read and agreed to the published version of the manuscript.

Funding: This research and APC were funded by the Ministry of Strategy and Finance in South Korea, grant number EH210009.

Institutional Review Board Statement: Not applicable.

Informed Consent Statement: Not applicable.

Conflicts of Interest: The authors declare no conflict of interest.

Appendix A

SEM images of varying amounts (0.43 wt.%, 0.62 wt.%, 2.94 wt.%, 3.66 wt.%, 4.29 wt.%, and 4.90 wt.%) of SWNT-incorporated composites are shown below.

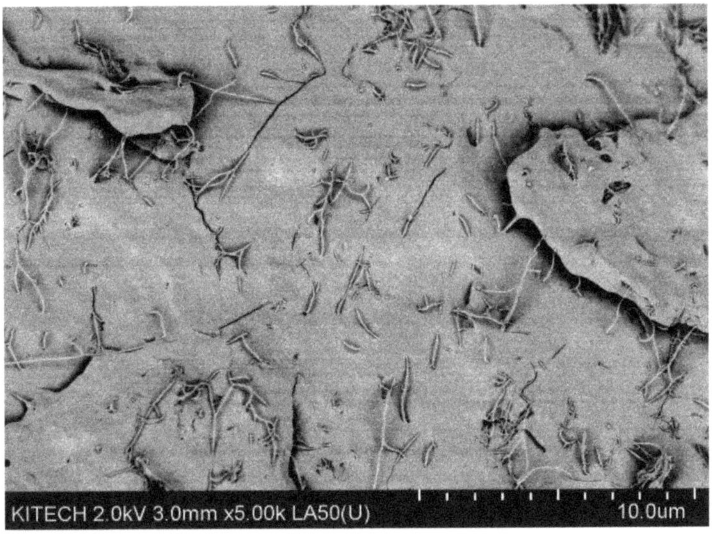

Figure A1. SEM image of the 0.43 wt.% SWNT/PP composite.

Figure A2. SEM image of the 0.62 wt.% SWNT/PP composite.

Figure A3. SEM image of the 2.94 wt.% SWNT/PP composite.

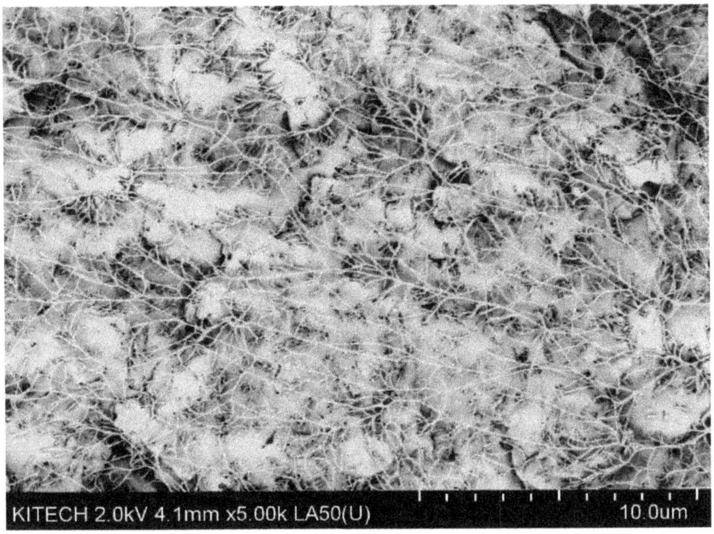

Figure A4. SEM image of the 3.66 wt.% SWNT/PP composite.

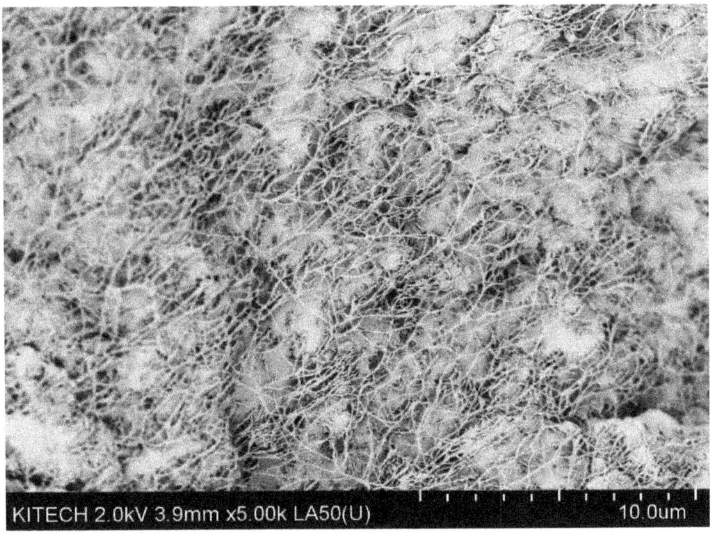

Figure A5. SEM image of the 4.29 wt.% SWNT/PP composite.

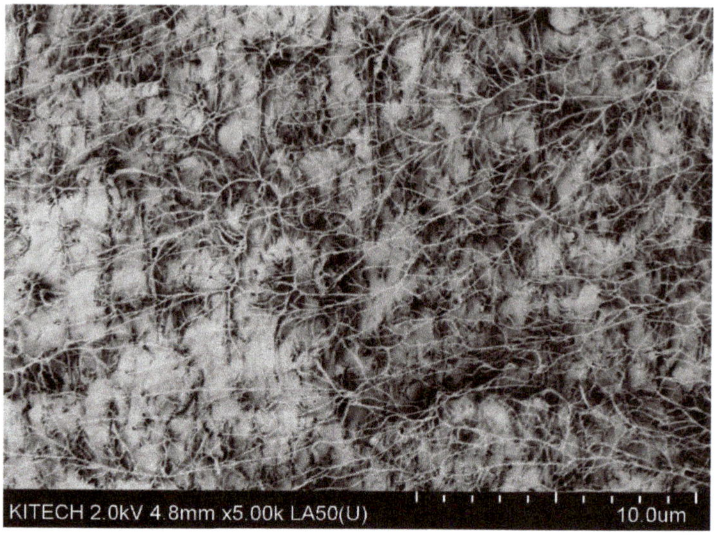

Figure A6. SEM image of the 4.90 wt.% SWNT/PP composite.

References

1. Villmow, T.; Kretzschmar, B.; Pötschke, P. Influence of screw configuration, residence time, and specific mechanical energy in twin-screw extrusion of polycaprolactone/multi-walled carbon nanotube composites. *Compos. Sci. Technol.* **2010**, *70*, 2045–2055. [CrossRef]
2. Iijima, S. Helical microtubules of graphitic carbon. *Nature* **1991**, *354*, 56–58. [CrossRef]
3. Charlier, J.-C.; Michenaud, J.-P. Energetics of multilayered carbon tubules. *Phys. Rev. Lett.* **1993**, *70*, 1858. [CrossRef]
4. Saito, R.; Dresselhaus, G.; Dresselhaus, M. Electronic structure of double-layer graphene tubules. *J. Appl. Phys.* **1993**, *73*, 494–500. [CrossRef]
5. White, C.; Robertson, D.; Mintmire, J. Helical and rotational symmetries of nanoscale graphitic tubules. *Phys. Rev. B* **1993**, *47*, 5485. [CrossRef] [PubMed]
6. Dresselhaus, M.; Dresselhaus, G.; Pimenta, M. The remarkable properties of carbon nanotubes as nanoclusters. *Eur. Phys. J. D At. Mol. Opt. Plasma Phys.* **1999**, *9*, 69–75.
7. Ebbesen, T.; Lezec, H.; Hiura, H.; Bennett, J.; Ghaemi, H.; Thio, T. Electrical conductivity of individual carbon nanotubes. *Nature* **1996**, *382*, 54–56. [CrossRef]
8. Dai, H.; Wong, E.W.; Lieber, C.M. Probing electrical transport in nanomaterials: Conductivity of individual carbon nanotubes. *Science* **1996**, *272*, 523–526. [CrossRef]
9. Hone, J.; Llaguno, M.; Biercuk, M.; Johnson, A.; Batlogg, B.; Benes, Z.; Fischer, J. Thermal properties of carbon nanotubes and nanotube-based materials. *Appl. Phys. A* **2002**, *74*, 339–343. [CrossRef]
10. Berber, S.; Kwon, Y.-K.; Tománek, D. Unusually high thermal conductivity of carbon nanotubes. *Phys. Rev. Lett.* **2000**, *84*, 4613. [CrossRef]
11. Che, J.; Cagin, T.; Goddard, W.A., III. Thermal conductivity of carbon nanotubes. *Nanotechnology* **2000**, *11*, 65. [CrossRef]
12. Ruoff, R.S.; Lorents, D.C. Mechanical and thermal properties of carbon nanotubes. *Carbon* **1995**, *33*, 925–930. [CrossRef]
13. Treacy, M.J.; Ebbesen, T.; Gibson, J. Exceptionally high Young's modulus observed for individual carbon nanotubes. *Nature* **1996**, *381*, 678–680. [CrossRef]
14. Li, F.; Cheng, H.; Bai, S.; Su, G.; Dresselhaus, M. Tensile strength of single-walled carbon nanotubes directly measured from their macroscopic ropes. *Appl. Phys. Lett.* **2000**, *77*, 3161–3163. [CrossRef]
15. Pan, Z.; Xie, S.; Chang, B.; Wang, C.; Lu, L.; Liu, W.; Zhou, W.; Li, W.; Qian, L. Very long carbon nanotubes. *Nature* **1998**, *394*, 631–632. [CrossRef]
16. Walters, D.; Ericson, L.; Casavant, M.; Liu, J.; Colbert, D.; Smith, K.; Smalley, R. Elastic strain of freely suspended single-wall carbon nanotube ropes. *Appl. Phys. Lett.* **1999**, *74*, 3803–3805. [CrossRef]
17. Salvetat, J.-P.; Bonard, J.-M.; Thomson, N.; Kulik, A.; Forro, L.; Benoit, W.; Zuppiroli, L. Mechanical properties of carbon nanotubes. *Appl. Phys. A* **1999**, *69*, 255–260. [CrossRef]
18. Demczyk, B.; Wang, Y.; Cumings, J.; Hetman, M.; Han, W.; Zettl, A.; Ritchie, R. Direct mechanical measurement of the tensile strength and elastic modulus of multiwalled carbon nanotubes. *Mater. Sci. Eng. A* **2002**, *334*, 173–178. [CrossRef]

19. El Assami, Y.; Drissi Habti, M.; Raman, V. Stiffening offshore composite wind-blades bonding joints by carbon nanotubes reinforced resin–a new concept. *J. Struct. Integr. Maint.* **2020**, *5*, 87–103.
20. De Volder, M.F.; Tawfick, S.H.; Baughman, R.H.; Hart, A.J. Carbon nanotubes: Present and future commercial applications. *Science* **2013**, *339*, 535–539. [CrossRef]
21. Kingston, C.; Zepp, R.; Andrady, A.; Boverhof, D.; Fehir, R.; Hawkins, D.; Roberts, J.; Sayre, P.; Shelton, B.; Sultan, Y. Release characteristics of selected carbon nanotube polymer composites. *Carbon* **2014**, *68*, 33–57. [CrossRef]
22. Chen, G.-X.; Li, Y.; Shimizu, H. Ultrahigh-shear processing for the preparation of polymer/carbon nanotube composites. *Carbon* **2007**, *45*, 2334–2340. [CrossRef]
23. Sen, R.; Zhao, B.; Perea, D.; Itkis, M.E.; Hu, H.; Love, J.; Bekyarova, E.; Haddon, R.C. Preparation of single-walled carbon nanotube reinforced polystyrene and polyurethane nanofibers and membranes by electrospinning. *Nano Lett.* **2004**, *4*, 459–464. [CrossRef]
24. Vigolo, B.; Poulin, P.; Lucas, M.; Launois, P.; Bernier, P. Improved structure and properties of single-wall carbon nanotube spun fibers. *Appl. Phys. Lett.* **2002**, *81*, 1210–1212. [CrossRef]
25. Chen, G.-X.; Kim, H.-S.; Park, B.-H.; Yoon, J.-S. Highly insulating silicone composites with a high carbon nanotube content. *Carbon* **2006**, *44*, 3373–3375. [CrossRef]
26. Regev, O.; ElKati, P.N.; Loos, J.; Koning, C.E. Preparation of conductive nanotube-polymer composites using latex technology. *Adv. Mater.* **2004**, *16*, 248–251. [CrossRef]
27. Kasaliwal, G.; Göldel, A.; Pötschke, P. Influence of processing conditions in small-scale melt mixing and compression molding on the resistivity and morphology of polycarbonate–MWNT composites. *J. Appl. Polym. Sci.* **2009**, *112*, 3494–3509. [CrossRef]
28. Krause, B.; Pötschke, P.; Häußler, L. Influence of small scale melt mixing conditions on electrical resistivity of carbon nanotube-polyamide composites. *Compos. Sci. Technol.* **2009**, *69*, 1505–1515. [CrossRef]
29. Pegel, S.; Pötschke, P.; Petzold, G.; Alig, I.; Dudkin, S.M.; Lellinger, D. Dispersion, agglomeration, and network formation of multiwalled carbon nanotubes in polycarbonate melts. *Polymer* **2008**, *49*, 974–984. [CrossRef]
30. Villmow, T.; Pegel, S.; Pötschke, P.; Wagenknecht, U. Influence of injection molding parameters on the electrical resistivity of polycarbonate filled with multi-walled carbon nanotubes. *Compos. Sci. Technol.* **2008**, *68*, 777–789. [CrossRef]
31. Lellinger, D.; Xu, D.; Ohneiser, A.; Skipa, T.; Alig, I. Influence of the injection moulding conditions on the in-line measured electrical conductivity of polymer–carbon nanotube composites. *Phys. Status Solidi-B-Basic Solid State Phys.* **2008**, *245*, 2268. [CrossRef]
32. Villmow, T.; Pötschke, P.; Pegel, S.; Häussler, L.; Kretzschmar, B. Influence of twin-screw extrusion conditions on the dispersion of multi-walled carbon nanotubes in a poly (lactic acid) matrix. *Polymer* **2008**, *49*, 3500–3509. [CrossRef]
33. Verma, P.; Saini, P.; Choudhary, V. Designing of carbon nanotube/polymer composites using melt recirculation approach: Effect of aspect ratio on mechanical, electrical and EMI shielding response. *Mater. Des.* **2015**, *88*, 269–277. [CrossRef]
34. Selke, S.E.; Hernandez, R.J.; Culter, J.D. *Plastics Packaging: Properties, Processing, Applications, and Regulations*, 2nd ed.; Hanser Publishers: Munich, Germany, 2004.
35. Bhattacharyya, A.R.; Sreekumar, T.; Liu, T.; Kumar, S.; Ericson, L.M.; Hauge, R.H.; Smalley, R.E. Crystallization and orientation studies in polypropylene/single wall carbon nanotube composite. *Polymer* **2003**, *44*, 2373–2377. [CrossRef]
36. Manchado, M.L.; Valentini, L.; Biagiotti, J.; Kenny, J. Thermal and mechanical properties of single-walled carbon nanotubes–polypropylene composites prepared by melt processing. *Carbon* **2005**, *43*, 1499–1505. [CrossRef]
37. Kashiwagi, T.; Grulke, E.; Hilding, J.; Groth, K.; Harris, R.; Butler, K.; Shields, J.; Kharchenko, S.; Douglas, J. Thermal and flammability properties of polypropylene/carbon nanotube nanocomposites. *Polymer* **2004**, *45*, 4227–4239. [CrossRef]
38. Assouline, E.; Lustiger, A.; Barber, A.; Cooper, C.; Klein, E.; Wachtel, E.; Wagner, H. Nucleation ability of multiwall carbon nanotubes in polypropylene composites. *J. Polym. Sci. Part B Polym. Phys.* **2003**, *41*, 520–527. [CrossRef]
39. Xia, H.; Wang, Q.; Li, K.; Hu, G.H. Preparation of polypropylene/carbon nanotube composite powder with a solid-state mechanochemical pulverization process. *J. Appl. Polym. Sci.* **2004**, *93*, 378–386. [CrossRef]
40. Coleman, J.N.; Cadek, M.; Blake, R.; Nicolosi, V.; Ryan, K.P.; Belton, C.; Fonseca, A.; Nagy, J.B.; Gun'ko, Y.K.; Blau, W.J. High performance nanotube-reinforced plastics: Understanding the mechanism of strength increase. *Adv. Funct. Mater.* **2004**, *14*, 791–798. [CrossRef]
41. Kasaliwal, G.; Villmow, T.; Pegel, S.; Pötschke, P. Influence of material and processing parameters on carbon nanotube dispersion in polymer melts. In *Polymer–Carbon Nanotube Composites*; Woodhead Publishing: Sawston, UK, 2011; pp. 92–132.
42. Alig, I.; Pötschke, P.; Lellinger, D.; Skipa, T.; Pegel, S.; Kasaliwal, G.R.; Villmow, T. Establishment, morphology and properties of carbon nanotube networks in polymer melts. *Polymer* **2012**, *53*, 4–28. [CrossRef]
43. Wang, T.; Song, B.; Qiao, K.; Huang, Y.; Wang, L. Effect of dimensions and agglomerations of carbon nanotubes on synchronous enhancement of mechanical and damping properties of epoxy nanocomposites. *Nanomaterials* **2018**, *8*, 996. [CrossRef] [PubMed]
44. Kasaliwal, G.R.; Göldel, A.; Pötschke, P.; Heinrich, G. Influences of polymer matrix melt viscosity and molecular weight on MWCNT agglomerate dispersion. *Polymer* **2011**, *52*, 1027–1036. [CrossRef]
45. Bauhofer, W.; Kovacs, J.Z. A review and analysis of electrical percolation in carbon nanotube polymer composites. *Compos. Sci. Technol.* **2009**, *69*, 1486–1498. [CrossRef]
46. Stauffer, D.; Aharony, A. *Introduction to Percolation Theory*; CRC Press: Boca Raton, FL, USA, 1994.
47. Sahini, M.; Sahimi, M. *Applications of Percolation Theory*; CRC Press: Boca Raton, FL, USA, 1994.

48. Seo, M.-K.; Park, S.-J. Electrical resistivity and rheological behaviors of carbon nanotubes-filled polypropylene composites. *Chem. Phys. Lett.* **2004**, *395*, 44–48. [CrossRef]
49. Pötschke, P.; Fornes, T.; Paul, D. Rheological behavior of multiwalled carbon nanotube/polycarbonate composites. *Polymer* **2002**, *43*, 3247–3255. [CrossRef]
50. Abdel-Goad, M.; Potschke, P.; Zhou, D.; Mark, J.E.; Heinrich, G. Preparation and rheological characterization of polymer nanocomposites based on expanded graphite. *J. Macromol. Sci. Part A Pure Appl. Chem.* **2007**, *44*, 591–598. [CrossRef]
51. Hemmati, M.; Rahimi, G.; Kaganj, A.; Sepehri, S.; Rashidi, A. Rheological and mechanical characterization of multi-walled carbon nanotubes/polypropylene nanocomposites. *J. Macromol. Sci. Part B* **2008**, *47*, 1176–1187. [CrossRef]
52. Chang, T.; Jensen, L.R.; Kisliuk, A.; Pipes, R.; Pyrz, R.; Sokolov, A. Microscopic mechanism of reinforcement in single-wall carbon nanotube/polypropylene nanocomposite. *Polymer* **2005**, *46*, 439–444. [CrossRef]
53. Andrews, R.; Jacques, D.; Minot, M.; Rantell, T. Fabrication of carbon multiwall nanotube/polymer composites by shear mixing. *Macromol. Mater. Eng.* **2002**, *287*, 395–403. [CrossRef]
54. Kharchenko, S.B.; Douglas, J.F.; Obrzut, J.; Grulke, E.A.; Migler, K.B. Flow-induced properties of nanotube-filled polymer materials. *Nat. Mater.* **2004**, *3*, 564–568. [CrossRef]
55. Tjong, S.C.; Liang, G.; Bao, S. Electrical behavior of polypropylene/multiwalled carbon nanotube nanocomposites with low percolation threshold. *Scr. Mater.* **2007**, *57*, 461–464. [CrossRef]
56. Gorrasi, G.; Romeo, V.; Sannino, D.; Sarno, M.; Ciambelli, P.; Vittoria, V.; De Vivo, B.; Tucci, V. Carbon nanotube induced structural and physical property transitions of syndiotactic polypropylene. *Nanotechnology* **2007**, *18*, 275703. [CrossRef]
57. Logakis, E.; Pollatos, E.; Pandis, C.; Peoglos, V.; Zuburtikudis, I.; Delides, C.; Vatalis, A.; Gjoka, M.; Syskakis, E.; Viras, K. Structure–property relationships in isotactic polypropylene/multi-walled carbon nanotubes nanocomposites. *Compos. Sci. Technol.* **2010**, *70*, 328–335. [CrossRef]

processes

Article

Ameliorated Electrical-Tree Resistant Characteristics of UV-Initiated Cross-Linked Polyethylene Nanocomposites with Surface-Functionalized Nanosilica

Yong-Qi Zhang [1], Ping-Lan Yu [2], Wei-Feng Sun [1,*] and Xuan Wang [1,*]

1 Key Laboratory of Engineering Dielectrics and Its Application, Ministry of Education, School of Electrical and Electronic Engineering, Harbin University of Science and Technology, Harbin 150080, China; jonegen@126.com

2 Chaozhou Power Supply Bureau, Guangdong Power Grid Co. Ltd., Chaozhou 521000, China; kingstel@163.com

* Correspondence: sunweifeng@hrbust.edu.cn (W.-F.S.); topix_xuan@sina.com (X.W.); Tel.: +86-158-4659-2798 (W.-F.S.)

Citation: Zhang, Y.-Q.; Yu, P.-L.; Sun, W.-F.; Wang, X. Ameliorated Electrical-Tree Resistant Characteristics of UV-Initiated Cross-Linked Polyethylene Nanocomposites with Surface-Functionalized Nanosilica. *Processes* **2021**, *9*, 313. https://doi.org/10.3390/pr9020313

Academic Editors: Shaghayegh Hamzehlou and M. Ali Aboudzadeh

Received: 23 January 2021
Accepted: 5 February 2021
Published: 8 February 2021

Abstract: Given the high interest in promoting crosslinking efficiency of ultraviolet-initiated crosslinking technique and ameliorating electrical resistance of crosslinked polyethylene (XLPE) materials, we have developed the funcionalized-SiO_2/XLPE nanocomposites by chemically grafting auxiliary crosslinkers onto nanosilica surfaces. Trimethylolpropane triacrylate (TMPTA) as an effective auxiliary crosslinker for polyethylene is grafted successfully on nanosilica surfaces through thiolene-click chemical reactions with coupling agents of sulfur silanes and 3-mercaptopropyl trimethoxy silane (MPTMS), as characterized by nuclear magnetic resonance and Fourier transform infrared spectroscopy. The functionalized SiO_2 nanoparticles could be dispersively filled into polyethylene matrix even at a high filling content that would generally produce agglomerations of neat SiO_2 nanofillers. Ultraviolet-initiated polyethylene crosslinking reactions are efficiently stimulated by TMPTA grafted onto surfaces of SiO_2 nanofillers, averting thermal migrations out of polyethylene matrix. Electrical-tree pathways and growth mechanism are specifically investigated by elucidating the microscopic tree-morphology with fractal dimension and simulating electric field distributions with finite-element method. Near nano-interfaces where the shielded-out electric fluxlines concentrate, the highly enhanced electric fields will stimulate partial discharging and thus lead to the electrical-trees being able to propagate along the routes between nanofillers. Surface-modified SiO_2 nanofillers evidently elongate the circuitous routes of electrical-tree growth to be restricted from directly developing toward ground electrode, which accounts for the larger fractal dimension and shorter length of electrical-trees in the functionlized-SiO_2/XLPE nanocomposite compared with XLPE and neat-SiO_2/XLPE nanocomposite. Polar-groups on the modified nanosilica surfaces inhibit electrical-tree growth and simultaneously introduce deep traps impeding charge injections, accounting for the significant improvements of electrical-tree resistance and dielectric breakdown strength. Combining surface functionalization and nanodielectric technology, we propose a strategy to develop XLPE materials with high electrical resistance.

Keywords: nanodielectrics; crosslinked polyethylene; auxiliary crosslinker; electrical tree; dielectric breakdown strength

1. Introduction

Due to the prospective advantages of high production rate, low raw material cost, and minimal radiation damage, ultraviolet (UV) crosslinking technology has attracted considerable focus in present years [1–3]. Related researches for UV-initiated crosslinking polyethylene (UV-XLPE) are now primarily focused on optimizing UV irradiation light sources and developing photon-initiation systems to improve electrical insulation performances [4,5]. In order to gain high photon-initiation yields in crosslinking process, effective photon-initiators are capable of efficiently absorbing UV irradiation at a specific wavelength range consistent with UV irradiation spectra. The photon-initiation system consisting of benzophenone (BP) and trimethylolpropane triacrylate (TMPTA) has be verified to efficiently expedite polyethylene crosslinking reactions for the industrial manufacture of high-voltage insulation cables [6]. However, given the low molecular weights and poor compatibility with polyethylene molecules, photon-initiator and auxiliary-crosslinker molecules can easily diffuse out of polymeric blends even at ambient temperature, leading to reductions of UV-initiation efficiency and XLPE crosslinking degree [7]. Meanwhile, photon-initiation systems of small molecules with high volatility are liable to be evaporated and deposited on a UV-source lampshade at an elevated temperature in UV-initiated crosslinking process, which will corrode irradiation instruments and abate UV-light transmission efficiency [8]. It is thereby urgent to develop a novel photon-initiation system or scheme with high polyethylene-compatibility and low thermal volatility for UV-initiation technology in the manufacture of XLPE high-voltage cables.

Inorganic nanoparticles incorporated into polymer matrix (polymer dielectric nanocomposites, nanodielectrics) can comprehensively present amelioration in space charge characteristics, electrical breakdown field, electrical-tree resistance, thermal conductivity, thermostability, and mechanical properties [9,10]. Accordingly, the surface modification technology of nanodielectrics provides a strategy for further improving the electrical, thermal, and mechanical properties of cable insulating materials [11]. Surface-modifications ensure a high dispersion of nanofillers maintaining a small size in polymer matrix, which can highly participate in dynamic exchange reactions, thus leading to improved mechanical properties of nanocomposites [12]. By modifying nano-capsules in geometry and size to adjust their interaction with electric field, the insulation defects caused by dielectric breakdown or electrical-trees can be well repaired, which has been widely used in self-healing insulating materials in recent years [13]. Attributed to the extensive source, large specific surface area, high reactivity, non-toxic and pollution-free chemistry, and visible light transparency, SiO_2 nanoparticles have been comprehensively utilized for developing UV-cured nanocomposites such as SiO_2/acrylic polyurethane, SiO_2/epoxy resin, and SiO_2/methacrylate to improve the mechanical and electrical properties of polymers [14–16].

In the present paper, the migration and volatilization of small molecular photon-crosslinking system are circumvented by utilizing SiO_2 nanoparticles with the surfaces being grafted by auxiliary crosslinker in combination with a large molecular photon-initiator to improve crosslinking degree and insulation performances. Conforming to thiolene-click chemistry and nano-surface modification [17,18], the auxiliary crosslinker TMPTA is chemically grafted onto nanosilica surfaces to acquire an auxiliary crosslinking function by using 3-mercaptopropyl trimethoxy silane (MPTMS). The dispersion of SiO_2 nanofillers in polyethylene matrix could thereby be improved due to the well compatibility of TMPTA with polyethylene molecules. Accordingly, functionalized-SiO_2/XLPE nanocomposites have been developed in a combination of UV-initiation, chemical surface modification, and nanodielectrics. After molecular and micro-structure characterizations of nuclear magnetic resonance, infrared spectroscopy, and scanning electron microscopy, the electrical-tree resistance and insulation performance of functionalized-SiO_2/XLPE nanocomposites are investigated together with finite-element electric field simulations and fractal dimension analyses to elucidate the mechanism of improving electrical resistance.

2. Experimental Schemes

2.1. Material Preparations

Pristine materials of preparing nanocomposites are used as follows: linear low density polyethylene (LLDPE, DFDA-7042, Petrochina Jilin Petrochemical Co. Ltd., Changchun, China) as the matrix material; triallyl isocyanurate (TMPTA, chemically pure, Macklin Biochemical Technology Co. Ltd., Shanghai, China) as the auxiliary crosslinker; 4-Hydroxyl benzophenone laurate (BPL, chemically pure, Harbin University of Science and Technology, Harbin, China) as the photon-initiator; pentaerythritol ester (Irganox1010, Shanyi Plastics Co. Ltd., Dongguan, China) as the antioxidant; 3-merraptnpropylt rimethnxysilane (MPTMS, chemically pure, Jiangsu Heyuan Chemical Co. Ltd., Nanjing, China) as the silane coupling-agent; dichloromethane (DCM), anhydrous ethanol (EtOH), triethylamine (TEA), and silica (SiO_2) 20 nm nanoparticles (Tianjin Fuyu Fine Chemical Co. Ltd., Tianjin, China) for the nanosilica surface modifications.

The surface modification process of functionalizing SiO_2 nanoparticles with an auxiliary crosslinker is schematically illustrated in Figure 1, and implemented as follows: 3.92 g MPTMS (0.02 mol), 0.23 g TEA, and 4.98 g TMPTA (0.02 mol) are dissolved into a 10 mL solvent of DCM in nitrogen protection and ice-water bath, and then warmed to room temperature, persisting for 24 h in thermal insulation to finally obtain the liquid product (MTMPTA) with 78% yield. Furthermore, the dried nanosilica of 10 g is homogeneously blended into a 100 mL ethanol solution (3:1) by ultrasonic dispersion treatment for 30 min, after which MTMPTA is instilled under pH = 4 with diluted hydrochloric acid. By filtering the suspension and washing with ethanol three times, surface-modified nanosilica (TMPTA-*s*-SiO_2) is eventually achieved after being hot-degassed for 3 h at 60 °C in vacuum.

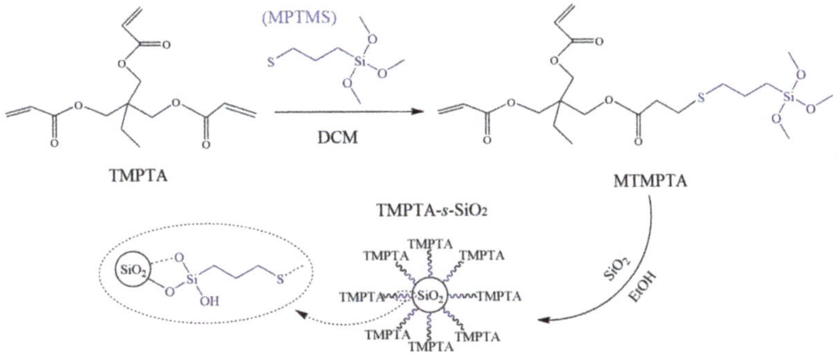

Figure 1. Schematic reactions of trimethylolpropane triacrylate (TMPTA) grafting onto the nanosilica surface.

According to the blending components listed in Table 1, four mixtures for preparing pure and composite XLPE materials are uniformly blended and pressed into film samples at 160 °C by a plate vulcanizer, in which pressure is increased from 0 to 15 MPa at a rate of 5 MPa/5 min to melt material. For photon-initiated crosslinking reactions, the obtained melting films are irradiated by a light source array of UV LED units (NVSU233A-U365, Riya Electronics Chemistry Co. Ltd., Shanghai, China) for 180 s under normal temperature and pressure [19], in which the emitting power and light-wavelength are controlled at 1.0 w and 365 nm, respectively, with a light-incidence of a 60° angle to the film plane [20]. After hot-degassing in short-circuit for 24 h at 80 °C in vacuum to eliminate residual chemical by-products and mechanical stresses, pure UV-XLPE and three kinds of UV-XLPE nanocomposites are finally prepared, as listed in Table 1.

Table 1. Blending components (wt%) for preparing XLPE nanocomposites.

Materials	LLDPE	BPL	TMPTA	TMPTA-*s*-SiO_2	SiO_2	Antioxidant
XLPE	96.7	2.0	1.0	0	0	0.3
0.5wt%TMPTA-*s*-SiO_2/XLPE	97.2	2.0	0	0.5	0	0.3
1.5wt%TMPTA-*s*-SiO_2/XLPE	96.7	2.0	0	1.5	0	0.3
1.5wt%SiO_2/XLPE	95.7	2.0	1.0	0	1.5	0.3

2.2. Material Characterization and Property Test

Nuclear magnetic resonance (Bruker-1, ^{1}H-NMR) is utilized to determine the hydrogen (H) content according to the type and quantity of H detected in samples. A Fourier transform infrared spectrometer (FTIR-6100, Jiasco Trading Co. Ltd., Shenyang, China) is utilized to characterize molecular structures of TMPTA-s-SiO_2 samples. According to ASTM-D 2675-2011 standard, the solvent-extracted gel contents of XLPE and nanocomposite materials are tested to evaluate the crosslinking degree of XLPE matrix.

Brittlely fractured cross-sections of XLPE nanocomposites are observed by an ultra-high-resolution scanning electron microscope (SEM, SU8020, Hitachi Co. Ltd., Tokyo, Japan) adjusted to a magnification of 75 k under an acceleration voltage of 5~10 kV. SEM samples need to be sputtered with thin gold layers covering all the fractured surfaces to conduct out the undesirable charges introduced by electron impacts.

For electrical-tree inception and propagation, the sample with an inserted tungsten needle (tip curvature radius = 3 μm) is immersed in silicone oil to avoid creepage discharge, and then a needle electrode is applied with alternating current (AC) high-voltage power at 50 Hz frequency. The step-up increasing voltage is adopted to reach 2.5 kV at 0.1 kV/s persisting for 15 min, and then continuously rise up to 0.1 kV/s until electrical-tree appearing. An optical system consisting of a charge-coupled device (CCD) camera and in situ optical microscope are accessed through the computer interface to observe electrical-tree morphology in time intervals of 5 min. The electrical-tree of 10 μm length is specified as identifying tree inception and 10 samples are tested for each material in statistical analyses with 2-parameter Weibull distribution. Electrical-tree propagation is further evaluated though finite-element electrical field simulations under 50 Hz AC voltage.

Thermally stimulated currents (TSC, Harbin University of Science and Technology, Harbin, China) are tested to evaluate the energy level distributions of charge traps. First, the film samples in 100 μm thickness are applied with an electric field of 40 kV/mm for 30 min at room temperature. The sample-electrode system in short-circuit is swiftly cooled down to −30 °C and stabilized for 10 min, after which TSC through short-circuit samples is consecutively tested with a microcurrent meter when temperature rising from −30 to 175 °C at a heating rate of 3 °C/min.

Circular film samples with a diameter and thickness of 80 mm and 0.1 mm, respectively, are fabricated to measure AC dielectric breakdown strength (DBS) with asymmetric columnar electrodes (25 mm and 75 mm in diameter for high-voltage and ground electrodes, respectively) at the same temperature as electrical-tree experiments. The applied electric field is continuously increased at a rate of 4 kV/s until attaining the possible maximum voltage that is recorded promptly just before electrical breakdown.

3. Results and Discussion

3.1. Material Characterization

As indicated from the ^{1}H-NMR spectrum of MTMPTA molecules in Figure 2a, the successfully synthesized MTMPTA is demonstrated by the peaks at δ = 5.82, 6.38, and 6.09 ppm identifying the chemical displacements of H atoms in –CH=CH_2 group, and δ = 2.74 and 2.61 ppm indicating the H atoms on –CH_2– adjacent to sulfur atom (S). FTIR spectra of the functionalized nanoparticles TMPTA-s-SiO_2 grafted with auxiliary crosslinkers are shown in Figure 2b. Infrared absorption peaks at 3438 cm^{-1}, 2952 cm^{-1}, 2841 cm^{-1}, and 2561 cm^{-1} indicate the atomic vibrations of Si–OH on nanosilica surfaces, methyl, methylene, and –SH in MPTMS molecules, respectively. Characteristic peaks at 1726 cm^{-1} and 1637 cm^{-1}, respectively, originate from stretching vibrations of double-bonds in –C=O and –CH=CH_2 groups of TMPTA molecules. The simultaneously arising peaks of methyl/methylene (2957 cm^{-1}), carbonyl (1717 cm^{-1}), and ethylene (1638 cm^{-1}) in the infrared spectrum of TMPTA-s-SiO_2 verify that TMPTA has been substantially grafted to nanosilica surfaces. The ^{1}H-NMR and FTIR spectra imply that the grafting rate of TMPTA on the surfaces of TMPTA-s-SiO_2 nanoparticles can be adjusted by modifying MTMPTA content in surface modification processes. Nevertheless, the density of TMPTA in TMPTA-s-SiO_2/XLPE nanocomposites can also be controlled by increasing the concentration of TMPTA-s-SiO_2 nanofillers to ameliorate the crosslinking degree and electrical resistance.

Thereby, 30 wt% dosage of MTMPTA is specialized in the synthesis of TMPTA-s-SiO_2 nanoparticles which are subsequently used to prepare XLPE nanocomposites.

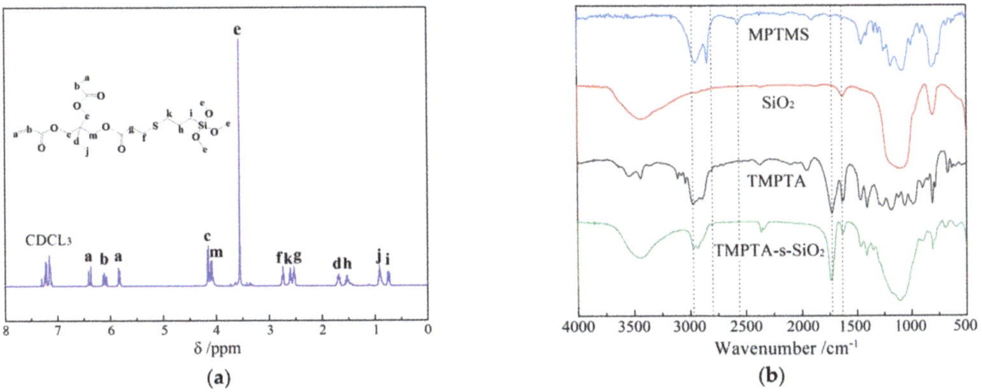

Figure 2. (**a**) ^{1}H-NMR spectrum of MTMPTA molecules and (**b**) Fourier transform infrared (FTIR) transmission spectra of MPTMS, neat nanosilica, TMPTA, and TMPTA-s-SiO_2 nanoparticles.

3.2. Crosslinking Degree

Identical photon-initiator content and irradiation time are applied for polyethylene crosslinking reactions in preparations of UV-initiated XLPE nanocomposites to investigate the effect of grafting TMPTA to nanosilica surface on the crosslinking degree of polyethylene matrix, as illustrated by gel content (a general indicator of crosslinking degree) shown in Figure 3. For the 0.5 wt% and 1.5 wt% filling content of TMPTA-s-SiO_2, crosslinking degrees are remarkably increased by 30% and 120%, respectively, compared with that of pure XLPE. Furthermore, the XLPE nanocomposite prepared by filling neat SiO_2 nanoparticles (SiO_2/XLPE nanocomposite) without TMPTA-functionalized surface shows a low crosslinking degree almost identical to that of pure XLPE. As shown in Table 1, the content of dissociative TMPTA added into polyethylene matrix to assist crosslinking reactions approaches 1.0 wt% for pure XLPE and SiO_2/XLPE nanocomposite, which is much higher than the actual content of TMPTA grafted onto surfaces of SiO_2 nanofillers in TMPTA-s-SiO_2/XLPE nanocomposites. Additionally, the surface functionalization of SiO_2 nanofillers by grafting TMPTA shows a great validity of fixing TMPTA molecules into XLPE matrix to avoid thermal migrations out of crosslinking reaction system and promote the auxiliary crosslinking efficiency. Moreover, the significantly further increment of crosslinking degree by increasing filling content from 0.5 wt% to 1.5 wt% of TMPTA-s-SiO_2/XLPE nanocomposites is also a manifestation of the auxiliary function of TMPTA in crosslinking reactions to improve crosslinking degree, which can be stably introduced into polyethylene crosslinking system through additional nanosurface functionalization in nanodielectric technology.

3.3. Microscopic Morphology of Nanocomposites

The outmost TMPTA layer cladding SiO_2 nanocore possesses a high compatibility with polyethylene molecular-chain and determine the dispersivity and size of nanofillers in XLPE matrix, which accordingly contributes to the acquired improvements in dielectric properties of composite materials. It is indicated from cross-sectional SEM images illustrated in Figure 4 that the surface-modified nanofillers in TMPTA-s-SiO_2/XLPE nanocomposites are majorly distributed in sizes of <80 nm with a favorable high dispersivity. It is impossible to completely prevent nanofillers from agglomerating in polyethylene matrix for the composite materials prepared with melting blend method, as shown in Figure 4c for 1.5 wt%SiO_2/XLPE nanocomposite with some SiO_2 fillers larger than 150 nm in diameter. Nevertheless, the surface-modified silica fillers remain in nanoscales even for 1.5 wt% filling content, as shown in Figure 4b, which means they fulfill successful nanodielectrics. Compared with 0.5 wt% filling content, a higher density of TMPTA-s-SiO_2 nanofillers with an almost identical size is uniformly dispersed in XLPE matrix without any nanoparticle agglomeration.

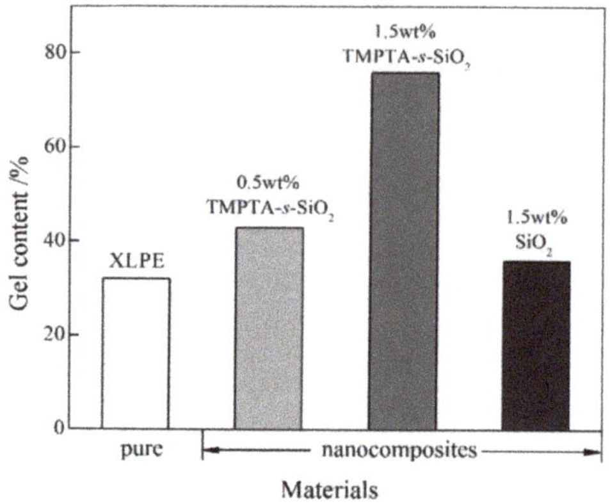

Figure 3. Gel contents of XLPE and its nanocomposites filled with TMPTA-s-SiO_2 or neat SiO_2 nanoparticles.

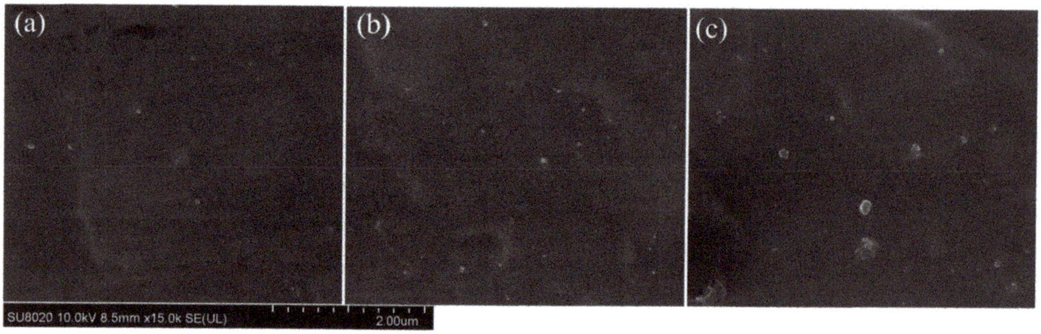

Figure 4. Cross-sectional SEM images: (**a**) 0.5 wt%TMPTA-s-SiO_2/XLPE, (**b**) 1.5 wt%TMPTA-s-SiO_2/XLPE, and (**c**) 1.5 wt%SiO_2/XLPE nanocomposites.

3.4. Electrical Tree Development

Electrical-tree inception voltages are fitted by 2-parameter Weibull statistics for XLPE and its nanocomposites, as shown in Figure 5. Characteristic inception voltages and shape parameters are listed in Table 2. Despite the almost same shape parameters for all the materials, the characteristic electrical-tree inception voltages of TMPTA-s-SiO_2/XLPE nanocomposites with 0.5 wt% and 1.5 wt% filling contents have been increased to 6.53 kV and 7.15 kV, respectively, which are 9% and 19% higher than that of pure XLPE, while SiO_2/XLPE nanocomposite represents an almost identical inception voltage (2% higher) to pure XLPE. Therefore, the evident improvement in electrical tree inception voltage is dominantly attributed to the TMPTA-grafting layer on nanosilica surface. It is suggested that the charge carriers (electrons or holes) injected from needle electrodes cannot be accelerated when transporting through TMPTA-s-SiO_2 nanofillers and will be scattered to form divergent discharging channels in high electric field regions. Accordingly, the carbonized discharging channels (tree like damages) will be inhibited from directly developing to the ground electrode plate, as manifested by the shorter electrical-tree length and higher inception voltage (judged by the same standard of tree length) as shown in Figure 5 and Table 2.

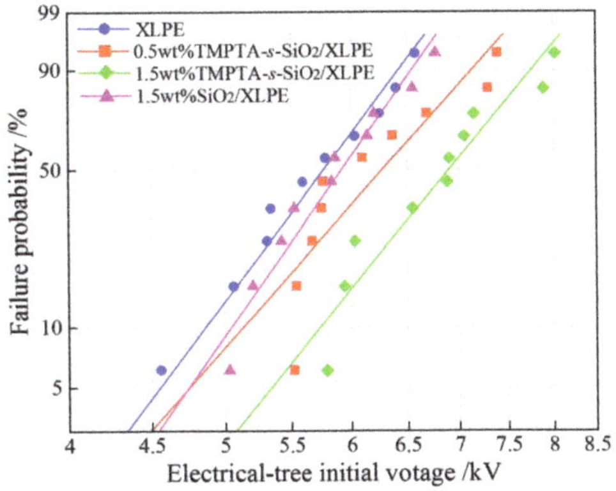

Figure 5. Weibull 2-parameter distributions of the electrical-tree inception voltages for XLPE and its nanocomposites.

Table 2. Characteristic electrical-tree inception voltages at 63.2% probability and shape parameters of 2-parameter Weibull statistics at 95% confidence interval.

Materials	Characteristic Inception Voltage/kV	Shape Parameter	Increment/%
XLPE	5.99	12.37	
0.5 wt%TMPTA-s-SiO_2/XLPE	6.53	9.349	9
1.5 wt%TMPTA-s-SiO_2/XLPE	7.15	10.17	19
1.5 wt%SiO_2/XLPE	6.10	11.79	2

By applying an AC voltage for 7200 s, after which nearly no change can be observed in tree morphology, the structure features of electrical-trees growing in XLPE and its nanocomposites are pictured by optical microscopy, as shown in Figure 6. TMPTA-*s*-SiO_2/XLPE nanocomposites present electrical-trees in shapes of disperse, dense, and short shrubbery, while the pure XLPE and SiO_2/XLPE nanocomposite show pine-branch like electrical-trees in the morphology of sparse branches concentrating around the long main trunks.

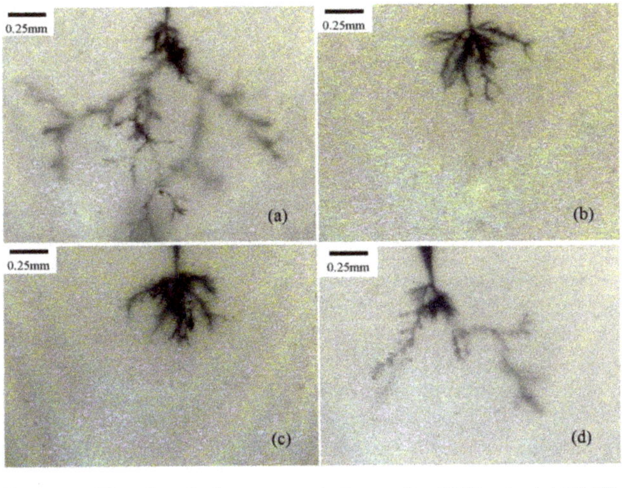

Figure 6. The electrical-tree morphology after 7200 s in (**a**) XLPE and the nanocomposites of (**b**) 0.5 wt%TMPTA-s-SiO_2/XLPE, (**c**) 1.5 wt%TMPTA-s-SiO_2/XLPE, and (**d**) 1.5 wt%SiO_2/XLPE.

Based on dielectric permittivity and electrical conductivity, the local electric field distributions around needle tip (or considered as the front of electrical-tree) are simulated with finite element method as implemented in COMSOL multi-physics software, with the results being shown in Figure 7. The triangular region in white color represents electrode needle-tip or tree-front. According to the nanofiller sizes observed from SEM images and the molecule dimension of MTMPTA, we model TMPTA-*s*-SiO_2 and neat SiO_2 nanofillers by the smaller shell-core structure and the larger neat sphere respectively, which are dispersively put into the uniform background of XLPE matrix to simulate nanocomposites, as shown in Figure 7b–d.

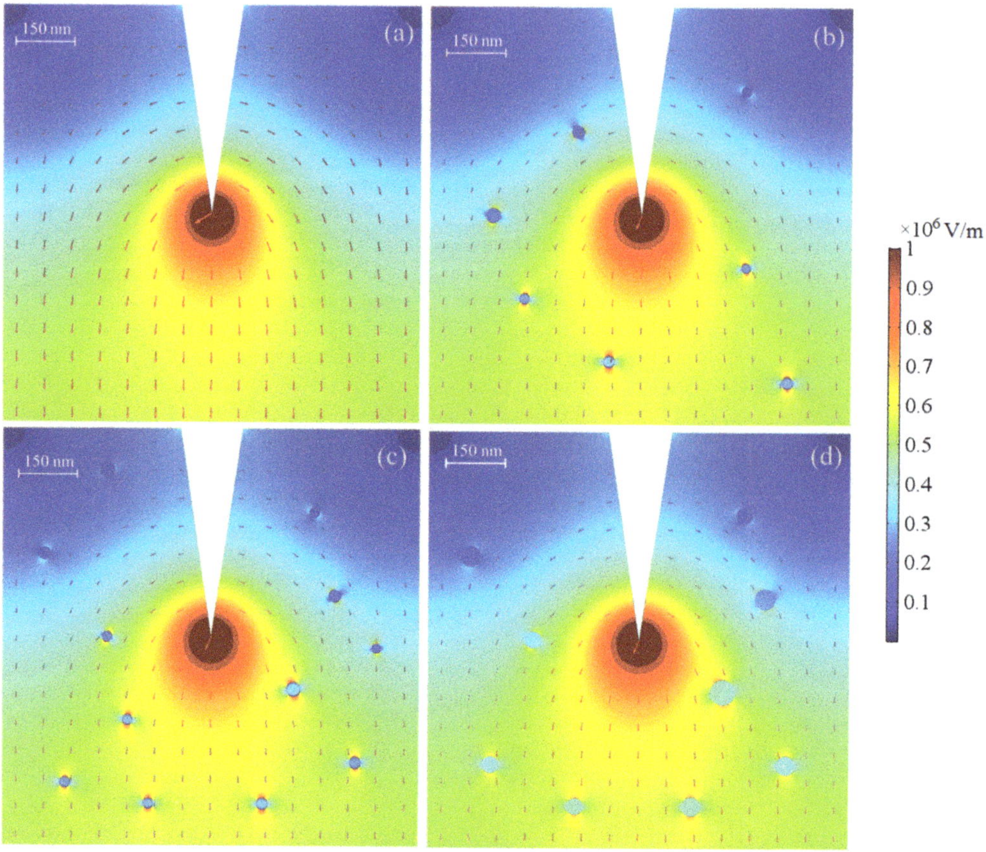

Figure 7. Electric field distributions around initiation tip (identical to tree front) simulated for (**a**) XLPE, (**b**) 0.5 wt%TMPTA-*s*-SiO_2/XLPE, (**c**) 1.5 wt%TMPTA-*s*-SiO_2/XLPE, and (**d**) 1.5 wt%SiO_2/XLPE nanocomposites.

The electric field in SiO_2 core of the surface-modified nanofiller is remarkably lower than that of XLPE matrix, leading to an evident concentration of electric fields near two sides of the outer surfaces crossing electric fluxlines between the two adjacent nanofillers as shown in Figure 7b,c. In contrast, the modeled particles of neat SiO_2 nanofillers in XLPE background show unconspicuous discrepancy of the electric fields inside and outside nanosurfaces, as shown in Figure 7d. The dielectric permittivity of the surface-modified MTMPTA shell is significantly higher than that of SiO_2 core and XLPE matrix, so as to expel the electric fluxlines out of nanofillers and accordingly intensify the electric fields (concentrate the continuous electric fluxlines) at the outer surfaces where the electric fluxlines enter and flow out through nanofillers. The much lower and higher electric fields in SiO_2 core and outside MTMPTA shell respectively than the background electric field of XLPE matrix could be comprehended as "dielectric shielding" and cannot be effectively presented by the neat SiO_2 nanofillers with a low dielectric permittivity as XLPE. It is reasonably suggested that the electrical-trees will be apt to form and propagate along the paths between TMPTA-s-SiO_2 nanofillers due to the highly increased spontaneous probability of partial discharge under high electric fields near the surfaces of TMPTA-s-SiO_2 nanofillers. Consequently, electrical-trees will be more possible to grow consecutively along the network pathways through TMPTA-s-SiO_2 nanofillers, which means a larger numbers of circuitous routes for electrical-tree developing from high voltage to ground, as manifested by the higher inception voltage and the larger fractal dimension with shorter tree-length that will be demonstrated in follows.

Fractal structure of electrical-trees can be evaluated from the growth length L along electric field and fractal dimension D_f, in which growth quantity Y and tree length L_B as a function of growing time are formularized as follows [21]:

$$\begin{aligned} Y &= (L/L_B)^{D_f}, \\ L &= L_B{}^{(1+1/D_f)}\left[\frac{\omega}{2\pi}\exp\left(\frac{\alpha l \pi \varepsilon E^2 - U_0}{kT}\right)\right]^{1/D_f} t^{1/D_f} \end{aligned} \tag{1}$$

where $\omega/2\pi$ symbolizes the intrinsic frequency of atomic vibrations in polymer materials; U_0 represents the activation energy of electrical-tree expansions; ε denotes dielectric permittivity; E signifies local electric field; α indicates thermodynamic parameter of atomic vibrations in polymer materials; l symbolizes the linear dimension of micro-cracks; and T and k are thermodynamic temperature and Boltzmann constant, respectively. Specifically for polyolefin materials such as XLPE, atomic-vibration frequency is predominately located on $\omega/2\pi = 6.25 \times 10^{12}$ Hz, and the molecular vibration parameter $\alpha = 6.68 \times 10^{-16}$ is determined from activation volume [22]. Meanwhile, the electric field should be calculated approximately as following [22]:

$$E = \frac{2U}{r\ln(1 + 4R/r)} \tag{2}$$

where U = 6 kV indicates the electrical voltage applied on needle electrode; r = 3 μm identifies the curvature radius of needle tip; and R = 3 mm denotes the distance from bottom plate to needle tip.

Fractal dimension can quantitatively depict the spatial topological characteristics for characterizing electrical-tree structure. As implemented with MATLAB programming, box counting is performed on the electrical-tree images from optical microscope observations to evaluate the fractal dimension of electrical-trees produced in pure XLPE and XLPE nanocomposites. Tree length versus fractal dimension of XLPE is calculated from Equations (1) and (2) in comparison with experimental results observed from electrical-tree morphology in pure XLPE and 1.5 wt%TMPTA-s-SiO_2/XLPE nanocomposite, as shown in Figure 8. In macroscopic scale, smaller fractal dimension of electrical-trees means a more tree-development along electric field. For 1.5 wt%TMPTA-s-SiO_2/XLPE nanocomposite, the absolute variation slope of electrical-tree length vs. fractal dimension is significantly smaller than theoretical values of XLPE (well consistent with experimental results). The electrical-tree development directly towards to ground electrode (tree length) is significantly reduced by the partial discharging routes in the network of high electric fields near the "dielectric shielding" surfaces of TMPTA-s-SiO_2 nanofillers, leading to the expedition of electrical-tree developing along the more circuitous trajectories in dispersive geometries, as indicated by the large fractal dimensions of electrical-trees produced in TMPTA-s-SiO_2/XLPE nanocomposites.

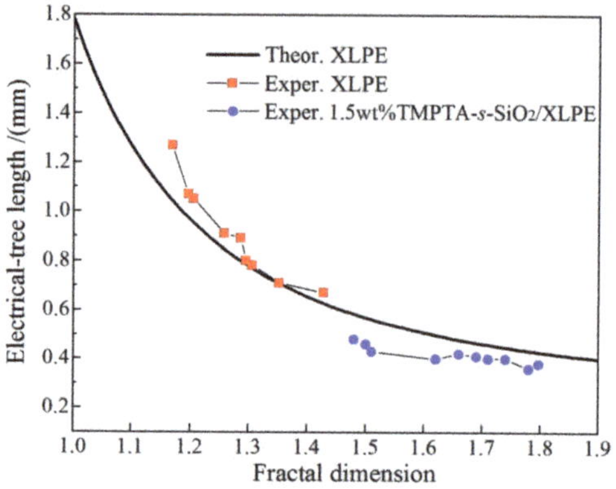

Figure 8. Growth length versus fractal dimension of electrical-trees in pure XLPE and 1.5 wt%TMPTA-s-SiO_2/XLPE nanocomposite.

3.5. Charge Trapping Characteristics

In order to investigate the deep-trapping mechanism in TMPTA-s-SiO_2/XLPE nanocomposites that can enhance dielectric breakdown strength, energy level distributions of charge traps are evaluated from TSC results [23]. TSC testing temperatures from −30 to 175 °C cover all of the required energies for thermal excitation of the charges captured in the deep traps introduced by nanosurface-grafting. The measured TSC temperature spectra and the derived trap densities as a function of trap energy level (trap level distributions) are shown in Figure 9. The general peaks arising in trap levels of 0.9−1.0 eV originate from the intrinsic shallow traps introduced by the structural defects between XLPE lamellae.

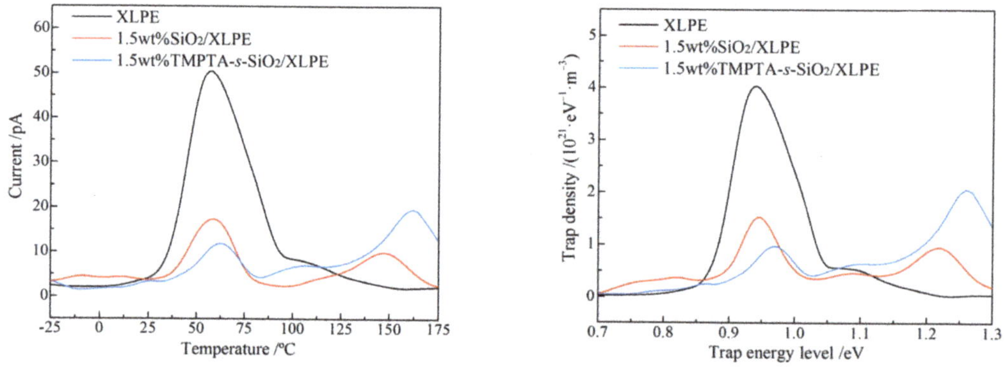

Figure 9. Thermally stimulated current (TSC) temperature spectra (**left**) and trap energy level distributions (**right**) of pure XLPE and the XLPE nanocomposites filled with TMPTA-s-SiO_2 and neat SiO_2 nanoparticles, respectively.

Due to the deep traps introduced by TMPTA grafted onto surfaces of SiO_2 nanofillers, a characteristic peak with lower amplitude and smaller integral area than XLPE intrinsic peak appears at 1.25 eV for TMPTA-s-SiO_2/XLPE nanocomposites, which is observably higher in peak intensity than that at 1.2eV for SiO_2/XLPE nanocomposite which derives from discharging traps at nano-SiO_2/XLPE interfaces. Because charge carriers prefer to be captured into deeper traps than shallower traps, no enough carriers have been trapped into intrinsic traps of XLPE matrix under high-voltage electric field in TMPTA-s-SiO_2/XLPE nanocomposites. Therefore, the amplitude and integral area of trap distributions at 0.9–1.0 eV are significantly lower than that of pure XLPE. It is proved by TSC tests that the surface functionalization of grafting TMPTA onto SiO_2 nanoparticles has achieved deeper and higher density of charger traps in TMPTA-s-SiO_2/XLPE nanocomposites than that in pure XLPE and SiO_2/XLPE nanocomposite, which will contribute to the improvement of electrical resistance.

3.6. Dielectric Breakdown Strength

Two-parameter Weibull statistics is utilized to analyze electrical breakdown fields under AC voltage, as shown in Figure 10. Scale parameter E_b characterizes the electrical breakdown field with a probability of 63.2%, while shape parameter β indicates the dispersivity of experimental data, as listed in Table 3.

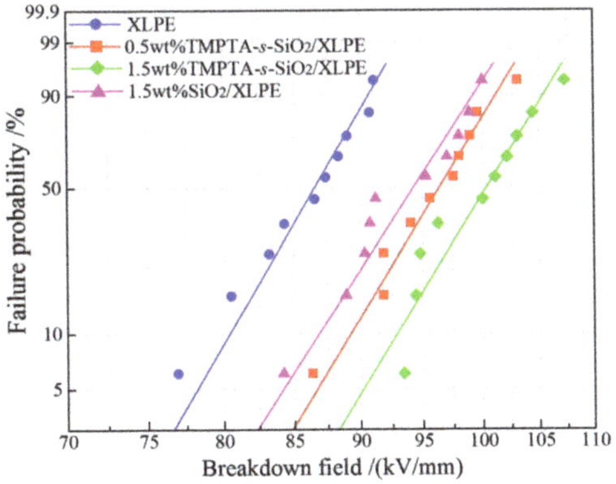

Figure 10. Electrical breakdown fields under AC voltage fitted with 2-parameter Weibull distribution for pure XLPE, TMPTA-s-SiO_2/XLPE and SiO_2/XLPE nanocomposites.

Table 3. Characteristic breakdown field E_b at 63.2% probability and shape parameter β obtained by fitting experimental results of electrical breakdown fields with 2-parameter Weibull distribution at 95% confidence interval.

Materials	E_b/(kV/mm)	β
XLPE	87.74	25.74
0.5 wt%TMPTA-s-SiO_2/XLPE	97.79	24.58
1.5 wt%TMPTA-s-SiO_2/XLPE	101.83	24.42
1.5 wt%SiO_2/XLPE	95.70	23.12

DBS results in Weibull distributions prove that electrical breakdown performances of TMPTA-s-SiO_2/XLPE nanocomposites are better than that of pure XLPE and SiO_2/XLPE nanocomposite. The relatively lower breakdown field of 1.5wt%SiO_2/XLPE nanocomposite than that of TMPTA-s-SiO_2/XLPE nanocomposites can be primarily attributed to the lack of "dielectric shielding" effects from nanosurface-modifications of grafting high polarizibility molecules to prevent electrical-trees developing directly towards ground electrode. The slightly increased breakdown field derives primiarily from the charge traps produced at interfaces between neat nanosilica and XLPE matrix [24], which however has been abated by the small specific surface area of the aggregated nanosilica caused by poor compatibility of SiO_2 with polyethylene molecules. Meanwhile, XLPE nanocomposites filled with neat SiO_2 nanoparticles will engender considerable particle agglomerations when filling content exceeds 1.5 wt%. Furthermore, TMPTA molecules with polar-groups grafted on nanosilica surface will not be evaporated in crosslinking reactions, thus retaining a high density of effective polar-groups acting as deep traps to efficiently improve DBS under the charge trapping and scattering mechanisms [6,19]. It is thus suggested a reasonable correlation of DBS improvement with TMPTA surface-functionalization of nanosilica for XLPE-based nanodielectrics. These results support the charge-trapping mechanism and demonstrate that TMPTA is an effective graft candidate to improve the electrical resistance performances of polyethylene insulating materials.

4. Conclusions

In order to simultaneously ameliorate the crosslinking degree and electrical resistance of UV-XLPE insulating materials, chemical modification and nanodielectric technology are coordinately exploited by grafting auxiliary crosslink agent (TMPTA) onto nanosilica surface. FTIR and ^{1}H-NMR characterizations demonstrate that TMPTA molecules have been successfully grafted onto nanosilica surfaces. TMPTA molecules being fixed on the surface of TMPTA-*s*-SiO_2 nanofillers disperse in polyethylene matrix acquire a high crosslinking efficiency without evaporation in crosslinking reaction, which could be further promoted by increasing the concentration of TMPTA-*s*-SiO_2 nanofillers. Furthermore, the larger sizes of neat SiO_2-nanofillers in XLPE matrix is also a manifestation of inhibiting particle agglomerations by functional surface-modifications. TMPTA-*s*-SiO_2/XLPE nanocomposites show a higher tree inception voltage by about 19% than pure XLPE and SiO_2/XLPE nanocomposite. The high polarizability layer of TMPTA covering SiO_2 core can eliminate electric field in SiO_2 core and highly increase electric field near outer surface of nanofillers, which can not be presented by neat SiO_2 nanofillers. Due to the highly probable discharge near TMPTA-*s*-SiO_2 nanofiller surface, the electrical-trees are produced and propagating with a fast speed along the long circuitous pathways between TMPTA-*s*-SiO_2 nanofillers, which accounts for the improvements in the fractal dimension and inception voltage of electrical-tree growth. In addition, polar-groups in TMPTA molecules grafted on nano-silica surfaces introduce effective deep traps in high density to further enhance DBS. TMPTA-*s*-SiO_2/XLPE nanocomposites achieve comprehensive significant improvements in crosslinking degree, electrical-tree resistance and AC electrical breakdown field.

Author Contributions: Conceptualization, X.W.; Data curation and formal analysis, Y.-Q.Z.; Methodology, P.-L.Y.; Investigation and writing, W.-F.S.; Project administration, X.W. All authors have read and agreed to the published version of the manuscript.

Funding: This research was funded by the National Natural Science Foundation of China (grant no. 51337002).

Institutional Review Board Statement: Not applicable.

Informed Consent Statement: Not applicable.

Data Availability Statement: Experimental methods and results are available from the authors.

Conflicts of Interest: The authors declare no conflict of interest.

References

1. Ouyang, B.; Ming, H.; Deng, X. A review about development of HV XLPE cable materials and processes. *Insul. Mater.* **2016**, *49*, 1–6.
2. Ye, G.D.; Zhou, H.; Yang, J.W.; Zeng, Z.H.; Chen, Y.L. Photoinitiating behavior of macrophotoinitiator containing aminoalkylphenone group. *J. Therm. Analy. Calor.* **2006**, *85*, 771–777. [CrossRef]
3. Lu, Y.; Tang, J.; Zhao, H.; Hao, G.; Huang, B. Study of ventilation cooling for an irradiation box of low-voltage cable ultraviolet cross-linking. *J. Harbin. Univ. Sci. Technol.* **2013**, *18*, 45–50.
4. Wu, Q.H.; Qu, B.J. Photoinitiating characteristics of benzophenone derivatives as new initiators in the photocrosslinking of polyethylene. *Polym. Eng. Sci.* **2001**, *41*, 1220–1226. [CrossRef]
5. Wu, Q.H.; Qu, B.J. Synthesis of di(4-hydroxyl benzophenone) sebacate and its usage as initiator in the photocrosslinking of polyethylene. *J. Appl. Poly. Sci.* **2002**, *85*, 1581–1586. [CrossRef]
6. Zhao, X.D.; Sun, W.F.; Zhao, H. Enhanced insulation performances of crosslinked polyethylene modified by chemically grafting chloroacetic acid allyl ester. *Polymers* **2019**, *11*, 592. [CrossRef]
7. Fu, Y.W.; Wang, X.; Wu, Q.H.; Zhao, H. Study of crosslinked and electrical characteristics for cable insulating material of new UV XLPE. *Trans. Chin. Electrotech. Soc.* **2018**, *33*, 178–186.
8. Chen, J.Q.; Zhao, H.; Zheng, H.F.; Chen, C.M.; Li, Y.; Sun, K. Research and design of electrodeless UV curing lamp with elliptic concentrator. *Electr. Mach. Contr.* **2017**, *21*, 109–113.

9. Danikas, M.G.; Tanaka, T. Nanocomposites-a review of electrical treeing and breakdown. *IEEE Electr. Insul. Mag.* **2009**, *25*, 19–25. [CrossRef]
10. Tanaka, T. Dielectric nanocomposites with insulating properties. *IEEE Trans. Dielectr. Electr. Insul.* **2005**, *12*, 914–928. [CrossRef]
11. Kango, S.; Kalia, S.; Celli, A.; Njuguna, J.; Habibi, Y.; Kumar, R. Surface modification of inorganic nanoparticles for development of organic–inorganic nanocomposites—A review. *Prog. Poly. Sci.* **2013**, *38*, 1232–1261. [CrossRef]
12. Wemyss, A.M.; Bowen, C.; Plesse, C.; Vancaeyzeele, C.; Nguyen, G.T.M.; Vidal, F.; Wan, C. Dynamic crosslinked rubbers for a green future: A material perspective. *Mater. Sci. Eng. R Rep.* **2020**, *141*, 100561. [CrossRef]
13. Zhang, Y.; Khanbareh, H.; Roscow, J.; Pan, M.; Bowen, C.; Wan, C. Self-healing of materials under high electrical stress. *Matter* **2020**, *3*, 989–1008. [CrossRef]
14. Xu, J.W.; Pang, W.M.; Shi, W.F. Synthesis of UV-curable organic–inorganic hybrid urethane acrylates and properties of cured films. *Thin Solid Films* **2006**, *514*, 69–75. [CrossRef]
15. Sangermano, M.; Malucelli, G.; Amerio, E.; Priola, A.; Billi, E.; Rizza, G. Photopolymerization of epoxy coatings containing silica nanoparticles. *Prog. Org. Coat.* **2005**, *54*, 134–138. [CrossRef]
16. Medda, S.K.; Kundu, D.; De, G. Inorganic–organic hybrid coatings on polycarbonate.: Spectroscopic studies on the simultaneous polymerizations of methacrylate and silica networks. *J. Non Cryst. Solids* **2003**, *318*, 149–156. [CrossRef]
17. Crucho, C.I.C.; Baleizão, C.; Farinha, J.P.S. Functional group coverage and conversion quantification in nanostructured silica by ^{1}H NMR. *Analyt. Chem.* **2016**, *89*, 681–687. [CrossRef] [PubMed]
18. Sangermano, M.; Colucci, G.; Fragale, M.; Rizza, G. Hybrid organic–inorganic coatings based on thiol-ene systems. *React. Funct. Poly.* **2009**, *69*, 719–723. [CrossRef]
19. Zhao, X.D.; Zhao, H.; Sun, W.F. Significantly improved electrical properties of crosslinked polyethylene modified by UV-initiated grafting MAH. *Polymers* **2020**, *12*, 62. [CrossRef]
20. Qiu, P.; Chen, J.Q.; Sun, W.F.; Zhao, H. Improved DC dielectric performance of photon-initiated crosslinking polyethylene with TMPTMA auxiliary agent. *Materials* **2019**, *12*, 3540. [CrossRef] [PubMed]
21. Ding, H.Z.; Xing, X.S.; Zhu, H.S. A kinetic model of time-dependent dielectric breakdown for polymers. *J. Phys. D Appl. Phys.* **1994**, *27*, 591.
22. Chi, X.H.; Gao, J.G.; Zheng, J.; Zhang, X.H. The mechanism of electrical treeing propagation in polypropylene. *Acta Phys. Sin.* **2014**, *63*, 177701–177708.
23. Tian, F.; Bu, W.; Shi, L.; Yang, C.; Wang, Y.; Lei, Q. Theory of modified thermally stimulated current and direct determination of trap level distribution. *J. Electrost.* **2011**, *69*, 7–10. [CrossRef]
24. Roy, M.; Nelson, J.K.; MacCrone, R.K.; Schadler, L.S.; Reed, C.W.; Keefe, R. Polymer nanocomposite dielectrics-the role of the interface. *IEEE Trans. Dielectr. Electr. Insul.* **2005**, *12*, 629–643. [CrossRef]

MDPI
St. Alban-Anlage 66
4052 Basel
Switzerland
Tel. +41 61 683 77 34
Fax +41 61 302 89 18
www.mdpi.com

Processes Editorial Office
E-mail: processes@mdpi.com
www.mdpi.com/journal/processes

www.ingramcontent.com/pod-product-compliance
Lightning Source LLC
LaVergne TN
LVHW071027170726
843515LV00006B/1219
* 9 7 8 3 0 3 6 5 2 0 3 4 6 *